The twentieth century has seen biology come of age as a conceptual and quantitative science. Biochemistry, cytology, and genetics have been unified into a common framework at the molecular level. However, cellular activity and development are regulated not by the interplay of molecules alone, but by interactions of molecules organized in complex arrays, subunits, and organelles. Emphasis on organization is, therefore, of increasing importance.

So it is too, at the other end of the scale. Organismic and population biology are developing new rigor in such established and emerging disciplines as ecology, evolution, and ethology, but again the accent is on interactions between individuals, populations, and societies. Advances in comparative biochemistry and physiology have given new impetus to studies of animal and plant diversity. Microbiology has matured, with the world of viruses and procaryotes assuming a major position. New connections are being forged with other disciplines outside biology—chemistry, physics, mathematics, geology, anthropology, and psychology provide us with new theories and experimental tools while at the same time are themselves being enriched by the biologists' new insights into the world of life. The need to preserve a habitable environment for future generations should encourage increasing collaboration between diverse disciplines.

The purpose of the Modern Biology Series is to introduce the college biology student—as well as the gifted secondary student and all interested readers—both to the concepts unifying the fields within biology and to the diversity that makes each field unique.

Since the series is open-ended, it will provide a greater number and variety of topics than can be accommodated in many introductory courses. It remains the responsibility of the instructor to make his selection, to arrange it in a logical order, and to develop a framework into which the individual units can best be fitted.

New titles will be added to the present list as new fields emerge, existing fields adv̲a̲n̲c̲e̲, *and new authors of abil̲i̲t̲*   *r.* *Only thus,*   *e* *explosion o̲*

James D. Eb *Howard A. S*

Modern Biology Series   Consulting Editors

James D. Ebert
Carnegie Institution of Washington

Howard A. Schneiderman
University of California, Irvine

Published Titles

Delevoryas	**Plant Diversification**
Ebert-Sussex	**Interacting Systems in Development,** *second edition*
Ehrenfeld	**Biological Conservation**
Fingerman	**Animal Diversity,** *second edition*
Griffin-Novick	**Animal Structure and Function,** *second edition*
Levine	**Genetics,** *second edition*
Novikoff-Holtzman	**Cells and Organelles,** *second edition*
Odum	**Ecology,** *second edition*
Ray	**The Living Plant,** *second edition*
Savage	**Evolution,** *second edition*
Sistrom	**Microbial Life,** *second edition*
Van der Kloot	**Behavior**

Cells and Organelles

second edition

Alex B. Novikoff

Albert Einstein College of Medicine
Yeshiva University

Eric Holtzman

Columbia University

Holt, Rinehart and Winston
New York   Chicago   San Francisco
Atlanta   Dallas   Montreal
Toronto   London   Sydney

To Professor Arthur W. Pollister
*from whom we both acquired much of our enthu-
siasm for cells. Between us we spanned 37 years of
his teaching histology and cytology at Columbia.*

Library of Congress Cataloging in Publication Data

*Novikoff, Alex Benjamin, 1913-
    Cells and organelles.*

> *(Modern biology series)
> Includes bibliographies and index.
>   1. Cytology.   I. Holtzman, Eric, 1939-
> joint author.   II. Title. [DNLM: 1. Cells.
> 2. Organoids. QH581 N943c]*
> QH581.2.N67   1976   574.8'7   75-40099

0-03-089721-1

Printed in the United States of America
89 006 9 8 7 6 5 4 3

Preface to the first edition

Cytologists study cells, all cells, and by many techniques. Because of their dependence on the microscope, they tend to analyze living systems in terms of visible structure, but in recent years cytologists have become increasingly concerned with biochemistry. *Cell biologists* start with a bias toward viewing cells in terms of molecules. Recently they have been much concerned with nucleic acids and proteins, the macromolecules which molecular biology has revealed to play primary roles in heredity. However, as studies progress, it is becoming more and more difficult to make a sharp differentiation among the various approaches to the cell—structural (classical cytology), physiological (biochemistry and biophysics), and molecular (cell biology). As distinctions between cell biology and cytology have become blurred, and perhaps outmoded, this book is about cytology and cell biology.

The book is divided into five parts. The first part introduces the major features of cells and the methods by which they are currently studied. In the second, we consider each of the organelles in turn, presenting structural and functional information. Then, in the third part, we discuss the diversity of cell types constructed from the same organelles and macromolecules. The fourth part presents major mechanisms by which cells reproduce, develop, and evolve. The final part is a brief look at the progress and the future of cell study.

Many of the cells used to illustrate the principles we discuss are from higher animals. This is not only because the authors have had more direct experience with such cells. It also reflects the historical development of cytology; the cells of higher animals have been best analyzed from the viewpoint of correlated structure and function. However, increasing attention is being focused on protozoa, higher plants, and bacteria and related cells. Wherever possible we have referred to studies of such organisms.

We begin the book and each of the parts with introductions outlining the major themes of the chapters that follow and commenting upon some topics that do not fit conveniently into one chapter. In the chapters we combine descriptive and experimental information to provide key portions of the evidence on which contemporary concepts are based. The illustrations have been obtained from leading students of the cell. The figure legends and suggested reading lists will familiarize the reader with the names of some of those who have contributed to the progress of cytology and cell biology.

We hope that we convey some of the excitement felt today by students of the cell. In addition, we hope that not only past achievements, but important problems awaiting future solution become evident to the reader.

New York, N.Y.   *A. B. N.*
January, 1970   *E. H.*

Preface to the second edition

Little need be added to what we have written above, regarding either the sense of excitement among students of the cell, or our general approach to presenting the material.

Note should be made that the 1974 Nobel Prize for Physiology or Medicine was awarded to three cell biologists "for their discoveries concerning the structural and functional organization of the cell": Albert Claude, Christian de Duve, and George E. Palade. Their contributions to cell fractionation procedures and the techniques of electron microscopy permitted them, and others, to lay the basis for much of the cell biology of today.

In planning the second edition we gave serious consideration to the suggestion from a number of teachers and students that we include references for the statements we make in the book. We resisted the temptation chiefly for two reasons: We did not wish to overly enlarge the book's size, or diminish its readability. The book is being utilized for courses at quite a variety of levels, and reference lists suitable for all would be unwieldy. In addition, in our experience as teachers we find reference lists to go out of date very rapidly. Even to support an "old" point, one tends to refer to newly appearing review articles and research reports in which recently appreciated nuances are covered. We do intend our "Further Reading" lists to provide access to the background literature, and we have revised and updated them accordingly.

We take this occasion to thank most sincerely those who read and reviewed our book in manuscript form, and those who have troubled to write us their views since publication of the first edition.

New York, N.Y.   *A. B. N.*
February, 1976   *E. H.*

Contents

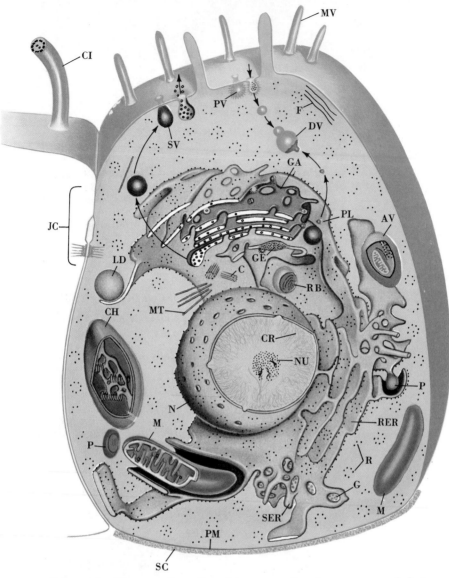

Schematic diagram of a cell and its organelles drawn to reveal their three-dimensional structure.

AV, autophagic vacuole; C, centriole; CH, chloroplast; CI, cilium; CR chromatin; DV, digestion vacuole; F, microfilaments; G, glycogen; GA, Golgi apparatus; GE, GERL; JC, junctional complex; LD, lipid droplet; M, mitochondrion; MT, microtubules; MV, microvillus; N, nucleus; NU, nucleolus; P, peroxisome; PL, primary lysosome; PM, plasma membrane; PV, pinocytic vesicle; R, ribosomes and polysomes; RB, residual body; RER, rough endoplasmic reticulum; SC, extracellular coat (as drawn, "basal lamina"); SER, smooth endoplasmic reticulum; SV, secretion vacuole

The organelles have been drawn only roughly to scale. Also, the sizes and relative amounts of different organelles can vary considerably from one cell type to another. For example, only plant cells show chloroplasts. A detailed enumeration of the organelle content of one cell type is presented in Chapter 2.12.

INTRODUCTION

The analogy between cells and atoms is a familiar one and like many familiar comparisons it is both useful and limited. Cells and atoms are units. Each is composed of simpler components which are integrated into a whole that exhibits special properties not found in any of the parts or in random mixtures of the parts. Both exhibit considerable variation in properties, based on different arrangements of components; the number of variations far exceeds the number of major components. Both serve as basic building blocks for more complex structures.

However, the analogy cannot be pressed too far; cells can reproduce themselves, whereas atoms cannot. The ability to utilize the non-living environment to make living matter is probably the most fundamental property of life, and cells are the simplest self-duplicating units. Duplication is based on DNA (deoxyribonucleic acid) which can be *replicated* to form perfect copies of itself. Thus the genetic information encoded in DNA is perpetuated from one cell generation to the next, sometimes without significant variation over vast periods of time. DNA is unique among macromolecules in its replication. Only in certain viruses has another macromolecule (RNA, ribonucleic acid) been shown to replace DNA in its central hereditary role; the replication of RNA in these viruses is based on the same principles as that of DNA.

Genetic information is expressed in cells by the mechanisms of *transcription* and *translation*. Transcription transfers the DNA-coded information to RNA molecules. Translation results in the formation of specific proteins whose properties are determined by the information carried by these RNA molecules. Among the proteins are *enzymes*, catalytic molecules that control most of the chemical reactions of cells. Enzymes differ in the kinds of molecules they affect (their *substrates*) and in the kinds of reactions they catalyze. Many are involved in the synthesis of the other cellular macromolecules, the nucleic acids (DNA and RNA), the lipids (fats and related compounds), and the polysaccharides (polymers made of many linked sugar molecules). Through this chain of transcription, translation, and enzymatic activities, DNA directs its own replication and controls as well the rest of *metabolism*, the sum total of all the chemical reactions that take place in cells. The chain is universal and thus all cells are made of the same classes of macromolecules (nucleic acids, proteins, lipids, polysaccharides) and smaller molecules such as water and salts. Duplication, and the presence in different cells of similar molecular and structural materials and mechanisms, are features of *cellular constancy*, one of the main themes of this book.

A second theme of the book is *cell diversity*. Cells may be classified into a large number of categories. Eucaryotic cells are distinguished from procaryotic cells, plant cells from animal cells, and muscle cells from gland cells. These distinctions derive from differences in morphology and metabolism. Eucaryotes differ from procaryotes in complexity of cellular organization. The unicellular protozoa, most algae, and the cells of multicellular plants and animals fall in the eucaryote category. In these cells, different specialized functions (such as respiration, photosynthesis, and DNA replication and transcription) are segregated into discrete cell regions which are often delimited from the rest of the cell by membranes. The cell's *organelles* reflect this segregation; they are subcellular structures of distinctive morphology and function. The most familiar of the organelles is the nucleus, which contains most of the DNA of the cell and enzymes involved in replication and transcription. The nucleus is separated by a surrounding membrane system from the rest of the cell, the cytoplasm. The cytoplasm contains many organelles including the *mitochondria*, the chief intracellular sites of respiratory enzymes; in plants, the cytoplasm contains *chloroplasts* in which are present the enzymes of *photosynthesis*, a metabolic process unique to plant cells. The mitochondria, chloroplasts, and a number of other cytoplasmic organelles are also delimited as discrete structures by surrounding membranes.

The procaryotes include the bacteria, the blue-green algae, and some other organisms. In contrast to the eucaryotes, they have relatively few membranes dividing the cell into separate compartments. This is not to say that all components are mixed together in a random fashion. The DNA, for example, does occupy a more or less separate nuclear region, but this is not delimited by a surrounding membrane. In fact, the traditional distinguishing feature of procaryotes is the absence of a membrane-enclosed nucleus (the suffix *caryote* refers to the nucleus). Respiratory and photosynthetic enzymes are not segregated into discrete mitochondria or chloroplasts, although, as will be seen, the enzymes are held in ordered arrangements within the cell.

The diversity of cell types owes its origin to evolution. By comparison with other macromolecules, DNA is remarkably stable. But *mutations* do occur at an appreciable, though low, frequency. Mutations alter the genetic information that is encoded in DNA and passed by a cell to its progeny; thus they can produce inherited changes in metabolism. Some result in a *selective advantage*, roughly defined as an increase in the number of viable offspring produced per lifetime by an organism. In the evolving population, organisms carrying such advantageous mutations will slowly replace organisms without them. The pattern of the spread of a mutation in a population depends on reproduction and thus, ultimately, on mechanisms of division of cells.

Usually the daughter cells resulting from division of *unicellular organisms* are essentially similar to the parent cells; the daughters contain replicates of the parent cells' DNA, and the DNA establishes the range of potential responses to the environment by specifying the available range of metabolic possibilities. Given a similar environment, there is little difference between parent and daughter. If the environment changes, parents and daughters will change in similar fashion and within genetically imposed limits. Diversity of cell types rests upon mutation.

In *multicellular organisms*, diversification of cell type without mutation is a regular feature of development. Most multicellular animals and plants start life as a single cell, a *zygote*, with a nucleus formed by the fusion of two parental nuclei. (Usually this results from fusion of sperm and egg or the equivalent.) The cell divides to produce daughter cells with identical DNA, but these *differentiate* into specialized cell types with different morphology and metabolism (for example, gland cells producing digestive enzymes or muscle cells rich in contractile proteins). Part of the mechanism for this is based on the fact that the immediate environment of a given cell is strongly influenced by the other cells of the organism. Cell interactions are of major importance in the development of multicellular organisms. They are key factors in determining what portion of its total genetic endowment will be expressed in a given cell. In different cell types, different portions of the DNA apparently are used in the transcription that underlies macromolecule synthesis. Presumably, for a given cell type only a particular part of the total genetic information is responsible for the cell's characteristics. Thus constancy in DNA coexists with diversity in metabolism and morphology of the cells which carry that DNA. Differentiation implies that cells are not mere aggregates of independent molecules or structures each "doing its own thing," and that DNA molecules are not autonomous rulers of subservient collections of other molecules.

As in development, the normal functioning of adult multicellular organisms also depends upon the interaction of neighboring cells and upon long-range cell-to-cell interactions mediated, for example, by hormones or nerve impulses. Cells are integrated into tissues, tissues into organs, and organs into an organism. Similarly, cells are themselves highly organized; molecules are built into structures in which they function in a coordinated and interrelated manner, and often show properties not found in a collection of the same molecules free in solution. The products of one organelle may be essential to the operation of another. Cell functions depend upon mutual interaction of parts. These interactions are elements of a network of mechanisms that regulate metabolism. *Cell organization* and the implications of organization for function are a third theme of the book.

Reproduction and constancy, evolutionary and developmental diversity, the intergration of cellular components into a functional whole—all are subjects for investigation in cytology and cell biology. The fourth theme of the book is the dependence of major biological findings upon the development of *new methods of study* and upon the *choice of the best organism* for the problem at hand. We will illustrate the kinds of experiments and approaches currently used in cytology and cell biology. The central tool is the microscope, but (as outlined in the Preface) microscopy is increasingly supplemented by chemical and physical studies. Descriptive and experimental approaches supplement each other. The great diversity of cells and organisms presents opportunity for choice of cell types especially well suited for analyses of new problems. Investigations of pathological material and of cells experimentally subjected to abnormal conditions provide valuable clues to normal functioning.

Study of the cell is progressing rapidly and the solution of many problems presently unsolved may be anticipated with confidence. Some of the unsolved problems are of practical importance. As our understanding of cells increases, so does our ability to control and modify them. This ability is crucial for medicine and agriculture. It also raises important ethical and social questions.

p a r t **1**

CYTOLOGY
OF
TODAY

About fifty years ago, the last edition of E. B. Wilson's great book, *The Cell in Development and Heredity*, was published. It was a summation and synthesis of a vast cytological literature. The work reflects the extraordinary ingenuity of early experimenters and the great excitement over what was then a recent appreciation of the roles of chromosomes in heredity. It is concerned mainly with eucaryotic cells. The nucleus had been extensively studied; some of the cytoplasmic organelles had been identified although clarification of their functions was only beginning. Biochemical analysis of cells was in its infancy.

Since that time, and especially in the last twenty-five years, there has been a remarkable development of techniques applied to the study of cells. The electron microscope has extended the investigation of structure down to the level of macromolecules. Biochemists have separated and analyzed cell molecules and organelles and determined their metabolic functions. Cytology and biochemistry have been combined to the extent that modern cytology is often referred to as *biochemical cytology*.

To illustrate current views of cell *organization*, we will begin with the rat *hepatocyte*, the major cell type of the liver. This cell type has many important functions, ranging from the secretion of blood proteins and the storage of carbohydrates to the destruction of toxic material produced elsewhere in the body. This variety of physiological functions is one reason that hepatocytes are widely studied by biochemists. Biochemical study is facilitated by the relative homogeneity of the organ; hepatocytes constitute over 60 percent of the cells and 90 percent of the weight of the liver in the rat. (The remainder consists of cells of blood vessels, ducts, and supporting tissues and specialized *phagocytes*, cells that engulf and remove from the blood a variety of materials such as some dead red blood cells.) Thus, constituents isolated from the liver come primarily from one cell type, the hepatocyte. Rats are readily available and they have large livers (almost 12 grams in an adult rat) from which relatively great quantities of cell constituents may be obtained. The hepatocytes are easily disrupted, to provide isolated organelles that can be studied by biochemical techniques. In addition, the liver is relatively easy to prepare for both light microscopy and electron microscopy.

To illustrate current views of *cellular metabolism*, we have chosen the metabolic pathway responsible for the formation of most of the ATP (*adenosine triphosphate*) of cells. ATP provides energy for virtually all cell functions: transport of substances into and out of the cell, chemical reactions within the cell, and integrated activities such as secretion, movement, and cell division.

c h a p t e r **1.1**

A PORTRAIT OF A CELL

The relations of hepatocytes to the architecture of the liver are shown schematically in Figure 1-1. Liver, like any other organ, must be nourished by molecules that enter the hepatocytes and other cells from the blood (for example, small molecules such as sugars, water, salts, and some macromolecules). The protein *albumin, lipoproteins* (complexes of lipid and protein), and other macromolecules are secreted by the hepatocyte into the blood, as are other materials including wastes (such as CO_2). Bile, which contains molecules formed in the breakdown of hemoglobin and of toxic substances, is excreted into small extracellular channels, the *bile canaliculi*, which lead to the bile duct; the latter in turn empties into the intestine where the bile facilitates fat digestion.

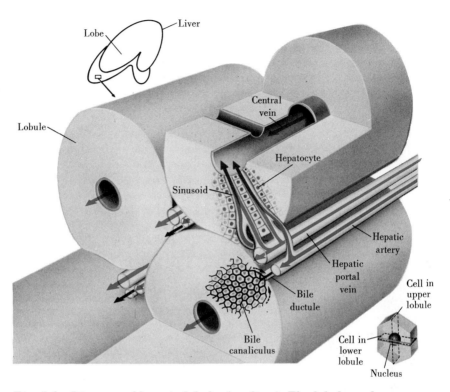

Fig. I-1 *Diagram of hepatic lobules (rat liver). The lobule at the upper right illustrates the relations of the liver cells (hepatocytes) to the blood; that at the lower right illustrates hepatocyte relations with the channels (bile canaliculi) into which the cells secrete bile. The two lobules diagram different views of the same cells as indicated at the lower right. Nutrient-rich blood, carried from the intestine by the hepatic portal vein, and oxygen-rich blood from the hepatic artery enter the sinusoids within each lobule. After exchanges have occurred with the hepatocytes arranged along the sinusoids, the blood enters the central veins of the lobules and is carried out of the liver.*

The flow of bile past the hepatocytes is in the opposite direction from the flow of blood. The hepatocytes are aligned along the blood and bile spaces, and the arrangement of organelles within the cell reflects the polarization of functions; large areas of the cell surface are involved in the exchange with the blood and other smaller areas in the secretion of bile. The key organelles are presented by a series of diagrams in Figures I-2 through I-5. The diagrams are based upon light microscopy (Fig. I-6), electron microscopy (Fig. I-7), and biochemistry. To convey the three-dimensional structure of the organelles, a hypothetical "generalized" cell is also included (as the frontispiece), preceding the

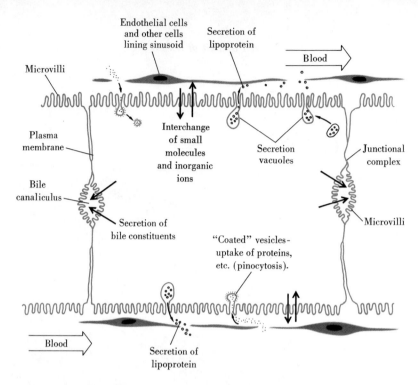

Fig. I-2 *Diagrammatic representation of a hepatocyte (from a rat) show-*
ing the plasma membrane *that surrounds the cell. This membrane and*
structures associated with it control entry and exit of material. Arrows
indicate interchanges of molecules between cells and blood. Mac-
romolecules entering the cell are shown as small dots, although generally
they are invisible by current techniques. The lipoprotein particles secreted
by the cell are seen as small spheres by electron microscopy. The junctional
complex, a specialized region of association between adjacent cells, restricts
movement of material between hepatocytes. For example, it prevents most
molecules from moving directly from the blood sinusoids into bile cana-
liculi. The complex also probably helps in holding adjacent cells together.

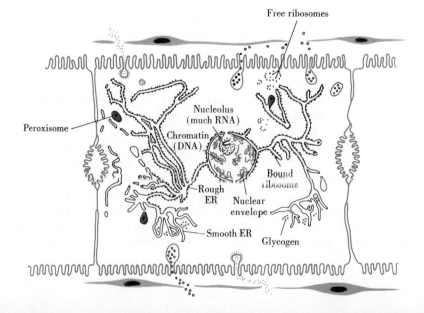

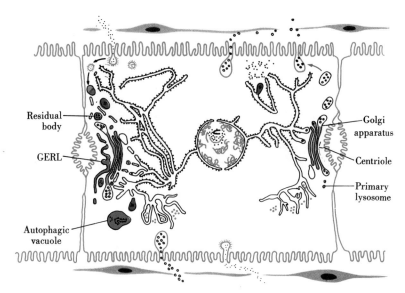

Residual body

GERL

Autophagic vacuole

Golgi apparatus

Centriole

Primary lysosome

Fig. I-4 *The Golgi apparatus, which has been added to the structures seen in Figure I-3, consists of flat sacs and vesicles of varying sizes. Its function includes concentrating of material produced in the ER and "packaging" of several classes of molecules into membrane-delimited vesicles. The large Golgi vesicles (often called vacuoles) are filled with secretion material such as lipoproteins (Fig. I-2). Among the small vesicles some probably are "primary" lysosomes involved in the transport of enzymes used in the intracellular digestion which takes place within lysosomes. Other lysosomes shown in the diagram are autophagic vacuoles and residual bodies.*

A pair of centrioles is found close to the Golgi apparatus; these organelles are of uncertain function in hepatocytes, but traditionally have been thought to be important in cell division. A specialized area of smooth ER, GERL, appears to participate in lysosome formation.

◀ **Fig. I-3** *The nucleus and other organelles have been added to the structures seen in Figure I-2. The chromatin of the nucleus contains most of the cell's DNA while the nucleolus is rich in RNA. The nuclear envelope is part of the endoplasmic reticulum (ER), a complex interconnected system of membrane-delimited channels (large sacs and smaller tubules) in which certain types of macromolecules are transported within the cell.*

Ribosomes (granules composed of RNA and protein), usually complexed with information-carrying mRNA (messenger-RNA) to form polysomes, are attached to part of the ER; other ("free") ribosomes are not attached to ER. ER with ribosomes attached is called rough ER. Proteins, synthesized on the polysomes of rough ER, enter the ER channels to be carried to other parts of the cell. The smooth ER lacks ribosomes; it is involved, for example, in the synthesis of steroids and some other lipids.

Stored carbohydrate, in the form of granules of glycogen, is found close to networks of smooth ER tubules. Peroxisomes are membrane-delimited structures containing enzymes that catalyze reactions many of which involve hydrogen peroxide; their cellular roles are incompletely understood.

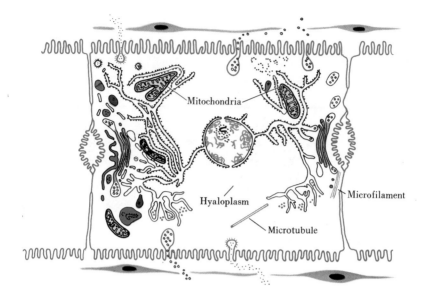

Fig. I-5   *The diagram now shows a few mitochondria (the average rat hepatocyte contains a thousand or more). The mitochondria are the chief sites of respiratory (oxygen-consuming) metabolism; they produce ATP (adenosine triphosphate, a key molecule in energy transfer and utilization) and many other important molecules. The* hyaloplasm *is the "background" cytoplasm which, by current microscopic techniques, appears structureless.* Microtubules *and* microfilaments *participate in maintaining the cell's shape and in intracellular motion.*

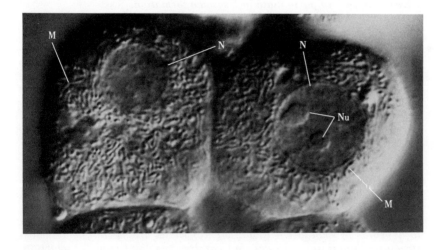

Introduction. Information on the number of different organelles is presented in Chapter 2.12. Two sets of definitions should be noted. *Intracellular* refers to things within the cell, *extracellular* to material outside the cell, and *intercellular* to components found between cells. *Vesicle* and *vacuole* are somewhat imprecise terms; both are used to designate a class of more or less spherical intracellular structures, each delimited by a membrane. A vesicle is a small vacuole. The definition of *hyaloplasm* will be considered in Chapter 1.2B.

The diagrams indicate the complexity of cell structure and function. Each of the different organelles has a distinctive morphology, biochemical composition, and function. Detailed discussions of each organelle are found later in this section and particularly in Part 2. We now turn our attention to the major techniques by which "portraits" of cells have become possible.

c h a p t e r **1.2**

METHODS OF BIOCHEMICAL CYTOLOGY

The microscopic study of cells is limited both by the microscope and by the manner of preparing the specimen for observation. In general, living cells and tissues are difficult to study directly with the ordinary light microscope. Multicellular tissues are usually too thick to permit penetration of light; single living cells are often transparent, with little visible internal detail. Thus one line of development of techniques for cell study is centered around improvements in microscopes and in methods for preparing and observing cells.

A second line of development is concerned with coordinating structural findings with biochemical information. Several methods have been devised which permit the direct use of the microscope in studying chemical and metabolic features of cells, and powerful, new nonmicroscopic techniques are being used in conjunction with them.

◄ *Fig. I-6   Two hepatocytes, still alive, isolated from the liver of a rat and viewed with a Nomarski differential interference microscope (Section 1.2.2). This microscope produces an image that conveys a strong three-dimensional impression. N indicates the cells' nuclei, Nu nucleoli, and M two of the many mitochondria visible in the cytoplasm. × 2,500. (From G. B. David, Improved Isolation Separation and Cytochemistry of Living Cells (Stuttgart: G. Fischer-Verlag.)*

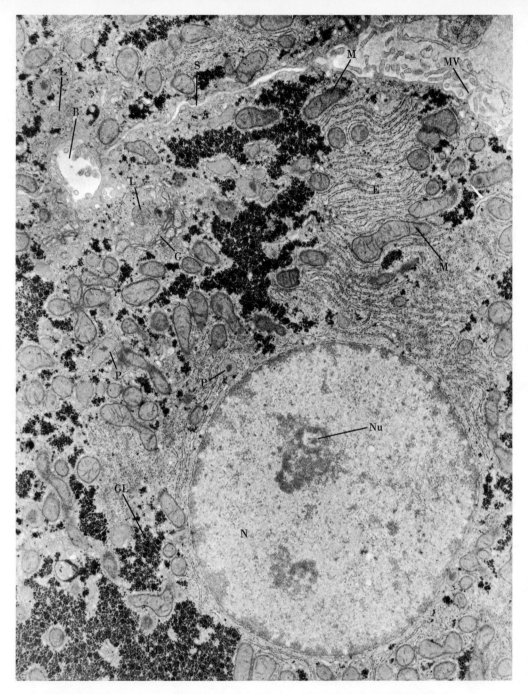

Fig. I-7 *Major organelles of a rat hepatocyte. This is a low-magnification electron micrograph that illustrates the overall appearance of the cell. B indicates a bile canaliculus and MV a microvillus at the sinusoid surface; many of the other microvilli near MV have been sectioned so as to appear unattached to the cell. The dense deposits at GL are masses of glycogen granules. M indicates mitochondria, N the nucleus, Nu a nucleolus, P peroxisomes, L lysosomes, G the Golgi apparatus, and S the intercellular space separating this cell from an adjacent hepatocyte. E indicates endoplasmic reticulum. × 25,000. (Courtesy of W.-Y. Shin.)*

c h a p t e r **1.2A**
MICROSCOPIC TECHNIQUES

During the latter part of the nineteenth and early part of the twentieth centuries, the light microscope approached the theoretical and practical limits of its performance as an optical instrument. Improvements in lenses and design resulted, by the beginning of this century, in instruments fundamentally similar to those in use today. During this period, the basic techniques of preparation of material for microscopy were also developed:

(1) *Fixing* of cells or tissue with agents that serve to kill and stabilize structure with lifelike appearance and prevent post-mortem disruptive changes known as *autolysis*.

(2) *Embedding* in hard materials that provide support of the tissue for *sectioning*, the preparation of thin slices. Sectioning tissue permits study of complex structures by providing an unimpeded view of deep layers.

(3) *Staining* of cells with dyes that color only certain organelles, thus providing contrast, for example, between nucleus and cytoplasm or between mitochondria and other cytoplasmic structures.

At first glance, it might seem improbable that tissue put through such elaborate procedures would bear any resemblance to living material. Certainly there is reason for caution in interpretation. However, several generations of investigators have provided increasingly trustworthy methods for tissue preparation. When comparison is possible, the results often compare remarkably well with direct observation on living cells and with information obtained by various indirect means. A variety of preparative methods are available for study of specific cell features.

In the present century, development of cytological techniques has proceeded chiefly in four directions: (1) invention of microscopes based on newly understood physical principles, notably the *electron* microscope which permits use of much higher magnifications; (2) development of new optical devices, such as the *phase contrast* microscope, and perfection of others, such as the *polarizing* microscope, that facilitate detailed study of living cells; (3) evolution of *cytochemical* methods for obtaining chemical information about microscopic preparations; and (4) development of techniques for the disruption of of organelles and other components for biochemical study.

1.2.1   ***ELECTRON***   The structures of interest to cytologists
          MICROSCOPY   range widely in size. In terms of the units
                          used for microscopic measurement (Fig. I-8),
the diameter of a mitochondrion may be about 0.5 μm, the length of
many bacteria roughly 1 μm, and the diameters of most mammalian
cells in the range of 5–50 μm. All microscopes are characterized by
limits of *resolution* which, in turn, determine the limits of useful magnification. These limits are inherent in the interaction of light with the
specimen and with the optical system. No matter how perfect the microscope, an image can never be a perfect representation of the object
(Fig. I-9). As a result, objects lying close to one another cannot be
distinguished (*resolved*) as separate objects if they lie closer than approximately one half the wavelength of the light being used. The
wavelengths of visible light determine color. Blue light has a wavelength
of 475 nm and red light of 650 nm; the average wavelength of white light
(for example, sunlight), which is a mixture of all colors, is about 550 nm.
Thus for light microscopes, the limit of resolution is about 0.25 μm. If
spaced apart by less than this limit, adjacent but separate objects appear in the light microscope as one object. This effect imposes an
absolute limit on the details that can be distinguished with the light
microscope, irrespective of magnification. *Magnification* refers to the
apparent size of objects, *resolution* refers to the clarity of the image.
Microscope images can be enlarged almost indefinitely by optical and
photographic means, but beyond a point the image appears increasingly
blurred and nothing is gained by further magnification. Thus light microscopes are rarely used at magnifications greater than 1000–1500 ×.

The development of the electron microscope as a practical instrument in the 1940s and 1950s made possible useful magnifications of
100,000 × or greater. Electrons may be regarded as an extremely

Unit	Symbol		Chief Use in Cytological Measurement
Centimeter	cm	= 0.4 inch	Macroscopic realm (naked eye) Giant egg cells
Millimeter	mm	= 0.1 cm	Macroscopic realm (naked eye) Very large cells
Micrometer (Micron)	μm (μ)	= 0.001 mm	Light microscopy Most cells and larger organelles
Nanometer (Millimicron)	nm (mμ)	— 0.001 μm	Electron microscopy Smaller organelles, largest macromolecules
Angstrom unit	Å	= 0.1 nm	Electron microscopy, X-ray methods Molecules and atoms

Fig. I-8   Units used in measuring the dimensions of cells and organelles.

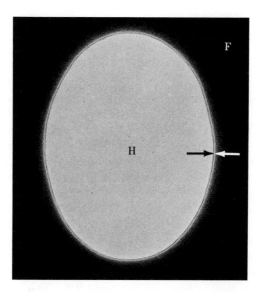

Fig. I-9   *An electron micrograph that illustrates one factor limiting resolution. A hole (H) in an electon-opaque film (F) was photographed under optical conditions that exaggerate the effect of diffraction, a phenomenon that occurs whenever light or electrons pass edges. The hole actually has a simple edge but the image of the edge shows a "diffraction fringe." The fringe consists of concentric bands, one dark and one bright (arrows). Under usual conditions of microscopy, the distorting effects of diffraction and other comparable optical phenomena are less severe but they cannot be totally eliminated.* × *120,000. (Courtesy of L. Biempica.)*

short-wavelength form of radiation. In practice, they can provide resolution of details separated by 1–5 Å, a thousandfold better than the light microscope. In the electron microscope, either magnetic or electrostatic lenses focus electrons in a manner analogous to the way glass lenses of ordinary microscopes focus light. The magnified image is viewed on a fluorescent screen like a TV screen and is also recorded on photographic plates. Most biological macromolecules have dimensions on the order of 10–100 Å. The diameter of a DNA molecule (not the length, which is very much greater) is 20 Å; many protein molecules have diameters of 50–100 Å. Thus macromolecules cannot be distinguished in the light microscope, but are well within the range of resolution of the electron microscope.

Fixation for light microscopy and for electron microscopy usually depends on the use of agents that precipitate macromolecules and keep them insoluble. *Formaldehyde* and related compounds such as *glutaraldehyde* are widely used. They react with protein molecules to produce linking of previously separate molecules and other changes promoting insolubility. *Osmium tetroxide* (osmic acid), commonly used in electron microscopy, probably also can link separate molecules; it appears to react with lipids and other components but its reactions are incompletely understood (see Section 2.1.2).

For light microscopy, fixed tissue usually is embedded in paraffin wax. The tissue is dehydrated and soaked in molten wax which penetrates the tissues and then is cooled to form a hard matrix. Embedded tissue is sectioned (on mechanical slicers called *microtomes*) with sharp steel knives to produce sections 2–10 μm or more in thickness. The sections are mounted on glass slides for study in the microscope. On occasion, tissue is not embedded, but is instead frozen solid and sectioned while frozen. Freezing makes tissue hard enough for cutting reasonably thin sections, but also tends to damage cells. The technique is used for quick preparation (during surgical operations, for example, when it is desirable to examine tissue samples for diagnostic purposes) and for other applications such as retention of certain cell lipids and enzymes. Several of the many staining procedures used for light microscopy are outlined in Sections 1.2.3 and 2.2.2.

For electron microscopy these techniques are inadequate. Electrons can pass only through very thin sections and are blocked entirely by glass. Thus to embed tissue, it is soaked in unpolymerized plastic which is hardened by the addition of appropriate catalysts that promote polymerization. The plastics provide support for cutting sections 500–1000 Å or less in thickness. To accomplish this, sharp glass or diamond edges are used, as steel does not readily attain or keep an edge that is sharp enough. In place of glass slides, sections are mounted on a screenlike metal mesh and observed through the empty spaces of the screen.

The stains used in electron microscopy are not the colored dyes of light microscopy. Rather, they contain heavy metal atoms, usually lead or uranium, in forms that combine more or less selectively with chemical groups characteristic of specific types of structures in the cell. Since the presence of such atoms permits fewer electrons to pass through the specimen and produce an image, "stained" structures appear darker than their surroundings. Conventionally, a structure that appears dark in the electron microscope is referred to as *electron-dense* or *electron-opaque*.

Sectioning imposes difficulties of interpretation. If the object being studied is, for example, a cylinder 1.0 μm long, ten to twenty sections 500–1000 Å thick can be cut in planes perpendicular to the cylinder axis. Each section cut from the cylinder will appear then as a flat disc (Fig. I-10). To ascertain the three-dimensional structure, it is best to cut *serial sections*, a series of sections made at the same angle. These can be used to reconstruct the shape of the object in three dimen-

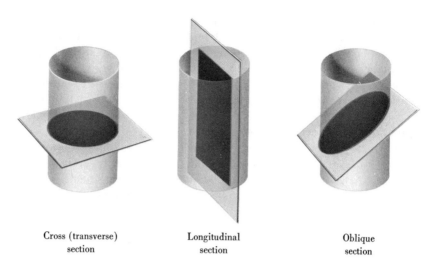

Cross (transverse)   Longitudinal   Oblique

section   section   section

Fig. I-10 *Sectioning a cylinder along different axes can produce the appearance of a circle, a rectangle, or an ellipse.*

sions. In our example, piling a series of discs one on top of another (mentally or by using cut photographs) would reproduce the original cylinder. However, serial sectioning for electron microscopy is difficult. Thus most microscopists take sections at random and compare the appearances at various angles to the axis (Fig. I-10).

Some specimens, particularly isolated particles of macromolecular dimensions, are suitable for viewing in the electron microscope without sectioning. Special techniques are used to make such materials stand out against the thin plastic film on which they are supported. Metals, such as platinum, may be evaporated onto them from an angle, which increases their electron density and can create a shadow-like effect that enhances the three-dimensional appearance. Structures in Figures II-43, III-15, and V-2 were prepared by such "shadowing" techniques. Another method, *negative staining*, is explained in Figure II-39. Sometimes the electron microscope images obtained with these procedures are electronically processed through image-intensifying devices which further increase the contrast between specimen and background, and thus can bring out fine details.

Innovations in techniques and instruments are continually being introduced in electron microscopy. Impressive results have been obtained with the "freeze fracture" method of specimen preparation. Fixation, embedding, and sectioning are avoided. Rather, the cells are quick-frozen and then broken open, exposing their interior structures; a fuller description is provided in Figure II-4. Recently, *scanning* electron microscopes have come into widespread use as a means for viewing surfaces of specimens too thick for conventional, "transmission" microscopes (see Fig. III-11). Scanning instruments of moderate resolving power are available (they can distinguish objects spaced 50-100 Å apart) and higher resolution instruments are being designed which may prove versatile for work at the macromolecular level. *High voltage* electron microscopes utilize voltages of 1000 to 3000 kilvolts (1 kv = 1000 volts) to accelerate electrons, rather than the 40–100 kv conventionally employed. This permits the electrons to penetrate much thicker sections, up to several micrometers; aspects of organelle geometry and interrelations may be made evident that are difficult to see in thinner sections. Already, a better appreciation of the three dimensional arrangement of microtubules, microfilaments and hyaloplasm has come from studying unsectioned, thin cultured cells. Some investigators hope that such microscopes will open the way for electron microscope observation of living cells, through use of special fluid-containing chambers.

1.2.2 **STUDY OF LIV-**   For the study of living cells, early light mi-
 ING CELLS   croscopists relied on a few large cell types
(egg cells, some plant cells, and protozoa) and developed microsurgical techniques to introduce material into single living cells and to alter living cell structures. Aside from these efforts, cytology has been based predominantly on the study of fixed, sectioned, and stained material. The constant improvement of methods has enabled cytologists to obtain increasingly reliable information from fixed preparations, although it remains true that the preparation of cells for microscopy may introduce *artifacts* (distorted and misleading appearances).

Living cells are dynamic entities, continually moving and changing; fixed and stained cells are stable and unchanging. Images from fixed material are at best only "snapshots" of dynamic cells. Often a sequence of events must be reconstructed by careful study of a large number of such "snapshots" taken at intervals.

Fortunately, ways are now available to the light microscopist for examining cells in action. The behavior and speed of light as it passes through a portion of a cell depend on the concentration, nature, and organization of the cellular constituents. For example, cell regions differ in *refractive index*, which can be thought of as a measure of the speed of light through matter. The refractive index depends in large part on the density of the region through which the light passes, that is, on the concentration of matter per unit volume. The higher the density, the higher the refractive index, and the slower the rate at which light passes. Thus light traveling through the nucleus will usually be affected differently from light traveling through the cytoplasm because of differences in the concentration of material. Differences in light paths due to factors of this type are not easily detectable in the ordinary microscope. The *phase contrast* and *interference* microscopes are light microscopes modified to translate refractive index differences into visible contrast, either in brightness or in color, by taking advantage of the phenomenon of interference. It is a well-known physical principle that two identical light waves can, for example be made to cancel each other, causing darkness rather than brightness when they arrive together at a given point. This may occur when one wave passes through a medium of refractive index different from that encountered by the other wave. By using such effects to manipulate the light passing through the cell rather than manipulating the cell itself, an "optical staining" is achieved so that various organelles now stand out in sharp contrast to their surroundings (Figs. I-6 and I-11). In addition, because of the influence of concentration on the optical properties of cells, the phase contrast and

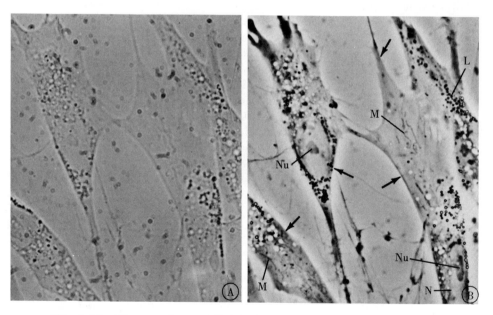

Fig. I-11   *The same living cells (chick fibroblasts grown in culture) photographed A, through an ordinary light microscope and B, through a phase-contrast microscope. The phase-contrast image clearly shows cell borders (arrows), nuclei (N), nucleoli (Nu), the elongate mitochondria (M), spherical lysosomes (L), and other details that are barely detectable in the ordinary microscope. × 800.*

interference microscopes can also be used to estimate the amounts of material present in different cells or cell regions; an example of such quantitative use is outlined in Section 2.11.2.

The *polarizing* microscope can detect regions in cells where constituents are disposed in highly ordered array. A *polarizer* built into the microscope produces *polarized light* which is passed through the specimen. Ordinary light can be pictured as a beam in which wavelike vibrations are occurring in all possible directions perpendicular to the direction in which the beam is proceeding. In polarized light the vibrations are confined to a single set of directions (for example, left and right, or up and down). Such light represents a sort of oriented field of electromagnetic energy that can interact in a distinctive fashion with specimen areas in which macromolecules or subunits are also arranged in a regular and oriented manner (for example, like a stack of coins or in parallel rows). Built into the microscope is an *analyzer*, an optical device that translates the behavior of polarized light that has passed through a specimen into visible contrast in the image. Ordered areas can be made to appear darker or brighter than areas of more random arrangement (Fig. IV-20); such areas are referred to as *birefringent*. The important point is that *order* can be detected in living cells, even when the oriented array is not visible by ordinary light microscopy.

The *ultraviolet* microscope makes use of the fact that certain substances in the cell, for example, nucleic acids, strongly absorb ultraviolet light (nucleic acids absorb light with a wavelength of 260 nm). Pictures taken in ultraviolet light show the locations of high concentrations of nucleic acids as regions darker than the rest of the cell.

In addition to improvements in optical techniques, the development of methods for growing cells in artificial conditions or *culture* has opened broad possibilities for studying living cells, both unicellular organisms and cells separated from tissues. An increasing variety of cell types can be maintained in artificial well-defined growth media. Chapter 3.11 will discuss some of the important results obtained by cell biologists with cultured cells.

Some activities of organelles in living cells, particularly in unicellular organisms and in cells in culture, can be vividly demonstrated by taking motion pictures through a phase-contrast, polarizing, or interference microscope. Such *microcinematography* also *records* these activities and facilitates their detailed study. Processes thus recorded may be speeded up photographically or may be viewed in slow motion. In Section 2.10.3 we refer to the study of the motion of cilia. Another striking example is the "degranulation" (merger of lysosomes and other granules with phagocytic vacuoles) of white blood cells when bacteria are engulfed by these cells (see Section 2.8.2).

1.2.3 CYTOCHEMISTRY The procedures of cytochemistry aim at producing a color or special contrast, such as enhanced brightness or darkness, at the sites of specific constituents within cells. Thus in the *Feulgen procedure*, sections are treated with dilute hydrochloric acid. The DNA, because of the unique sugar it contains (deoxyribose), is the only constituent changed in the right way (aldehyde groups form) to react with the *Schiff reagent*, a colorless form of pararosaniline dye. The modified DNA reacts with this reagent to restore the bright red color of the dye. This Feulgen reaction generally stains only the nuclei, where almost all cellular DNA is found. Figure I-12 shows a Feulgen-stained nucleus; the DNA is seen to be localized in the *chromatin*, of which chromosomes are composed. The small amount of DNA present in some cytoplasmic organelles usually does not show,

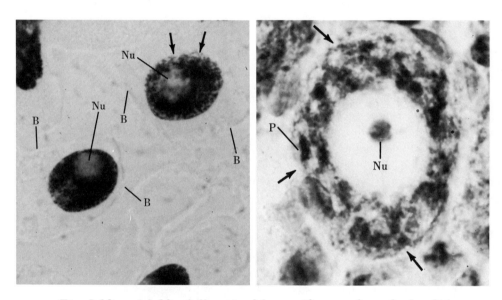

Fig. I-12 and I-13 *Cells stained by specific procedures for localizing nucleic acids. Figure I-12 shows two cells from a root tip of* Tulbaghia, *stained with the Feulgen method for DNA (see Section 1.2.3). The cytoplasm is unstained and therefore the cell borders (B) are barely visible. Most of the nuclear volume is occupied by intensely stained chromatin and the arrows indicate points at which the tangled, thread-like character of the chromatin can be seen. Nucleoli (Nu) are unstained.* × 1000. *Figure I-13 shows a neuron (nerve cell) from a rat ganglion stained with the dye pyronine, which stains basophilic structures such as those rich in RNA (see Section 2.2.2). The red color of the dye shows as black when green light is used for photography, as it was in this figure. In the nucleus, the nucleolus (Nu) stains intensely and in the cytoplasm, basophilic patches (P) are stained. Arrows indicate the cell's borders.* × 1600.

presumably because its concentration is below the limit needed to produce visible color.

If tissue sections are exposed to dyes that are soluble only in specific types of lipids, the cells will absorb the stain only in the areas that contain these lipids. Stains for RNA are also available. They show, for example, the localization of much RNA within a discrete intranuclear body, the *nucleolus* (Fig. I-13). Stain also appears in areas of the cytoplasm, which when viewed in the electron microscope, are seen to contain many small RNA-protein granules, the *ribosomes*.

Such staining techniques may give quantitative information. For instance, the amount of dye taken up may be directly proportional to the amount of stained component, and methods (*microspectrophotometry*) have been devised to measure through the microscope the amount of dye present at a given site within a cell. This quantitative approach using the Feulgen stain was useful in establishing that the amount of DNA in cell nuclei was characteristic of the given species and constant from cell to cell in a given organism, historical steps in establishing the role of DNA.

As mentioned in the Introduction, enzymes vary in the compounds they affect (their substrates) and in the nature of the reactions they catalyze. Under specific conditions the actions of some enzymes can generate insoluble products visible in the microscope. This permits direct microscopic study of enzyme distributions in cells. For example, *phosphatases*, enzymes which split the phosphate group from specific substrates, are localized by incubating the section with the appropriate substrate in the presence of a metal ion such as lead. The lead ions combine with the phosphates liberated at the sites of the enzyme. Lead phosphate is insoluble and precipitates as it forms. The resulting deposits at the enzyme sites are visible by electron microscopy. For light microscopy, the lead phosphate is converted to lead sulfide which has a deep brown color. In Figures I-14 through I-17 the plasma membrane, endoplasmic reticulum, Golgi apparatus, and lysosomes are shown to contain specific phosphatases which split different substrates. Figures I-18 and I-19 show two other organelles, mitochondria and peroxisomes, stained through effects of enzymes within the organelles. Thus far very few of the many enzymes known to biochemists have proved suitable for cytochemistry—these survive fixation and other steps in tissue preparation and are capable of yielding insoluble visible reaction products. Nevertheless, enzyme stains are useful in selectively staining organelles, as Figures I-14 through I-19 demonstrate.

Another staining approach, *immunohistochemistry*, utilizes the ability of certain macromolecules to bind to other molecules. For example, as we will discuss in Section 3.6.4, when higher vertebrate organisms are exposed to microorganisms, to proteins from other species, or to other foreign molecules, they synthesize antibodies, proteins that bind selectively to the foreign materials that evoked their synthesis. Through simple chemical reactions, antibodies can be coupled with fluorescent dyes for light microscopy, or with tracer molecules detectable in the electron microscope. Thus the locations of some molecules in cells and tissues have been determined by obtaining labeled antibodies specific for these molecules and using them as stains (Figs. II-26B and II-70).

Tissues are sometimes studied by microscopes based not on differences in refractive indices of materials but on other properties. Acoustic microscopes detect differences in "elasticity"—various materials reflect sound waves to varying degrees. X ray microscopes detect concentrations of dense materials.

1.2.4   *AUTORADIOG-* A valuable technique for tracing events in
RAPHY cells is *autoradiography*. Small *precursor* (metabolic forerunner) molecules that the cell uses as building blocks in the synthesis of macromolecules are made radioactive by incorporation of radioactive isotopes. The most widely used isotope is tritium (^{3}H), a radioactive form of hydrogen. Carbon^{14}, phosphorus^{32}, and sulfur^{35} are also employed. Probably the most frequently used radioactive precursor is *tritiated thymidine* (^{3}H-thymidine); this is thymidine (used by the cell to synthesize DNA) made radioactive by the substitution of tritium atoms for some of its hydrogen atoms. Typically, radioactive precursors are presented to cells which are then fixed at successive intervals and subsequently sectioned. The sections are coated with photographic emulsion similar to ordinary camera film (Fig. I-20). Exposure of camera film to light followed by photographic development leads to reduction of exposed silver salts in the film and to production of metallic silver grains. These grains form the image in the negative. Similarly, exposure to radioactivity and subsequent development produces grains in the autoradiographic emulsion. By examining the sections microscopically, both the underlying structure and the small grains in the emulsion are seen (Figs. I-21 and II-32). For example, if radioactive precursors of RNA are given to cells that are fixed only a few minutes later, almost all the grains are seen over the nuclei, suggesting that it is here that RNA is made from precursor molecules. If the interval between the exposure to radioactive precursors and the fixation is lengthened to an hour or two, fewer and fewer grains will appear over the nuclei and more and more over the cyto-

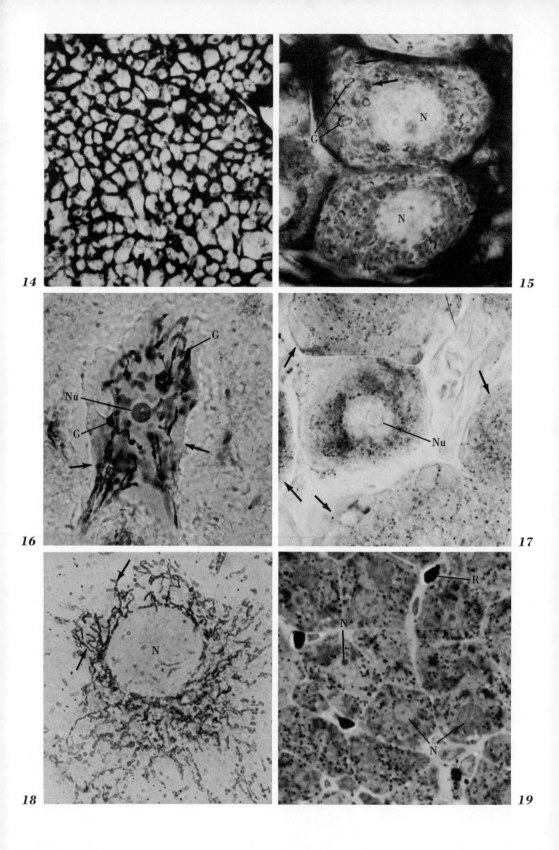

Fig. I-14 to I-19 *are light micrographs of preparations incubated to reveal sites of six different enzymes. In each case, "staining" is due to the deposition of a reaction product resulting from the action of the enzyme on a specific substrate. In Figures I-14 to I-17 the enzymes are* phosphatases *hydrolyzing different phosphate-containing substrates.*

Fig. I-14 *Staining due to splitting of ATP by a* plasma membrane *enzyme (or enzymes) of cells in a liver tumor of a rat. Each light polygonal area is a cell outlined by the dark reaction product on its surface.* × *500.*

Fig. I-15 *Staining due to splitting of the* diphosphate of inosine *(IDP) by enzymes at two sites in rat neurons (nerve cells). One site is the* Golgi *apparatus (G; see Fig. I-16). The other is the membrane systems of the* endoplasmic reticulum. *This results in staining of patches (arrows) in the cytoplasm comparable to the patches in Figure I-13 where it is the RNA of the ribosomes bound to the reticulum that is stained. (N) indicates nuclei.* × *1000.*

Fig. I-16 *Staining of the* Golgi *apparatus in a rat neuron due to the activity of an enzyme that splits* thiamine pyrophosphate *(TPP) and therefore is called* thiamine pyrophosphatase *(TPPase). The Golgi apparatus is the network indicated by G. Cell borders are seen at the arrows.* Nu *indicates a nucleolus.* × *880.*

Fig. I-17 *Staining of the* lysosomes *of rat neurons due to the presence of the enzyme,* acid phosphatase. *Portions of several neurons are seen in this micrograph.* Nu *indicates the nucleolus within the nucleus of the one in the center of the field and the arrows indicate the edges of several others. The lysosomes are the dark granules scattered in the cytoplasm of the neurons.* × *800.*

Fig. I-18 *Staining of the many* mitochondria *(two are indicated by arrows) in a mouse connective tissue cell grown in tissue culture. The enzyme responsible for staining transfers hydrogens of NADH (p. 34) to a soluble "tetrazolium" dye and thus converts the tetrazolium to an insoluble blue "formazan." The nucleus (N) is unstained.* × *1000. Compare this figure with the photograph of mitochondria in a living cell in Figure I-11.*

Fig. I-19 *Staining of* peroxisomes *in rat hepatocytes due to a reaction in which a soluble colorless compound,* diaminobenzidine, *is converted to an insoluble brown reaction product. The reaction is dependent on the enzyme* catalase, *within peroxisomes which are the dark cytoplasmic granules surrounding the unstained nuclei (N). Hemoglobin can also catalyze the reaction with the result that the red blood cells (R) in the hepatic circulation also stain.* × *800.*

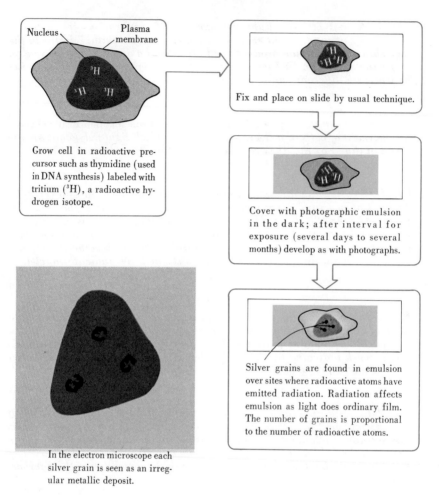

Nucleus

Plasma membrane

³H

³H   ³H

Grow cell in radioactive precursor such as thymidine (used in DNA synthesis) labeled with tritium (³H), a radioactive hydrogen isotope.

Fix and place on slide by usual technique.

Cover with photographic emulsion in the dark; after interval for exposure (several days to several months) develop as with photographs.

Silver grains are found in emulsion over sites where radioactive atoms have emitted radiation. Radiation affects emulsion as light does ordinary film. The number of grains is proportional to the number of radioactive atoms.

In the electron microscope each silver grain is seen as an irregular metallic deposit.

Fig. I-20   *Procedures of autoradiography.*

plasm (Figs. I-21 and I-22). These observations suggest that once RNA has been made in the nucleus, it moves out into the cytoplasm. Such use of "snapshots" taken at successive time intervals illustrates one of the most commonly used methods of studying dynamic biochemical events in cells.

The grains produced in the emulsion are similar, irrespective of the precursor used, that is, radioactive RNA, protein, or DNA all "look" alike. Interpretation of autoradiograms depends on the knowledge of biochemical pathways; precursors are carefully chosen so that they are used by the cell to build only one kind of molecule. Further, "control"

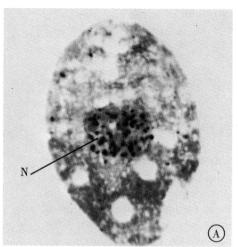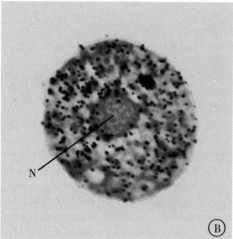

Fig. I-21 *Light microscope autoradiograms of* Tetrahymena, *a ciliated protozoan. The cells in* **A** *and* **B** *were both grown for approximately 10 minutes in tritiated cytidine (³H-C), a radioactive precursor of RNA. The cell in* **A** *was fixed almost immediately thereafter. The cell in* **B** *was allowed to grow in nonradioactive solutions for 90 minutes after exposure to ³H-C. In* **A** *almost all grains are seen over the nucleus (N) whereas in* **B** *most are over the cytoplasm. Experiments of this type are known as* pulse-chase *experiments. A component is labeled during a brief exposure to radioactive precursor (a pulse), and the subsequent behavior of the radioactive molecules is followed in the chase, the period of growth in a nonradioactive medium. In the present case RNA molecules are labeled in the nucleus during the pulse. During the chase they migrate to the cytoplasm and are replaced in the nucleus by newly made (unlabeled) molecules. Approx. ×1000. (Courtesy of D. Prescott, G. Stone, and I. Cameron.)*

preparations are employed to verify the interpretation. For example, after exposure to RNA precursors and incorporation of radioactivity, tissue sections can be treated with a solution of purified RNase (ribonuclease). This enzyme specifically breaks down RNA and removes it from the tissues. If such treatment removes the radioactivity from the mate-

	Percentage of grains over		
	Nucleolus	Rest of nucleus	Cytoplasm
30 minutes	54	41	5
30 minutes + 4 hours	14	19	67

Fig. I-22 *Mouse cells (L-strain fibroblasts) grown in culture were exposed for 30 minutes to radioactive RNA precursor (³H-cytidine) then either fixed immediately or grown for an additional 4 hours in nonradioactive medium before fixation and autoradiographic study. "Rest of nucleus" refers to the nonnucleolar region of the nucleus. (From the work of R. P. Perry.)*

rial being studied (so that no grains are produced in the emulsion), it is reasonable to conclude that the precursor had been incorporated into RNA.

c h a p t e r **1.2B**
CELL FRACTIONATION

Cell fractionation is an important and versatile technique for studying cell chemistry. The different subcellular structures are separated by centrifugation of homogenates. The latter are prepared by disrupting cells of liver or other organs in media that preserve the organelles. Solutions of sucrose or other sugars are generally used because they can be readily adjusted to maintain the integrity of organelles and to counteract the tendency of organelles freed from the cell to clump together. Some organelles, such as mitochondria and plastids, remain essentially intact. On the other hand, the endoplasmic reticulum, which in the living cell is an interconnected network of membrane-bounded cavities, is broken into separate pieces that form rounded vesicles (Fig. I-23). Similarly, the plasma membrane is fragmented into pieces of variable sizes in most cells, but methods are available for obtaining plasma membranes largely intact from some cells.

The behavior of a particle in a centrifugal field depends chiefly upon its weight and upon the resistance it encounters in moving through the suspension medium. Thus size, density, and shape influence the movement of a particle in a centrifugal field. The most widely used method for separating cell organelles is called *differential centrifugation*. It is based on differences in the speed with which structures sediment to the bottom of a centrifuge tube. At a given centrifugal force, structures that are relatively large, dense, and heavy sediment most rapidly. Thus using comparatively low forces (slow speed of centrifuge), nuclei, large fragments of plasma membrane, and some mitochondria may be sedimented as a pellet (Fig. I-23), while the remaining cell structures remain in suspension. With somewhat higher centrifugal

Fig. I-23 *Cell fractionation. Cells can be disrupted (homogenized) by a variety of procedures. The one illustrated involves a homogenizer employing movement of a close-fitting glass or plastic pestle within a tube. The space between pestle and tube wall can be varied to achieve breakage of cells and organelles with minimum damage. The fragments of ER seal off to form closed vesicles isolable as "microsomes."*

  Modern ultracentrifuges are capable of rotating at 50–100,000 revolutions per minute. This imposes forces in excess of 100,000 times the force of gravity and permits sedimentation of the smallest cell structures and of many macromolecules.

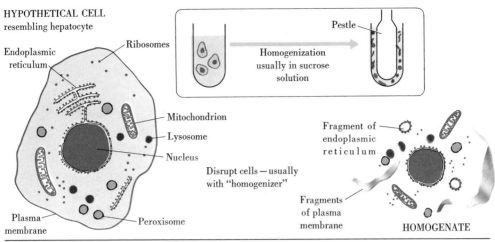

HYPOTHETICAL CELL
resembling hepatocyte

Endoplasmic reticulum

Ribosomes

Pestle

Homogenization usually in sucrose solution

Mitochondrion

Lysosome

Nucleus

Fragment of endoplasmic reticulum

Disrupt cells — usually with "homogenizer"

Fragments of plasma membrane

Plasma membrane

Peroxisome

HOMOGENATE

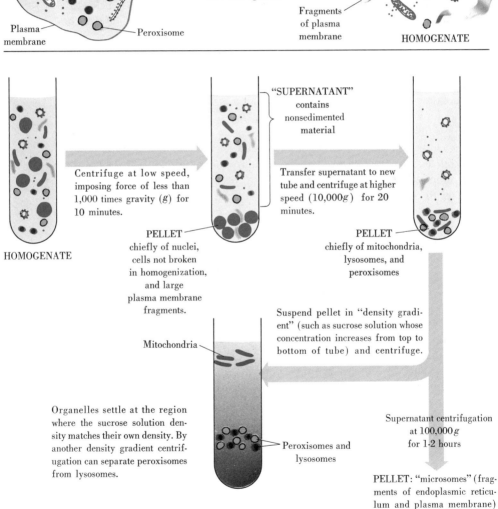

"SUPERNATANT" contains nonsedimented material

Centrifuge at low speed, imposing force of less than 1,000 times gravity (g) for 10 minutes.

Transfer supernatant to new tube and centrifuge at higher speed (10,000g) for 20 minutes.

HOMOGENATE

PELLET chiefly of nuclei, cells not broken in homogenization, and large plasma membrane fragments.

PELLET chiefly of mitochondria, lysosomes, and peroxisomes

Suspend pellet in "density gradient" (such as sucrose solution whose concentration increases from top to bottom of tube) and centrifuge.

Mitochondria

Organelles settle at the region where the sucrose solution density matches their own density. By another density gradient centrifugation can separate peroxisomes from lysosomes.

Peroxisomes and lysosomes

Supernatant centrifugation at 100,000g for 1-2 hours

PELLET: "microsomes" (fragments of endoplasmic reticulum and plasma membrane)

SUPERNATANT: ribosomes not bound to membranes and soluble molecules.

force, a second fraction containing most of the mitochondria is separated. Most lysosomes and peroxisomes are sedimented in the next fraction. The so-called microsomes, which consist mainly of fragmented endoplasmic reticulum but also of some small fragments of plasma membrane, come down with still greater force (note that there are no "microsomes" *as such* in the cell). Free ribosomes (those not bound to membranes) often remain unsedimented, together with other small structures and soluble molecules. The ribosomes can be brought down by prolonged use of high centrifugal forces.

The solution remaining after the ribosomes have been sedimented is referred to as the *soluble fraction* or *cytosol*. This fraction usually contains some material lost from the organelles in the course of homogenization and centrifugation. However, it also contains other molecules, including some enzymes, that are not known to be part of the intracellular structures identifiable in the microscope. Such molecules are usually thought to derive from the *hyaloplasm* (also called ground substance, cell matrix, or cell sap), which is defined by microscopists as the apparently structureless medium that occupies the space between the visible organelles. As will be evident in subsequent chapters, with the improvement of techniques less and less of the cell appears structureless. Thus the number of components thought of as truly part of the hyaloplasm is diminishing. However, at present it still remains appropriate to regard the hyaloplasm as a solution, or perhaps an almost fluid gel, containing material in transit from one cell region to another. Probably it also includes a number of dissolved or suspended enzymes that function as individual molecules or perhaps as parts of small groups of molecules rather than as components of large microscopically recognizable organelles.

Centrifugation through gradients of increasing concentration of sucrose or other substances can yield separations on the basis of density; the higher the concentration of dissolved sucrose, the greater the density of the solution. Organelles situate themselves at the levels in the tube where the densities of the solution match the densities of the organelles. Density gradient centrifugation (Fig. I-23) often enables separation of particles of roughly the same size but different density. Thus peroxisomes may be separated from lysosomes, or microsomes may be separated into those fractions in which the membranes are studded with ribosomes and those in which ribosomes are absent. Plasma membrane fragments also have been isolated in this manner. Modified procedures of gradient centrifugation can be used to achieve separations of macromolecules, such as RNA and DNA.

Isolated fractions are examined by electron microscopy to determine the integrity of the organelles and the purity of the fractions. This step is of importance since fractions thought to contain only one organelle may prove to be contaminated with others. For example, some lysosomes and peroxisomes may sediment with the mitochondria and some mitochondria sediment with the lysosomes and peroxisomes. Fractions that consist exclusively of one organelle are difficult to obtain, so that most work is done with fractions that are less than pure.

Fractions can be disrupted by chemical or physical means and subfractions can then be separated by additonal centrifugation. Thus fragments of rough endoplasmic reticulum (isolated as microsomes) can be treated with detergents such as deoxycholate that solubilize the membranes and free the ribosomes.

Isolated fractions can be subjected to the full range of biochemical analysis to learn the chemical composition, enzyme activities, and metabolic capabilities of subcellular structures. With the use of radioactive precursors and by isolating fractions at different time intervals, insights may be gained into temporal aspects of metabolic events. If cells are exposed, for instance, to radioactive amino acids, and fractions are isolated early after exposure, radioactive proteins are found only in the fractions that contain ribosomes, suggesting that ribosomes are the sites of protein synthesis. As the interval between exposure to radioactive amino acids and isolation of fractions is lengthened, other cell structures are found to contain radioactive protein. Thus it is possible to trace the transport of the protein, initially synthesized on the ribosomes, through its subsequent transport in the cell.

Biochemical techniques give information about the *average* properties of the material in a fraction. Often much is gained when they are supplemented by cytochemical and autoradiographic studies of intact cells or of isolated material; autoradiography can focus on single organelles and thus reveal the distribution of properties in the individuals underlying the average. Similarly, autoradiographic work is greatly strengthened by parallel biochemical investigations.

c h a p t e r **1.3**
A MAJOR METABOLIC PATHWAY

Figure I-24 outlines major metabolic sequences responsible for the breakdown of glucose. Full details are given in A. L. Lehninger, *Biochemistry* and in other references listed at the end of this part. Emphasis here will concern defining some major terms and summarizing certain features especially relevant to cell organization and useful for subsequent discussions.

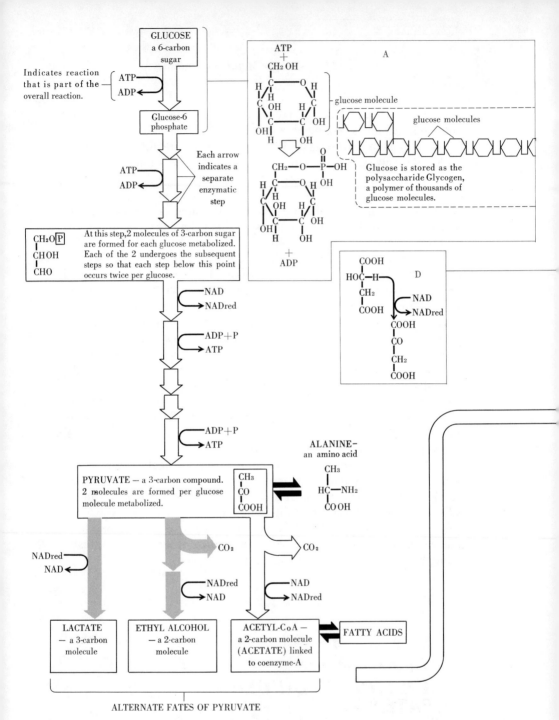

GLUCOSE
a 6-carbon sugar

Indicates reaction that is part of the overall reaction.

ATP
ADP

Glucose-6 phosphate

ATP
ADP

Each arrow indicates a separate enzymatic step

CH₂O P
|
CHOH
|
CHO

At this step, 2 molecules of 3-carbon sugar are formed for each glucose metabolized. Each of the 2 undergoes the subsequent steps so that each step below this point occurs twice per glucose.

NAD
NADred

ADP+P
ATP

ADP+P
ATP

A

ATP
+
CH₂OH

glucose molecule

glucose molecules

Glucose is stored as the polysaccharide Glycogen, a polymer of thousands of glucose molecules.

+
ADP

D

COOH
|
HOC—H
|
CH₂
|
COOH

NAD
NADred

COOH
|
CO
|
CH₂
|
COOH

PYRUVATE — a 3-carbon compound. 2 molecules are formed per glucose molecule metabolized.

CH₃
|
CO
|
COOH

ALANINE– an amino acid

CH₃
|
HC—NH₂
|
COOH

CO₂   CO₂

NADred
NAD

NADred
NAD

NAD
NADred

LACTATE
— a 3-carbon molecule

ETHYL ALCOHOL
— a 2-carbon molecule

ACETYL·CoA —
a 2-carbon molecule (ACETATE) linked to coenzyme-A

FATTY ACIDS

ALTERNATE FATES OF PYRUVATE

Fig I.24 *Some features of the reaction sequences in glucose metabolism. The chemical structures of several key molecules are indicated. A–D illustrate important biochemical reactions: A shows the phosphorylation of glucose; B, the condensation of two molecules; C, a decarboxylation (the removal of a carboxyl group (COOH) by its conversion to CO₂); and C and D, the transfer of hydrogen atoms to coenzymes (the dehydrogenation or oxidation of a substrate molecule coupled with the reduction of a coenzyme molecule). Also illustrated are some of the many pathways that intersect with the glucose pathway.*

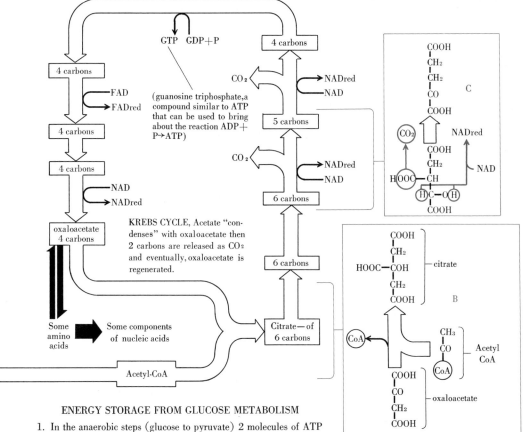

GTP   GDP+P

4 carbons

4 carbons

FAD
FADred

(guanosine triphosphate, a compound similar to ATP that can be used to bring about the reaction ADP+ P→ATP)

CO_2

4 carbons

4 carbons

NADred
NAD

5 carbons

CO_2

4 carbons

NADred
NAD

NAD
NADred

6 carbons

oxaloacetate
4 carbons

KREBS CYCLE, Acetate "condenses" with oxaloacetate then 2 carbons are released as CO_2 and eventually, oxaloacetate is regenerated.

6 carbons

Some amino acids

Some components of nucleic acids

Citrate— of 6 carbons

Acetyl-CoA

```
       COOH
       |
       CH2
       |
       CH2
       |                    C
       CO
       |
       COOH
```

CO_2 NADred

```
       COOH
       |
       CH2                   NAD
       |
HOOC— CH
       |
   HC—OH
       |
       COOH
```

```
       COOH
       |
       CH2
       |
HOOC—COH         citrate
       |
       CH2
       |
       COOH              B
```

```
       CH3
       |
       CO         Acetyl
       CoA         CoA
```

CoA

```
       COOH
       |
       CO
       |              oxaloacetate
       CH2
       |
       COOH
```

ENERGY STORAGE FROM GLUCOSE METABOLISM

1. In the anaerobic steps (glucose to pyruvate) 2 molecules of ATP are consumed and 4 ATPs are formed per glucose broken down.

2. In the Krebs cycle starting with pyruvate, the equivalent of one ATP (a GTP) is formed per pyruvate, or two per glucose. In addition, reduced coenzymes (NADred, FADred) are formed. Via electron transport and oxidative phosphorylation the reduced coenzymes can give rise to a total of 14 ATP's per pyruvate (28 per glucose).

3. Two NADred molecules per glucose are formed in the glucose to pyruvate steps. Their metabolic fate may differ somewhat from the Krebs cycle coenzymes but they represent the equivalent of 6 ATP's that could be formed by electron transport and oxidative phosphorylation.

4. TOTALS: Overall, 38 molecules of ATP formed per glucose.
    "Aerobic" steps: 30 ATPs from Krebs cycle and associated
                           pathways.
                      6 ATPs from electron transport based on
                           coenzymes reduced in glucose to
                           pyruvate steps.

   "Anaerobic" steps: 2 ATPs from glucose to pyruvate steps. (In
       anaerobic metabolism, the reduced coenzymes from the glu-
       cose to pyruvate steps can be used in formation of *ethyl
       alcohol* or *lactate* rather than entering electron transport).

Terminology: *NAD was once known as* diphosphopyriding nucleotide *and thus is often referred to as* DPN. *Sometimes, for convenience, the glucose-to-pyruvate steps are also referred to as* glycolysis *although strictly speaking, the term designates the path from glucose to lactate. "Aerobic" metabolism requires oxygen, "anaerobic" metabolism does not.*

c h a p t e r **1.3A**

METABOLIC BREAKDOWN OF GLUCOSE

1.3.1   **AEROBIC AND**
ANAEROBIC
PATHWAYS; SOME
DEFINITIONS

In hepatocytes, glucose (a 6-carbon sugar) is broken down first to two molecules of the 3-carbon compound, *pyruvic acid* (pyruvate). As seen in Figure I-24, this breakdown occurs in a series of steps, each step catalyzed by a specific enzyme. The product of each step serves as the substrate for the next enzyme; the products are referred to as *intermediates*.

The further metabolism of pyruvate to form carbon dioxide and water depends on a second series of reactions known as the *Krebs cycle*. Each pyruvate molecule is first converted to a 2-carbon *acetate* molecule. The acetate undergoes the reactions shown in Figure I-24. Each molecule is combined with a 4-carbon molecule (oxaloacetate) to produce a 6-carbon product. This 6-carbon molecule is broken down by a series of steps that ultimately regenerate an oxaloacetate molecule, which can then combine with another molecule of acetate and go through another turn of the cycle. At two steps in the series of reactions, CO_2 is released; these CO_2 molecules plus the one CO_2 released in the pyruvate-to-acetate conversion account for all three carbons of the pyruvate molecule.

In five of the Krebs cycle reactions, hydrogen atoms are transferred to *coenzymes*, either NAD (nicotinamide adenine dinucleotide) or FAD (flavin adenine dinucleotide). Coenzymes constitute a class of small nonprotein molecules that function, together with enzymes, in many metabolic reactions. In the reactions we are considering, the coenzymes accept hydrogen atoms enzymatically released from substrates and are thus *reduced*. From *reduced NAD* (often $NADH_2$ or $NADH^{\pm}$ is used as a shorthand designation) and *reduced FAD* ($FADH_2$), these hydrogen atoms are transferred eventually to oxygen.

The steps from glucose to pyruvate are *anaerobic*—they require no oxygen. Virtually all organisms are capable of degrading sugars by similar anaerobic pathways. In some microorganisms, the pyruvate molecules are further metabolized (anaerobically) to produce ethyl alcohol molecules; such production of alcohol from sugar is known as *fermentation*. Other cells instead convert pyruvate molecules anaerobically to lactic acid (lactate) molecules; the path from glucose to lactate is called *glycolysis*. Some bacteria, yeasts, and other microorganisms metabolize exclusively by fermentation, glycolysis, and other anaerobic pathways. Cells of most organisms have enzymes both for anaerobic pathways (glycolysis and others) and the Krebs cycle and related aerobic (oxygen-dependent) pathways. As Figure I-24 makes evident, the aerobic metabolism of glucose via the Krebs cycle makes use of the same glucose-to-pyruvate steps as glycolysis.

1.3.2   ***GLUCOSE***   Generally, metabolism of glucose or re-
METABOLISM   lated sugars and other carbohydrates is the
AND ATP   major source of cellular ATP. ATP is usu-
ally formed by the addition of a phosphate group to ADP (adenosine diphosphate). The chemical bond formed in this reaction is called a high-energy bond; its formation requires an input of energy derived from metabolism. When such a bond is broken, energy is made available. ATP is the major molecular form in which energy is stored and transferred to sites of use in cell functions.

Reduced coenzymes also are carriers of metabolic energy. In glucose metabolism, coenzymes are reduced in several reactions. These reduced coenzymes may be used in a special sequence of reactions that generate ATP. The ATP-generating reactions will be discussed in greater detail in Part 2. In brief they are as follows:

(1) The electrons from H atoms carried by the coenzymes enter a sequence of reactions known as *electron transport*. This involves a set of enzymes known as the *respiratory chain*.

(2) The terminal reaction of the series results in the transfer of the electrons plus the protons (H^+ ions) from hydrogen atoms to oxygen. This produces H_2O.

(3) Much ATP is generated in the course of electron transport. The process of formation of ATP is referred to as *oxidative phosphorylation*; it links oxygen consumption to the addition of phosphate to ADP.

Two conclusions from Figure I-24 should be noted. First, the aerobic pathways generate much more ATP than the anaerobic pathways; this reflects chiefly the production of reduced coenzymes in the Krebs cycle. In some anaerobic organisms mechanisms have evolved which supplement sugar-based ATP production with other anaerobic pathways that store energy. This facilitates survival of these organisms despite the absence of aerobic pathways. Second, the linking of glucose metabolism to ATP production by aerobic cells involves three more or less readily distinguishable sequences of reactions: the formation of pyruvate, the degradation of pyruvate molecules in the Krebs cycle, electron transport, and oxidative phosphorylation. The relations of this conceptual separation of glucose metabolism to actual cell organization will be discussed shortly.

c h a p t e r **1.3B**

METABOLIC INTERRELATIONS

1.3.3   ***CARBOHY-***   As with many other metabolic sequences,
DRATES, the glucose pathways intersect with other
PROTEINS, pathways (Figure I-24). For example, pyru-
AND FATS vate can be formed from the amino acid
alanine by a reaction that is reversible, so
that (if appropriate nitrogen sources are present) alanine can also be
formed from pyruvate. By such reactions, carbohydrates can be used to
synthesize amino acids which are combined to form proteins, or amino
acids can be metabolized via the Krebs cycle to generate ATP. Fatty
acids, components of fats, can be converted to acetate and degraded by
the Krebs cycle, or fatty acids can be built up from acetate units derived
from carbohydrates.

Reduced NAD is used in a variety of metabolic pathways as a
reducing agent (a source of hydrogens or electrons). ATP participates in
a very large number of reactions.

The controls that determine which sequence within the complex
network of possible pathways an individual molecule will follow are
largely unknown. The overall metabolic pattern responds to the availability in the cells of ATP, NAD, oxygen, various intermediates, and
such agents as hormones. For example, in very active muscle, much
glucose is rapidly broken down to pyruvate, but since the oxygen supply
is inadequate for immediate breakdown of all the pyruvate to CO_2 and
water, some pyruvate is converted anaerobically to lactate. This conversion is reversible, and the lactate is converted back to pyruvate
either in the muscle itself during periods of rest, or in other tissues
which the lactate reaches via the blood stream. More will be said later
about metabolic controls (see, for example, Sections 2.1.6 and 3.2.5).

1.3.4   ***METABOLIC***   If hepatocytes or other eucaryotic cells are
ORGANIZATION homogenized and cell fractions prepared by
AND INTERAC- centrifugation, the following distribution of
TIONS OF enzymes is observed.
ORGANELLES IN
EUCARYOTIC   (1) Enzymes of the glycolytic pathway
CELLS are present chiefly in the unsedimented
supernatant fluid.

(2) Krebs cycle, oxidative phosphorylation, and respiratory chain
enzymes are largely confined to the mitochondria.

(3) If the mitochondria are disrupted (for example, by exposure to *ultrasonic vibration*, very high frequency sound waves) most Krebs cycle enzymes are released as soluble material. Centrifugation of a preparation of disrupted mitochondria results in a pellet containing the membranous remnants of mitochondrial structure to which are attached electron transport and oxidative phosphorylation enzymes. Most Krebs cycle enzymes remain in the supernatant.

The generally accepted conclusion from these observations is that the glycolytic enzymes are not firmly bound to any of the visible cellu-

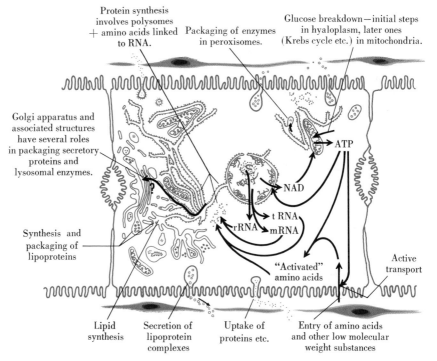

Fig. I-25   *A few examples of the probable interrelations of organelle functions in important metabolic sequences, the synthesis, transport, and packaging of proteins and lipoproteins. Protein synthesis depends upon several categories of RNAs (Fig. II-15) produced in the nucleus, on ATP synthesized by the mitochondria, and it utilizes amino acids that are transported through the plasma membrane. The proteins are made on polysomes (complexes of mRNA and ribosomes) and those molecules synthesized on polysomes bound to the ER enter the ER and are transported to other cell regions. Some of the proteins, made in the ER, complex with lipids that also are made in the ER (probably chiefly in the smooth regions). The lipoprotein complexes formed in this way accumulate in membrane-delimited secretory packages; structures associated with the Golgi apparatus may play a role in this process. The secretions are released by fusion of the membrane that surrounds them with the plasma membrane.*

lar organelles. Probably they are in the hyaloplasm, the supposedly structureless medium in which the organelles are suspended (see Chapter 1.2B). However, conceivably they are so loosely bound to intracellular structures that they are readily freed during homogenization. In contrast, the enzymes of the Krebs cycle, electron transport, and oxidative phosphorylation are located in the mitochondria. By centrifuging disrupted mitochondria, further compartmentalization is demonstrated; the electron transport and oxidative phosphorylation enzymes are firmly attached to the mitochondrial membranes, whereas most Krebs cycle enzymes are not.

Such observations form the basis of Figure I-25, a final diagram of the hepatocyte. In many metabolic pathways there is interaction among several different organelles. Even our discussion of the glucose pathway could easily be extended to include other organelles. For example, in hepatocytes the *synthesis* of NAD molecules (as distinct from the addition of H atoms to NAD) depends on an enzyme apparently localized in the nucleus. Figure I-25 illustrates only a tiny fraction of the known metabolic interrelations.

FURTHER READING

De Robertis, E.D.P, F.A. Saez, and E.M.F. De Robertis, *Cell Biology* 6th ed. Philadelphia: Saunders Co., 1975, 615 pp. A general text.

DuPraw, E.J. *Cell and Molecular Biology*. New York: Academic Press, 1968, 739 pp. A general text.

Dyson, R.D. *Cell Biology: A Molecular Approach*. Boston: Allyn and Bacon, 1974, 701 pp. A textbook discussing many aspects of cellular structure and biochemistry. Good reference lists.

Gray, P., Ed. *Encyclopedia of Microscopy and Microtechnique*. New York: Van Nostrand-Reinhold, 1973, 638 pp. A collection of useful concise discussions of major microscopic principles and techniques.

Hughes, A. *A History of Cytology*. New York: Abelard Schuman, 1959, 158 pp. A concise history of the field.

Lehninger, A. L. *Bioenergetics*, 2nd ed. New York: Benjamin, 1971. An introduction to key aspects of metabolism, especially as related to the formation and use of ATP.

Lehninger, A.L. *Biochemistry*, 2nd ed. New York: Worth Co., 1975, 1104 pp. A readable and comprehensive text.

Locke, M., Ed. *Ultrastructure of Cell and Organisms: A Series of Atlases with Text*. New York: Academic Press. A series of short monographs, issued at intervals, on topics of interest to cell biologists.

Loewy, A.G., and P. Siekevitz. *Cell Structure and Function*, 2nd ed. New York: Holt, Rinehart and Winston, 1969, 512 pp. A wide-ranging discussion of cells, stressing their biochemistry.

Porter, K.R., and A.B. Novikoff. "The Nobel Prize for Physiology or Medicine." *Science*, **186**:516-520, 1974. A concise summary of the main contributions of three Nobel laureates, Albert Claude, Christian de Duve, and George E. Palade.

Prescott, D.M., Ed. *Methods in Cell Biology*. New York: Academic Press, A series of volumes, issued at intervals since 1966, describing methods of cell study.

Tedeschi, H., *Cell Physiology: Molecular Dynamics*. New York: Academic Press, 1974, 724 pp. A textbook stressing cellular biochemistry and physiology. Good reference lists.

Weissmann, G., and R. Claiborne, Eds. *Cell Membranes: Biochemistry, Cell Biology and Pathology*. New York: Hospital Practice, 1975, 283 pp. A collection of articles on cell membranes and organelles that originally appeared in *Hospital Practice*.

Wilson, E.B. *The Cell in Development and Heredity*, 3rd ed. New York: Macmillan, 1925, 1232 pp. Outstanding summary of cytology of its period; problems discussed in a manner still of importance to students of cytology and cell biology.

Wolfe, S.L. *Biology of the Cell*. Belmont, Calif.: Wadsworth, 1972, 545 pp. A textbook of cell biology with particular emphasis on the nucleus and related topics. Good background reference lists.

Many articles of interest appear in the monthly, *Scientific American*.

McGraw-Hill Book Company publishes sets of "Biocore Units" comprising introductory articles by research specialists. Several of the articles concern cell biological topics.

The Bobbs-Merrill reprint series (Indianapolis: Howard W. Sams and Co.) includes a collection of reprints of research articles on cell structure and function (Biology, category C).

Many professional journals contain articles of interest to cytologists and cell biologists. These include, among many others, the *Journal of Cell Biology, Cell, Journal of Cell Science, Journal of Molecular Biology, Journal of Biological Chemistry, Chromosoma, Journal of Ultrastructure Research, Proceedings of the National Academy of Sciences of the United States, Subcellular Biochemistry*, and *Journal of Histochemistry and Cytochemistry*. In addition there are publications appearing annually or at other intervals that contain review articles summarizing recent progress in various fields. Among these are the *Annual Review of Biochemistry, Annual Review of Plant Physiology, Advances in Cell Biology, Advances in Cell and Molecular Biology, International Review of Cytology*, and the *Cell Biology* Monograph Series (formally known as *Protoplasmologica*) published by Springer-Verlag, Vienna.

p a r t **2**

CELL ORGANELLES

This part discusses major features of the organelles of eucaryotic cells; Part 3 will consider the procaryotes and will include additional information on eucaryote organelles as specific cell types are discussed. Almost all of Part 4 is devoted to further consideration of the nucleus, and Part 5 contains a summary of some recent information about nucleoli.

The membranes of eucaryotic cells divide the cell into compartments that are distinctive in both morphology and metabolism. Membranes surround the cell as a whole and the *membrane-bounded* or *membrane-delimited* organelles: the nucleus, mitochondria, chloroplasts, lysosomes, peroxisomes, and the cavities of the endoplasmic reticulum and Golgi apparatus. The membranes are not simple mechanical barriers. They are highly ordered arrays of molecules, chiefly lipid and proteins, in which enzymes are integrated; they participate in diverse activities. Membranes provide selective barriers that control the

40

amount and nature of substances that can pass between the cell and its environment and between intracellular compartments. Within several of the organelles, notably mitochondria and chloroplasts, the arrangement of enzymes on membranes provides remarkable efficiency of function, since the different enzymes involved in sequential reactions of metabolic pathways are held in close relationship with one another.

But must a structure be membrane-bounded to qualify as an organelle? As techniques improve, less and less of the cell appears unstructured, but only some of the organization involves membranes. Nucleoli, chromosomes, ribosomes, centrioles, and microtubules all are distinctively structured and have specialized roles in the cell, but no membrane surrounds them. The nonmembrane-bounded organelles grade down in size and complexity to protein filaments composed of a few hundred molecules. At the lower end of the size spectrum, the distinctions between organelle and macromolecule become difficult to define and perhaps meaningless. Should a multienzyme complex, in which a few or a few dozen enzymatically active protein molecules are complexed as a functional unit (Section 3.2.4), be called an organelle or a molecular aggregate? Are nucleoli to be considered organelles despite their being contained in other organelles (nuclei)? The decision appears to be a matter of arbitrary definition.

Organelles are not static units of unchanging size and shape, with fixed relations to other organelles and with inflexible function. All show movement within most cells. Some can grow and produce duplicates. Most exhibit evidence of *turnover*. For example, if the mitochondria in a hepatocyte are labeled by the incorporation of radioactive amino acids into their proteins, by five to eight days later, half of the label has been lost from the mitochondria. In number and mass, the population of mitochondria has remained the same, but despite this "steady state," half the organelles have "turned over." Does this mean that new mitochondria are continually arising as others are destroyed? Or does each mitochondrion add new material to its structure while simultaneously losing an equivalent amount of older constituents? Or do both processes take place? The answers are not yet known. The amount of a given component present in a cell at a given time depends on the rate of formation and on the rate of degradation. Both formation and degradation must be subject to precise cellular controls, and the relevant mechanisms are being sought. Turnover is a general phenomenon affecting virtually all components of an organism; the major exception is the DNA of the nucleus. In evolutionary terms turnover may partly reflect the fact that enzymes and other macromolecules are subject to spontaneous changes that permanently inactivate them—hence processes for their degradation and replacement seem advantageous for the cell's economy.

The information that will be reviewed in this part suggests that cytologists and cell biologists may, before long, be able to specify the particular molecular arrangements in each organelle and the detailed chemical reactions that occur within it. The analysis of several organelles has passed from the stage of mere identification of the structure and of the molecular constituents to the point where the normal modes of formation and duplication of the organelles are being studied in detail, and functional parts of organelles are being reconstituted from simpler components in the test tube.

c h a p t e r **2.1**
THE PLASMA MEMBRANE

A plasma membrane is present at the surface of all cells. Although some other closely associated structures play important roles, this membrane is the primary barrier that determines what can enter or leave the cell. Its properties strongly influence the formation of multicellular aggregates (such as tissues), as well as the passage of material between closely associated cells. Specialized plasma membrane regions are often present. In Figure I-2 the *microvilli* were mentioned; these are tubular projections that increase the surface area of hepatocytes and of many other cell types. These and other specializations of the plasma membrane will be discussed in Part 3. There we will also consider the extracellular *cell wall* surrounding plant cells. At this point our concern is with some general properties of plasma membrane structure and function.

2.1.1   ***MICROSCOPIC*** Light microscopy cannot reveal the pre-
STRUCTURE sence of the plasma membrane directly, since the thickness of the membrane is well below the resolving power of the light microscope. However, before the advent of electron microscopy, a great deal of indirect information about the membrane was accumulated from physiological experiments. Artificial systems were devised, in which two solutions containing different concentrations of given substances dissolved in water were separated by a membrane that resembled the plasma membrane in being semipermeable (permitting only some types of molecules to pass). If a given component can pass readily through the membrane, a net movement or *diffusion* of the component will take place from the solution of higher concentration to that of lower concentration ("down the concen-

tration gradient"), until the concentrations on both sides of the membrane are equal. If the membrane is impermeable or only slightly permeable to the dissolved components, but if, like the plasma membrane, it is moderately permeable to water, then water will pass rapidly into the more concentrated solution. This net movement of water in the direction that tends to equalize the concentrations of dissolved materials on each side of the membrane, is referred to as *osmosis*. In early experiments, cells were suspended in solutions containing different molecules dissolved in water, and the volume changes of cells were measured. The details of behavior of a cell in a given solution will depend on the total concentrations of all molecules dissolved inside the cell and in the surrounding solution as well as on the relative rate of passage through the membrane of dissolved molecules as compared with water; to a first approximation, at equilibrium, the total number of molecules of all dissolved material per unit volume inside and outside will be equal as will the concentration of each individual component that can diffuse through the membrane. Under some experimental conditions, extensive volume changes can be taken to indicate relative impermeability to the particular dissolved molecules and consequent influx or efflux of water; cells may shrink drastically or swell to bursting, depending on the concentration of the solutions. Studies of this type, and more sophisticated experiments in which concentrations of substances in cells are measured directly, have made it clear that a plasma membrane exists and exhibits great selectivity as to what can pass through. Gases move across the membrane with little difficulty. Water and other small molecules can pass through more readily than larger molecules with comparable chemical properties. Material that is soluble in lipids generally enters the cell more rapidly than nonlipid-soluble substances. Some substances pass through by passive diffusion, others only with the expenditure of energy by the cell.

This selectivity and other indirect observations suggest that the membrane is a highly organized and structurally complex entity. However, the electron microscope usually shows a relatively simple structure, and the basis of complex membrane functions is still not well understood. In conventional electron microscope preparations the plasma membrane appears to be a series of three layers with a total thickness of 75–100 Å (Fig. II-1). This three-layered structure has long been referred to as a *unit membrane* but this term has misleading connotations. Thus we will use the simple descriptive term, *three-layered membrane*. The microscope conveys little of the dynamic features of membrane structure—the movement of molecules within the membrane and of membranes within the cell.

It has become increasingly evident that the actual arrangement of molecules in many membranes is more complex than the three-layered

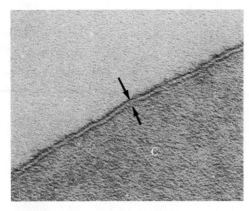

Fig. II-1 An electron micrograph of a portion of a human red blood cell (C). The arrows indicate the three-layered (dense–light–dense) structure of the plasma membrane. × 250,000. (Courtesy of J. D. Robertson.)

structure seems to imply. In the first flush of excitement among microscopists at actually seeing membranes, it was widely believed that most cellular membranes were quite similar to one another, and that extensive interconnections existed among different membrane-delimited cell compartments. Diagrams from the early days of electron microscopy often show the plasma membrane continuous with the endoplasmic reticulum (ER), and include other membrane continuities among organelles so that the cell appears permeated by membrane-bounded channels continuous with the extracellular space. As work has progressed it has become clear that the three-layered structure is an oversimplified representation for most membranes, and that while membranes of different organelles do have key features in common, they also differ markedly in composition and in important facets of their organization. In addition, on closer examination with improved techniques, some apparent continuities, most notably those between the endoplasmic reticulum (ER; Fig. I-3) and the plasma membrane, turned out to represent a kind of optical artifact. The images of adjacent separate objects sectioned in the proper plane can overlap so as to create an impression of continuity. True structural continuity between the ER and the plasma membrane is observed only under most unusual circumstances. Some of the other seeming connections among organelles are probably better interpreted in terms of dynamic movement of membranes from one to the other, than as static channels. Discussion of this problem will continue below.

2.1.2   **A MOLECULAR MODEL OF MEMBRANES; LIPIDS**   From a variety of correlated chemical and structural studies, a series of models thought to apply to the plasma membrane and to other cellular membranes has been proposed as conceptual frameworks for further investigation. Although it is evident that the models are over-simplified they have stimulated a great deal of experimentation and discussion.

Cellular membranes are rich in lipids. This has been concluded both from direct chemical analysis and from the permeability characteristics of the plasma membrane which provide indirect evidence for the presence of lipids. Because proteins are present as the other major membrane constituent, the membranes are referred to as *lipoprotein membranes*. Plasma membranes (and other membranes) of different cell types usually contain from one to four times as much protein as lipid, although a few cellular membranes contain more lipid than protein. This estimate is on a *weight* basis. Since lipid molecules are usually much smaller than protein molecules (lipids have molecular weights on the order of 1000 whereas proteins range from 10,000 to over 100,000) there are usually far more lipid than protein molecules in membranes. The ratio may approach 10 to 100 lipid molecules per protein molecule in some cases. In very rare cases, such as the membranes surrounding the gas vacuoles of certain procaryotes, membranous structures are found that seem to contain much protein but little lipid. These membranes have distinctive microscopic appearances, but are poorly understood.

Figure II-2 illustrates three of the major types of lipids found in nature: fats, phospholipids, and steroids. Fats consist of *fatty acids*, long hydrocarbon chains, linked to a "backbone" three carbons long. The backbone is formed from the compound *glycerol*. When all three glycerol carbons are attached to fatty acids, the resultant lipid is known as a *triglyceride*, the major component of most fats. (The term *triacylglycerides* is coming into use instead of triglycerides.) Mono- and diglycerides are constructed, as their names suggest, with one or two fatty acids linked to glycerol. Most phospholipids have a structure similar to triglycerides, except that in place of one of the fatty acids they have a more complex chain including phosphate and nitrogen-containing groups. Steroids have an architecture that is quite different; they are relatives of *cholesterol*, a molecule containing interconnected rings of carbon atoms. *Glycolipids* (lipids with sugar groups attached) also are found in cellular membranes.

The phospholipids and steroids are polar molecules; different ends of the molecules have different properties. One end is *hydrophobic*, that is, it tends to be insoluble in water. However, groups such as the charged phosphate and nitrogen groups at one end of a phospholipid molecule or the hydroxyl (OH) group found at one end of the cholesterol molecule, confer *hydrophilic* properties on the regions of the molecules in which they are present. The result is that these regions have a great affinity for water.

Plasma membranes and other cellular membranes are rich in polar lipids, and the models being discussed make use of this fact. The mod-

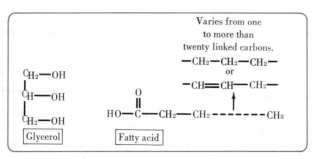

Varies from one
to more than
twenty linked carbons.

$-CH_2-CH_2-CH_2-$
or
$-CH=CH-CH_2-$

CH_2-OH
$CH-OH$
CH_2-OH
Glycerol

$HO-\overset{\overset{O}{\|}}{C}-CH_2-CH_2------CH_3$
Fatty acid

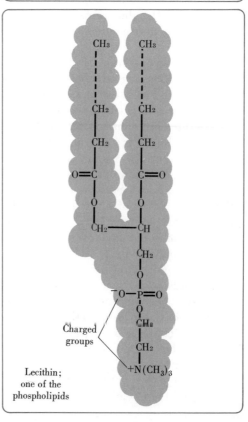

Glycerol
carbons

$CH_2-O-\overset{\overset{O}{\|}}{C}-CH_2-CH_2-CH_2---CH_3$

$CH-O-\overset{\overset{O}{\|}}{C}-CH_2-CH=CH----CH_3$

$CH_2-O-\overset{\overset{O}{\|}}{C}-CH_2-CH_2-CH_2---CH_3$

Fatty acid

Triglyceride

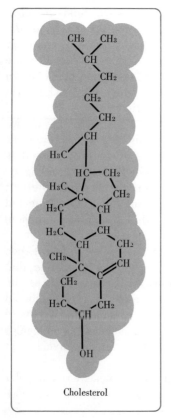

Cholesterol

Charged
groups

Lecithin;
one of the
phospholipids

◄ *Fig. II-2   Structures of lipid molecules. Common natural fatty acids have 16–20 carbons in the region indicated by the dotted lines. The nitrogen-containing compound attached to the phosphate on the third glycerol-derived carbon in lecithin is choline. The phospholipids of membranes are roughly 30–35 Å long; the steroids are somewhat shorter. (After J. B. Finean and others.)*

els suggest that the central region of membranes consists of two layers of lipid molecules, largely phospholipids and steroids (Fig. II-3). Each layer is one molecule thick. The layers are arranged with the hydrophobic ends of the molecules in one layer associated with the hydrophobic ends of the molecules in the other layer. As will be outlined later (Section 4.1.1), hydrophobic regions of molecules have considerable affinity for one another and, as a result, lipids will often associate spontaneously in ordered groups, including such layered arrays. The hydrophilic portions of the membrane lipids face outward toward the surfaces of the "bimolecular" layer of lipid. Proteins often contain many hydrophilic groups and associate readily with water. Thus the earlier versions of the models being discussed suggested that the membrane proteins were simply associated with the hydrophilic faces of the lipid layers (Fig. II-3A).

These views are attractive since they agree with the usual electron microscope image; presumably the light central layer of the three

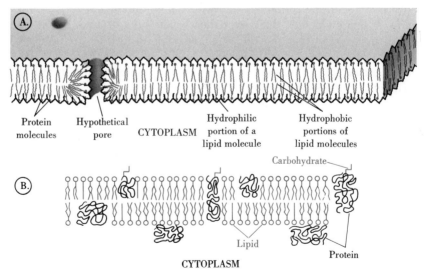

Protein molecules   Hypothetical pore   CYTOPLASM   Hydrophilic portion of a lipid molecule   Hydrophobic portions of lipid molecules

Carbohydrate

Lipid   Protein

CYTOPLASM

Fig. II-3   Models of the plasma membrane. (After H.A. Davson, J.F. Danielli, S. J. Singer, and others.)

would correspond to the hydrophobic portions of the lipids, while the two dense lines would represent the proteins and the hydrophilic regions of the lipids. Unfortunately, this correspondence by itself is only conjectural. Interpretation of the electron microscope image of a membrane in chemical terms is difficult. Electron density (Section 1.2.1) in most preparations depends largely on the location of osmium atoms from the osmium tetroxide used in fixation and on the locations of other heavy metal atoms used as stains. On chemical grounds, osmium tetroxide should react strongly with some of the hydrophobic portions of the membrane lipids (specifically, with double bonds, $C=C$, in the fatty acids); if the model is correct, this would be expected to produce a dense central layer rather than the two outer dense layers actually observed. On the other hand, when artificial mixtures with known arrangements of lipids or lipids and proteins are fixed with osmium tetroxide or stained with heavy metal stains, most electron density is produced in the regions where hydrophilic groups are expected; this empirical finding has been used to argue that the electron dense lines in the unit membrane are in the position that the model predicts they should be.

The plasma membrane of the red blood cell (Fig. II-1) has been studied more intensively than any other. Red cells are readily obtained in large quantities. In mammals they lack nuclei; the plasma membrane surrounds a cytoplasm consisting chiefly of one protein, hemoglobin. The absence of a nucleus and of cytoplasmic membranes facilitates plasma membrane isolation and purification. It has been demonstrated that the membrane contains roughly the amount of lipid needed to make a layer two molecules thick over the surface area of a red blood cell. This is an important line of evidence supporting the proposal that plasma membrane structure is based on a bimolecular lipid layer. The arrangement of proteins, however, is more complex than Figure II-3A indicates. In the first place, most of the protein molecules in membranes are probably folded into specific compact three-dimensional structures (Fig. II-3B). Given the quantities of protein present in plasma membranes, the proteins probably do not coat the entire surface of the lipid bilayer. Further, the different proteins present in the membrane vary markedly in the ease with which they can be detached. Some are relatively readily solubilized; others can be removed only by drastic treatments that disrupt the membrane and grossly perturb the lipid arrangement. This is explained by the proposition that different proteins have different associations with the lipids, as illustrated in Figure II-3B. The terms *integral* and *peripheral* or *intrinsic* and *extrinsic)* are sometimes used to distinguish proteins that are tightly bound to membranes from those with looser, more superficial association. Proteins possess both hydrophilic and hydrophobic groups; the latter can be arranged so that in some proteins extensive hydrophobic regions occur which permit close interaction with hydrophobic lipid regions. Integral proteins often contain a hydrophobic portion which penetrates, more or less deeply, into the lipid bilayer and for some proteins spans the membrane (Fig. II-3B). Protein molecules in the membrane probably interact and associate with one another in various ways.

As Figure II-3B suggests, the specific proteins and groups (such as carbohydrates) attached to proteins probably differ at the two membrane surfaces. There may also be differences in lipid composition between the cytoplasmic and extracellular faces of the plasma membrane.

The idea that certain proteins may actually span the membrane, stretching from one surface to the other, helps explain some of the functional properties of plasma membranes outlined below. Evidence for such organization comes from experiments in which membranes are exposed to agents that react with proteins, but cannot penetrate the membranes. (For example, the enzyme *lactoperoxidase*, too large to pass through the plasma membrane, can be used to catalyze the addition of radioactive iodine atoms to proteins.) Red blood cells are exposed to these agents either as intact cells, or after suitable disruption that exposes the inner (cytoplasmic) surface of their plasma membranes. It is sometimes even possible to make membrane fragments which seal off into inside-out vesicles, whose outer surface is the plasma membrane surface that originally faced the cytoplasm. The differences in the populations of proteins that react under different conditions confirm, for example, that certain proteins are restricted to the cytoplasmic surface of the membrane; these react only in disrupted cells. Since some proteins are accessible to the agents from either side of the membrane, they probably stretch across the width of the structure.

Present views of membrane organization have also been profoundly influenced by the advent of *freeze-fracture* and *freeze-etch* techniques (Figs. II-4 and II-5), an approach that avoids fixation and embedding (hence minimizes some possibilities for artifact) and provides face views of the membrane, rather than the cross-sectioned views that dominate conventional electron microscope images.

As Figure II-5 illustrates, freeze-fracture preparations of plasma membranes show numerous globular structures, roughly 75 Å in diameter, present in the central zone of the membrane; that is, inserted in the hydrophobic lipid region. These particles are big enough to span the membrane width, and most investigators believe they are proteins. This belief is supported by observations such as those indicating that artificial membranes made only of phospholipids show no particles, whereas the particles are present if the protein *rhodopsin* (Section 3.8.1) is mixed with the lipids.

The freeze-fracture procedure has permitted some initial dem-

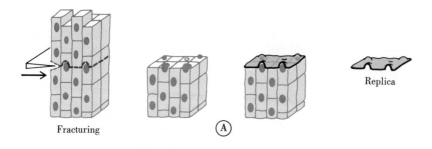

Fracturing   (A)   Replica

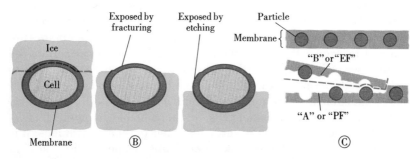

Fig. II-4.   *Freeze-fracturing and freeze-etching. (See also Fig. II-5.)*

A. Rapidly frozen tissue is fractured, usually with a sharp knife edge. A thin layer of carbon and heavy metal (often platinum) is evaporated onto the exposed surface. This produces a thin, detailed replica of the surface and it is the replica rather than the surface itself that is viewed in the electron microscope. The replicas provide face views of the exposed surfaces.

B. Fracturing often exposes the interior of a membrane, separating the two lipid layers (see Fig. II-3). After fracturing, the preparation can be "etched" before making the replica: In etching additional ice is removed by evaporation (sublimation). This can expose the outer surface of the membrane (see Fig. II-5).

C. Fracturing can reveal particles within a membrane. (Based on work by D. Branton, H. Moor, and others.) Often, most of the particles remain with one of the two surfaces produced by the fracture plane; the other surface shows what often seems to be a complementary pattern of pits or depressions. Such fracturing can yield views of the inner face of the outer half of the plasma membrane and views of the outer face of the inner half of the plasma membrane. The latter face is called A or +, the other face is called B or −. If the membrane shown here were a plasma membrane the cytoplasm would be below, and the extracellular space, above. Often the A face shows more particles than the B face. Recently, workers with freeze-etch techniques have suggested a new labeling system: the surface of the plasma membrane facing the extracellular space will be called ES (extracellular surface), the two faces exposed by fracturing will be EF and PF (extracellular face and protoplasmic face), the surface facing the cytoplasm will be PS (protoplasmic surface).

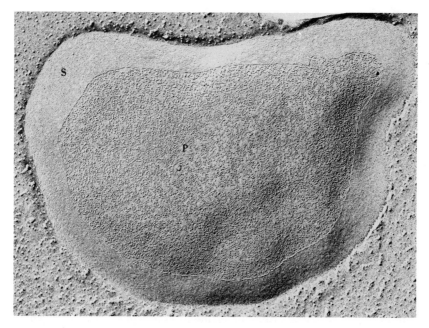

Fig. II-5   *Surface of a human red blood cell prepared by the freeze-etch procedure (Fig. II-4). The area at P is the interior of the membrane, exposed by fracturing. Numerous small particles are visible. The area at S is the outer surface of the membrane, exposed by etching. This surface lacks the particles. × 40,000. (From V.T. Marchesi and T.W. Tillack, Journal of Cell Biology, 45: 649.)*

onstrations of differences in the molecular organization of different portions of the plasma membrane of a given cell. Thus we will see (Section 3.5.2, Fig. III-21) that highly ordered arrangements of intramembrane particles characterize regions where adjacent cells form special junctional structures with one another. But while more or less stable specialized membrane regions do exist, the membrane is not a rigidly organized structure. With varying experimental conditions, some of the intramembrane particles can show marked changes in distribution, suggesting that they can move in the plane of the membrane. This possibility is also suggested by experiments in which cells are induced to fuse with one another (Chapter 3.11), and the distribution of certain surface proteins is followed by use of immunohistochemical techniques (p. 22). The surface proteins of two fused cells rapidly intermingle. Additional experiments, using biophysical techniques such as electron spin resonance, nuclear magnetic resonance, and other methods for looking at molecular interactions, buttress the view that proteins and lipids can move laterally within membranes. By comparison, however, movement of such molecules from one membrane surface to the other is quite rare. These observations are first steps toward the analysis of more subtle changes in membrane structure, important for function.

2.1.3   ***TRANSPORT*** How can molecules in aqueous solution
    ACROSS THE (that is, dissolved in water) outside the cell
    PLASMA pass through the lipoprotein plasma mem-
    MEMBRANE brane and gain access to the aqueous solu-
tion inside the cell? The model of the plasma membrane as a bimolecular layer of lipid accounts for the ability of small lipid-soluble molecules to enter the cell. Such material could pass down concentration gradients from the "outside" solution into the lipid layers and then into the "inside" solution. However, it is not obvious that water, inorganic ions [for example, sodium ions (Na^+) or potassium ions (K^+)], or a variety of other substances that are not readily soluble in lipids but enter the cell could easily pass a barrier of lipid layers. Some molecules require enzyme activity to pass the membrane, and the presence of only a few molecules of the appropriate specific enzyme per square micrometer of membrane would suffice to explain their ability to enter the cell. Others, however, appear to pass through without enzymatic intervention. To explain the passive diffusion of some molecules into the cell, and also to explain some restrictions on the sizes of molecules that can readily enter, it is often hypothesized that special channels pass through the membrane, interrupting the lipid layers. Usually these channels are referred to as *pores*, although one need not think literally in terms of holes. Rather, the channels are presumed to be arrangements of membrane molecules or of portions of molecules, such that there extends from one surface of the membrane to the other a continuous zone of hydrophilic groups (for example, portions of proteins or charged groups of lipids) and perhaps even of water molecules loosely bound to membrane components. Such a hypothetical pore is diagrammed in Figure II-3A. Probably a variety of nonlipid-soluble molecules could diffuse through such channels far more rapidly than through continuous layers of hydrophobic groups. For example, inorganic ions associate readily with groups bearing charges opposite to their own and with water; in solution they are surrounded by a loose "shell" of associated water molecules. The pores are usually thought of as having diameters of a few angstroms so that passage through them would be restricted to water or material such as inorganic ions and small molecules with diameters less than that of the pore.

Several alternative types of channels have been proposed. The molecules in the membrane are in constant motion; such movement might result in a transient realignment of appropriate portions of molecules that creates a temporary channel, a *dynamic pore*. Or perhaps more permanent channels exist, but their small size, irregular structure, or poor preservation during fixation prohibit identification by current microscopic techniques. Possibly membrane proteins of the "integral" type are involved.

Proteins, including enzymes, aid the passage of various molecules across the plasma membrane. For example, glucose enters some animal cells by "facilitated" or "mediated" transport mechanisms in which membrane-associated proteins take part. Proteins are thought to be implicated in sugar transport because sugars with only slightly different chemical structure may enter cells at very different rates, even if they are of similar sizes, lipid solubilities, and other features that govern passive movement through membranes. This argues for highly selective sites, presumably on proteins, that bind the sugar molecules as part of the transport mechanism. Membrane enzymes are involved in energy-dependent transport, termed *active transport*. Interruption of energy production by cells (for example, by exposure to low temperature or by treatment with metabolic poisons such as iodoacetic acid which interrupts glycolysis) can strongly affect the movement in or out of the cell of several important types of molecules, including inorganic ions.

Energy-dependent transport can operate against concentration gradients; that is, it can result in concentrations within the cell that are considerably higher or lower than concentrations outside the cell, for materials that would tend to be at equal concentration inside and out were active transport not operating. One important transport of this kind involves sodium ions (Na^+). Intracellular sodium concentration in most cells of higher organisms is kept much lower than extracellular concentration. When radioactive sodium ions are introduced into a solution containing nonradioactive cells, some of the radioactive ions rapidly enter the cells and replace nonradioactive sodium ions that leave. This "tracer" experiment indicates that sodium normally continually enters and leaves the cell and that the low intracellular concentration is not due to an inability of the ions to cross the membrane. Rather, an active extrusion mechanism (a sodium "pump") balances the entry of sodium ions and maintains the low intracellular level. Potassium ions (K^+) have the reverse distribution; they are more concentrated within the cell than outside. The relevance of these ion distributions to nerve transmission will be discussed in Section 3.7.3. For the moment it should be noted that a potential difference, comparable to the differences between poles of a battery (or, more accurately, between surfaces of a capacitor), exists across the membrane; the inside of the cell is normally electrically negative in comparison to the outside. Physical chemical theory indicates that the existence of this potential reflects the asymmetries in ion distribution and the fact that some ions penetrate the membrane more readily than others.

An ATPase that splits ATP at a much accelerated rate if Na^+ and

K$^+$ are present in the medium, has been found in many plasma membranes, and it is considered to participate in Na$^+$ transport. It is thought that the enzyme picks up sodium ions from inside the cell and exchanges them for potassium ions outside; the K$^+$ ions are then carried to the inside and exchanged for additional Na$^+$ ions. In the course of these events, a phosphate derived from ATP is temporarily linked to the enzyme, somehow providing energy for the transport. That the ATPase is involved in such ion transport is indicated (1) by the required presence of both Na$^+$ and K$^+$ for rapid activity in splitting ATP, and (2) by the fact that the drug *ouabain* inhibits both the splitting of ATP by the enzyme and the transport of Na$^+$ out of the cell. How the enzyme is able to act "asymmetrically," treating Na$^+$ and K$^+$ differently on the two sides of the membrane, is not known.

How could proteins "aid" other components to cross a membrane? If, as seems possible, some pertinent proteins span the membrane, perhaps slight changes in their configurations can open transient pores. Usually, active or facilitated transport is attributed to (hypothetical) carrier molecules which are presumed to attach to materials being transported at one surface of the membrane, carry them to the other surface, and release them. Carriers might be portions of larger molecules such as proteins which themselves remain in place, but rotate or change their shapes so that the carrier portions move. Or perhaps there also are carrier molecules which are small enough and so constructed that a complex between the entire carrier and the molecule being transported can cross the lipid bilayer. Or a transport system might involve multiple fixed sites in the membrane, with a transported component being passed from one to another.

Although much remains to be learned about the ways in which molecules cross the plasma membrane, new information is coming from many directions. Methods for isolating and purifying the plasma membrane are being improved. Another approach of importance is the study of artificial membranes made from purified phospholipids and other components. For example, membranes have been constructed that possess the simple lipid bilayer structure proposed in the model discussed in Section 2.1.2. Surprisingly, water can pass through these membranes fairly rapidly; the observed rates are roughly comparable to the rates of osmotic movement of water across plasma membranes of cells. Evidently a lipid bilayer is not as effective a barrier to water as might be thought. Of interest is the fact that the permeability to water of artificial membranes varies inversely with the proportion of cholesterol added to the phospholipids, a finding that may have important significance in explaining differences among natural membranes. The effects of cholesterol may be due to its altering the movements and spacing of membrane lipids. Some features of the electrical behavior of cell surfaces

(Section 3.7.3) are mimicked if a specific mixture of proteins is added to the artificial membrane. Inorganic ions (Na^+, K^+, and so forth) usually pass through the artificial membranes far more slowly than they do through natural membranes. However, if an antibiotic known as *valinomycin* is present, there is a great increase in the rate at which Na^+ and K^+ can diffuse through the artificial membrane. Furthermore, with valinomycin K^+ passes through artificial membranes at a faster rate than Na^+, an important selectivity feature that is characteristic of the passive movement of these ions through many natural membranes. Valinomycin is a circular molecule with a hydrophobic exterior surrounding a hydrophilic interior in which it can selectively bind ions like K^+, substituting for the "hydration shell" of loosely associated water molecules that normally surrounds the ions. The hydrophobic exterior apparently permits valinomycin to traverse the membrane's lipid layer, and thus to act as a carrier. The valinomycin effect may provide a model for natural "carrier" mechanisms. Other antibiotics are thought to form pores in artificial membranes, and these may aid in the search for natural pores.

2.1.4 BULK TRANSPORT: PINOCYTOSIS AND PHAGOCYTOSIS (ENDOCYTOSIS)

With a few special exceptions (Section 3.1.1) materials of macromolecular dimensions or larger cannot cross membranes. When they enter cells, they do so in membrane-delimited compartments formed from the plasma membrane; the pertinent processes are referred to collectively as *endocytosis*. *Phagocytosis* of bacteria or other large structures, by protozoans or by phagocytic cells of higher animals, involves the incorporation of these particles into intracellular vacuoles that originate by the folding of the plasma membrane around the material being engulfed. *Pinocytosis* is a similar phenomenon, except that here the particles taken up are of molecular or macromolecular dimensions. Proteins or other molecules are adsorbed to sites on the plasma membrane, and the membrane folds in to form small vacuoles or vesicles that move into the cell, carrying drops of the medium. This may be seen by the light microscope in living amebae because the vacuoles are quite large. In most cells of multicellular organisms, the vesicles are too small for light microscopy but can readily be seen by electron microscopy. In many cases the occurrence of pinocytosis can be demonstrated by introducing tracer molecules into the organism or the media in which cells are grown. One such tracer is the iron-containing protein *ferritin*, which is visible in the electron microscope because the large number of iron atoms in each molecule scatter electrons much more effectively than the cell components do. Another tracer is the enzyme *peroxidase* isolated from horseradish; many cells will take up this protein by pinocytosis. If cells are then fixed and incubated with the enzyme's substrate, the sites of peroxidase activity will be marked by a dense product (Fig. II-6). During the incubation many molecules of the reaction product are formed by the action of each molecule of peroxidase taken up by the cell. Thus the method is very sensitive and sites of only a few peroxidase molecules can be detected.

An interesting feature of many pinocytosis vesicles (Fig. III-33) is the fuzzy "coating" they show on the membrane surface facing the cytoplasm. Similar coating is sometimes seen on the surface of vesicles not involved in pinocytosis (for example, some vesicles produced by the

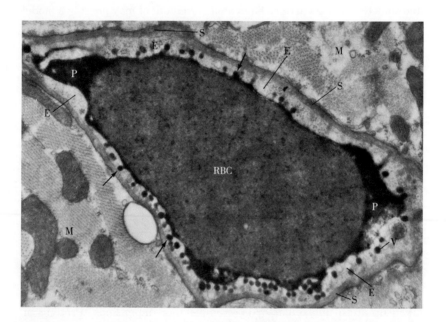

Fig. II-6 *A capillary (cut transversely) in the heart muscle of a rat that had been injected intravenously with the enzyme* peroxidase *5 minutes prior to fixation. The tissue was treated as mentioned in Section 2.1.4 so that peroxidase sites are marked by a dense reaction product. Most of the space within the capillary is occupied by a red blood cell (RBC). Surrounding this is a peroxidase-filled region, the plasma (P) suspending the blood cells. The capillary wall consists of thin endothelial cells (E) outside of which a space (S) separates the vessel from the muscle cells (M). Many pinocytosis vesicles (V) are present in the endothelial cells; some (arrows) are seen still to be connected to the cell surface (see also Fig. I-2).* × *25,000. (Courtesy of M. J. Karnovsky.)*

Golgi apparatus or associated systems, Fig. II-30). While the nature of the coat is not known, a reasonable hypothesis is that it represents a special organization of cytoplasm involved in the pinching off of the small membrane-enclosed vesicles from larger membrane structures.

As will be seen in Section 2.8.2, the macromolecules inside endocytic vacuoles often are degraded by enzymes to which they are exposed when the vacuoles fuse with lysosomes. A most important exception to this occurs in the cells lining capillaries (Fig. II-6). The pinocytic vesicles in these cells participate in the transport of molecules across the capillary wall. Having formed at one surface of the cell, they may fuse with the other surface, releasing their contents. Recent evidence also suggests that in some capillaries, vesicles still continuous with the cell surface may fuse with one another and with intracellular vesicles in such patterns that transient extracellular pathways are opened for direct movement of materials across the wall.

When pinocytosis and phagocytosis proceed rapidly, large amounts of plasma membrane are removed from the surface and taken into the cell as the membranes of the vacuoles. How the surface is replenished is still uncertain; we will discuss pertinent aspects of membrane formation in Section 2.5.4.

Endocytosis is often selective. It is likely that membrane receptors are involved, some of which bind specific molecules and thus concentrate them from the surrounding medium. Rates of endocytosis also are responsive to certain "inducing" molecules. (Fig. II-53).

2.1.5 CARBOHYDRATES AND PLASMA MEMBRANES

Polysaccharides, as a class, are polymers of carbohydrate molecules. *Glycogen*, composed of chains of glucose units (Fig. I-24), is the major intracellular storage form of sugar in animals. A similar polysaccharide, *starch*, is used in sugar storage by plants. Complex polysaccharides (for example, the *acid mucopolysaccharides*) contain groups such as sulfate groups and various small organic molecules linked to the sugars. Many polysaccharides are found complexed with proteins.

Animal cells often have a layer of protein and polysaccharide at their outer surface. Sometimes this forms a distinct coat or "fuzz" (Figs. II-7 and III-20). Even in compact tissues such as liver, the plasma membranes of adjacent cells are separated almost always by a space of 100–200 Å. [Only at special junctional membrane regions do adjacent cells approach much closer than this (Section 3.5.2).] The space between adjacent cells is often occupied by a matrix of proteins and polysaccharides that helps to cement cells together. Carbohydrates are also built into some of the molecules of plasma membranes. Many of the

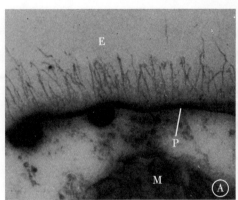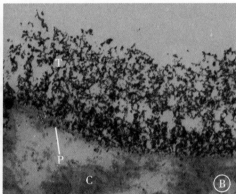

***Fig. II-7   A.** Portion of the surface of an ameba. Numerous fine filaments are seen attached to the extracellular surface of the plasma membrane (P). The tip of a mitochondrion is present at M. E indicates the space outside the cell. × 55,000. **B.** Portion of a similar cell that had been exposed to thorium dioxide particles in suspension. The electron-dense metal-containing particles are absorbed to the filaments (T). Such adsorption is the first step in the uptake of material through pinocytosis. C indicates the cell cytoplasm. × 35,000. (Courtesy of G. Pappas and P. Brandt.)*

key proteins present at the extracellular surface of red blood cell plasma membranes, and responsible for important features of membrane function and specificity, are *glycoproteins*; short chains of sugars are attached to certain of their amino acids. A negatively-charged carbohydrate derivative, *sialic acid (N*-acetylneuraminic acid), often is included in these chains. Some membrane lipids (glycolipids) such as the *gangliosides* also have sialic acids or sugars attached. Digestion of cell surfaces with the enzyme *neuraminidase*, which removes sialic acids, can considerably alter the properties of cells. For example, it removes some sites at which proteins and other molecules can be bound prior to endocytic uptake and it affects the electrical charge of the cell surface.

A number of animal cells and tissues have extensive special extracellular polysaccharide material. *Chondroitin sulfate* is a major component of cartilage, and *hyaluronic acid* is present in sites such as joints, the lower layer of the skin, and also in the external coats that surround some egg cells. The cell walls of plant cells are made largely of the polysaccharide *cellulose*, as will be discussed in Section 3.4.1.

2.1.6   ***MEMBRANE***   All cellular membranes are composed of
        SPECIFICITIES   the same classes of molecules, and it is
        AND RECEPTORS   widely assumed that the same basic structural principles apply to many, if not all. Differences in detail are being sought to help explain the special charac-

teristics of different membranes. Most of what is known along these lines is confined to cases where the differences are dramatic. For example, the plasma membranes of cells that line the bladder of mammals are notably thicker (12 nm) than is true for most cells. They show a number of other specialized structural features, and they are unusually rich in a category of glycolipids known as *cerebrosides*. These characteristics are associated with an unusually low permeability to water and to solutes of the types concentrated in urine. The plasma membranes that make up the myelin sheath that surrounds the axons of some nerve cells (Section 3.7.4) are poor in protein and in intramembraneous particles visible with freeze-fracture techniques. These features presumably relate to the fact that myelin carries out few transport or other enzymatic activities; its primary role is as a "passive" insulator. Special features of intracellular membranes will be especially evident in our discussions of chloroplasts and mitochondria (Sections 2.6.1 and 2.7.3; see also the comments on "purple membranes" on p. 200).

The particular set of proteins, lipids, and carbohydrates present at a given cell surface are responsible for a variety of specific properties of that cell. Thus cells have different selective binding sites at which they bind extracellular molecules in pinocytosis, or enzyme mediated transport. The factors responsible for blood type are localized at the surfaces of red blood cells. One indication of the existence of a complex cell-specific surface architecture is seen in surface specificity during cell association. Embryonic cells from different organs can be separated from one another by appropriate mild means (usually reduction of the calcium content of the medium and gentle proteolytic digestion are involved) and the cells of two organs, intermixed. The cells from each organ will "recognize" one another on contact. In time, cells of a given type separate from the mixture and constitute aggregates. These aggregates may even develop some features of the original organ. Such aggregation presumably reflects the presence of specific components or organization at the surfaces of different cells. Some investigators hold that glycoproteins are involved and are seeking to isolate the particular ones responsible.

Some of the agents that affect cells can enter them and interact directly with the nucleus or cytoplasm. Cells also respond to many extracellular molecules that cannot penetrate their plasma membranes. Often these latter responses are based upon binding of the molecules to specific receptor sites in the cell surface, whose activation serves to trigger a sequence of subsequent events. For example, when a nerve cell stimulates another nerve cell, or a gland cell or muscle fiber, it does so by releasing agents that alter the ionic permeability of the responding cell's surface (Section 3.7.5), and this in turn leads to other processes (nerve impulses, Section 3.7.5; muscle contraction, Section 3.9.1; secretion, Section 3.6.3). Various cells respond to hormones, circulating in the blood stream, through the production of intracellular "second messengers" by enzymes located at their surfaces. The best known of these messengers is cyclic adenosine monophosphate (cyclic AMP; cAMP). Thus when the hormone epinephrine (adrenaline) is released to the blood stream by the adrenal gland, as it is during stress, it is found that liver cells respond by increasing their release of glucose from its stored form, glycogen. The hormone binds to cell surface receptors on hepatocytes and by so doing, it stimulates the activity of a plasma membrane enzyme, *adenyl cyclase*. This enzyme generates cAMP from ATP *intracellularly*. The cAMP then binds to specific enzymes (protein kinases) which respond by catalyzing the addition of phosphate groups (phosphorylation) to other enzymes, among them key participants in the control of glycogen storage. The net effect of the phosphorylations is a decrease in the formation of glycogen and an increase in its conversion to glucose. (The cells also contain phosphodiesterase enzymes that cleave cyclic AMP, so the increased levels resulting from a given stimulus are transient and the cells can return to normal). Different cells possess different batteries of cell surface receptors, kinases, and so forth, so that permeability changes or cyclic AMP and other second messengers (such as cyclic guanosine monophosphate, cGMP, whose effects often are opposite to those of cAMP) are produced in response to varying events and agents, and call forth differing cell-specific responses.

An interesting group of derivatives of fatty acids (Fig. II-2) can produce effects on cells that often involve alterations in levels of cAMP. These are the *prostaglandins*, a type of compound with 20 carbon atoms five of which form a specific ring structure. They cause a vast array of physiological responses. Unlike the animal hormones produced by endocrine glands (thyroid, adrenal, and so on) prostaglandins are produced by many cell types, perhaps most cells in mammals.

c h a p t e r **2.2**
THE NUCLEUS

Cells without nuclei have very limited futures. The only common animal cell type without a nucleus, the mammalian red blood cell, lives only a few months; aside from its role in oxygen transport, it is extremely restricted in its metabolic activities. Egg cells from which nuclei have been experimentally removed may divide for a while, but the products of division never differentiate into specialized cell types, and eventually they die. Fragments without a nucleus, cut from such large unicellular organisms as amebae or the alga *Acetabularia*, survive temporarily, but ultimately they die unless nuclei from other cells are transplanted into them. Thus the nucleus is essential to long-term continuation of metabolism and to the ability of cells to alter significantly their structure and function (as in differentiation). In part, this reflects the primary role of the nucleus in producing the RNA required for protein synthesis. When cells change, their new functions and structures require new proteins. Even cells that are constant in metabolism and structure show continual replacement (*turnover*) of macromolecules and probably of organelles, including portions of the cytoplasmic protein-synthesizing machinery.

The nucleus is of central importance in cell heredity; it determines key morphological and metabolic features of a cell. For example, if a fragment containing the nucleus is cut from an *Acetabularia* of one species, characterized by a given morphology, the fragment will regenerate a whole cell of that species. This regenerative ability permits experiments of the type illustrated in Figure II-8, in which nuclei of one species are combined with cytoplasms from different species. The conclusion drawn from such experiments is that the nucleus produces material that enters the cytoplasm and participates in the control of cell growth and cell morphology. The crucial finding is that the morphology of the regenerated cells eventually becomes like that of the species from which the nucleus is taken. In the hybrid fragments with the nucleus from one species and most of the cytoplasm from the other species, old cytoplasmic material persists for a while and may influence cell form. Eventually, however, this is depleted and replaced by newly produced material from the nucleus.

Our concern in this chapter and the next will be with general aspects of nuclear structure and function, and with the roles of the nucleus in providing the nucleic acid molecules essential for protein synthesis. Parts 4 and 5 will deal with the details of chromosome architecture, cell division, and facets of the regulation of genetic activity.

2.2.1   DNA

The approximate composition of rat hepatocyte nuclei is 10–15 percent DNA, 80 percent protein, 5 percent RNA, 3 percent lipid. (This is on a percent of dry weight basis; water is a primary constituent of all cells and organelles and accounts for 70 percent of hepatocyte weight.) An early cytochemical finding was the presence in nuclei of virtually all the DNA of cells; only later was the presence of small but significant amounts of DNA in some cytoplasmic organelles recognized. The DNA content of cells of different organisms varies greatly. However, quantitative cytochemical studies of cells in multicellular animal and plants established that the DNA content of most nuclei in a given organism is twice $(2 \times)$ that of the sperm or egg cell nuclei (or, for some cells, a multiple of $2 \times$ based on continued doubling, such as $4 \times$, $8 \times$, or $16 \times$). This is as

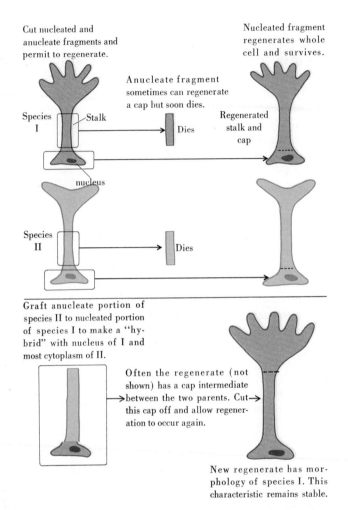

Cut nucleated and anucleate fragments and permit to regenerate.

Nucleated fragment regenerates whole cell and survives.

Anucleate fragment sometimes can regenerate a cap but soon dies.

Species I

Stalk

Dies

Regenerated stalk and cap

nucleus

Species II

Dies

Graft anucleate portion of species II to nucleated portion of species I to make a "hybrid" with nucleus of I and most cytoplasm of II.

Often the regenerate (not shown) has a cap intermediate between the two parents. Cut this cap off and allow regeneration to occur again.

New regenerate has morphology of species I. This characteristic remains stable.

Fig. II-8 *Experiments with the large single-celled alga* Acetabularia. *The intermediate caps sometimes formed in the initial "hybrids" probably reflect the fact that some time elapses before "old" material in the cytoplasm is depleted and replaced by new material from the nucleus. (After the work of Gibor, Hammerling, and others.)*

expected for the genetic material: each gamete contributes an equal amount of DNA to the zygote nucleus which by duplicating gives rise to all the nuclei in the cells of the organism. These considerations will be discussed further in Part 4.

DNA molecules are composed of two strands coiled together in a double helix (Fig. II-9). Each strand is a chain of nucleotides (a *polynucleotide* strand); all DNA nucleotides consist of a 5-carbon sugar (deoxyribose) with a phosphate group attached at one end and a

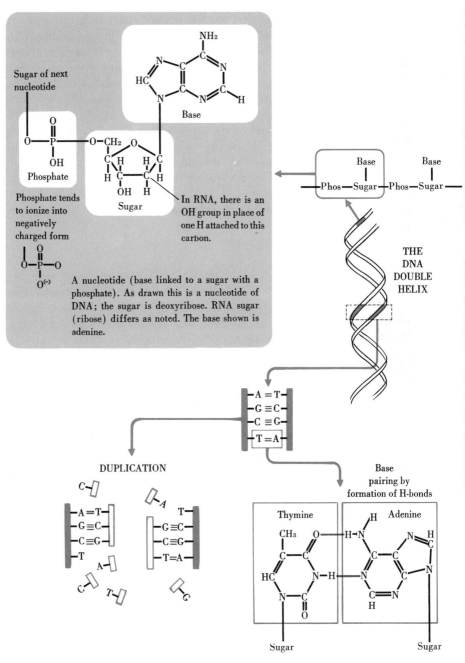

Sugar of next
nucleotide

O

O—P—O—CH₂

OH

Phosphate

Phosphate tends
to ionize into
negatively
charged form

O

O—P—O

O⁽⁻⁾

NH₂

Base

Sugar

In RNA, there is an
OH group in place of
one H attached to this
carbon.

A nucleotide (base linked to a sugar with a
phosphate). As drawn this is a nucleotide of
DNA; the sugar is deoxyribose. RNA sugar
(ribose) differs as noted. The base shown is
adenine.

Base   Base

—Phos—Sugar—Phos—Sugar—

THE
DNA
DOUBLE
HELIX

—A═T—
—G≡C—
—C≡G—
—T═A—

DUPLICATION

Base
pairing by
formation of H-bonds

Thymine   Adenine

Fig. II-9   *Some features of nucleic acids.*

nitrogen-containing ring compound (known as a *base*) at the other. Four different bases are present on different nucleotides: adenine (A), guanine (G), cytosine (C), and thymine (T). The two strands of nucleotides are aligned according to simple pairing rules. Every A on one strand pairs with a T on the other, while every G pairs with a C. The paired bases are held together by molecular interactions, chiefly *hydrogen bonds*, in which a hydrogen atom that is part of one base is also attracted to and loosely held by the second base. The strands are referred to as *complementary*, since the sequence of bases on one strand exactly dictates that of the other. It is the base sequences that carry the hereditary information. The replication of DNA results in the duplication of the base sequences (Fig. II-9), and this underlies the role of DNA in the transmission of hereditary information, as will be discussed in Part 4.

2.2.2   **PROTEINS**   In the nucleus, DNA is complexed with proteins among which are the *histones*. The phosphate groups of the DNA are negatively charged. When cells are stained with basic dyes (those dyes that are positively charged), the DNA- and RNA-containing regions stain because their negatively charged phosphate groups attract the oppositely charged dye molecules. Such regions are referred to as *basophilic*. The histones are basic proteins, that is, they have a high content of the basic amino acids lysine, arginine, and histidine which contain derivatives of amino (NH_2) groups; amino groups can assume a positively charged (NH_3^+) form. Histones are *acidophilic*; their positively charged amino group derivatives attract acid (negatively charged) dyes. In nuclei, however, histones do not stain strongly with acid dyes unless the DNA is first removed by chemical or enzymatic means. This suggests that many of the DNA phosphate groups bind to the oppositely charged histone amino groups, thus neutralizing the histones' charge and minimizing the attraction for acid dye molecules. When DNA is removed, the histone groups are free to bind dye. DNA and histone are usually present in nuclei in approximately equal amounts.

In addition to the histones, nuclei usually contain even larger amounts of other nonbasic proteins. These include enzymes and a heterogeneous group of proteins referred to as "nonhistone" proteins. Some of the nonhistone proteins may be closely associated with the histones and DNA, and there is growing suspicion that these associations are important in the control of nuclear functions (Section 4.4.6). Very recently the proteins *actin* and, perhaps, *myosin* have been reported to be present in some nuclei. This is still highly controversial, but since these proteins are generally thought of as agents of cellular motion (Section 2.11.2, and 2.11.3), there has been speculation that they might participate in the movement of nuclear components in cell division, in passage of material between the nucleus and the cytoplasm, or in the maintenance and alteration of nuclear structure.

The contribution of the nucleus to total cell mass may vary from approximately 5 percent (some muscle cells), to 5–10 percent in hepatocytes, and up to 50 percent or more in cells of the thymus gland and in other rapidly dividing cells such as plant root-tip cells and cancer cells. Variations in the ratio of nuclear volume to cell volume fall within the same range. Measurements on the nuclei of different cell types of a given organism indicate that the DNA per nucleus usually is essentially constant from cell to cell or varies as a multiple of the 2 × amount; the histone content generally parallels the DNA. In contrast, the content of nonbasic proteins and of RNA varies considerably. Sperm cell nuclei, which are metabolically inactive, have virtually no RNA or nonbasic protein; DNA may account for one third to one half of the nuclear dry weight. In cells with metabolically active nuclei, such as hepatocytes, DNA accounts for only 10–20 percent of the nuclear dry weight, and much nonbasic protein and RNA are present.

2.2.3   ***RNA***   A most important aspect of DNA function is the production (transcription) of RNA. Autoradiographic evidence outlined earlier (Section 1.2.4) indicates that most of the RNA of cells is produced in the nucleus in close association with DNA. DNA and RNA are similar in that both are polynucleotide chains, but RNA differs from DNA in several ways. RNA is single-stranded, except in certain viruses where, like DNA, it is double-stranded. The RNA 5-carbon sugar is *ribose* instead of deoxyribose. Finally, in RNA the base *uracil* replaces the thymine of DNA (see Fig. II-9).

The base sequences of RNA molecules are determined by DNA. RNA nucleotides align by base pairing along one DNA strand. (What determines which of the two strands is copied remains unclear.) The A base of RNA pairs with T of DNA, G of RNA with C of DNA. A transcription enzyme, RNA *polymerase*, joins the nucleotides to form an RNA strand that is complementary to a DNA strand. By this mechanism, DNA acts as a *template* for RNA synthesis.

The complementary relationship of RNA with DNA is the basis of an extremely valuable technique called *molecular hybridization*, or *RNA-DNA hybridization*, as shown in Figure II-10. If test-tube "hybrids" can form between particular DNAs and RNAs, then the two probably contain many complementary base sequences. Several different classes of RNA are present in cells; they differ in size and in role, as will be seen in the next chapter. Using hybridization, it can be shown that all major types of RNA are complementary with DNA from the nucleus (with the probable exception of the nucleic acids associated

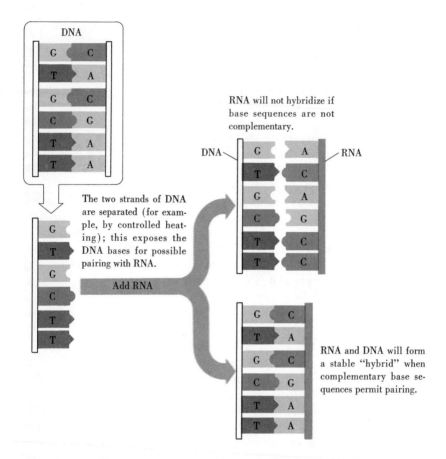

Fig. II-10 *Hybridization of nucleic acids. Purified nucleic acid molecules are mixed in the test tube. The formation of hybrids can be detected in several ways. For example, hybrids between radioactive single-stranded RNA and nonradioactive DNA will be double-stranded (one RNA strand, one DNA strand) and radioactive. Their radioactivity distinguishes them from the original double-stranded DNA while their double-strandedness produces differences (for example, in density) which permit them to be separated (by centrifugation, absorption to special gels or filters, or other methods) from the original single-stranded RNA molecules.*

with mitochondria and plastids). This strongly suggests that all are synthesized in the nucleus.

Once the RNA is made in the nucleus, much of it moves rapidly into the cytoplasm (Figs. I-21 and I-22). In rat hepatocytes, less than 10 percent of the total cellular RNA is present in the nucleus at any given moment; much of this is probably in process of synthesis or in transit to the cytoplasm.

2.2.4   ***NUCLEAR***   Nuclei have enzymes (polymerases) thought
                     ENZYMES   to participate in the production of DNA and
                                      RNA. There are at least 3 distinct forms of
RNA polymerase, apparently responsible for transcribing different
types of RNA. For example *RNA polymerase I*, present in nucleoli,
transcribes the RNA molecules destined for inclusion in ribosomes
(Section 2.3.4). Other enzymes are present as well, but these have been
studied only to a limited extent. Previously (p. 38) allusion was made to
one such enzyme involved in the synthesis of NAD in hepatocytes.
Nuclei of some cells may generate ATP by a pathway still not fully
elucidated. Whether nuclei synthesize their own proteins or receive
them from the cytoplasm is of current interest. The nuclear proteins
that have been studied most carefully such as the histones originate in
the cytoplasm, and migrate to the nucleus. However, several
laboratories report that preparations containing isolated nuclei, at least
from some cell types, can synthesize some proteins. Although it is
sometimes difficult to be sure that the isolated nuclei have been completely freed from cytoplasm, there is a possibility that different nuclear
proteins may originate in different parts of the cell, or that cell types
vary in the ability of their nuclei to synthesize proteins.

Even these fragmentary findings make it evident that nuclei cannot be regarded as mere libraries of DNA sequences that periodically
extrude information in the form of RNA. However, most of the details of
nuclear metabolism remain to be unraveled by future studies.

2.2.5   ***STRUCTURE***   As seen in the light microscope, the nucleus
                                        is bounded by a nuclear membrane. Within
the nucleus there is a heterogeneous collection of fibrils and of dense
areas that include *euchromation, heterochromation*, and *nucleoli* (Figs.
I-11, I-12, II-11, and II-12). Chromatin refers to the DNA-containing
structures in the nucleus. The finest structures seen in the electron
microscope are fibrils 20–40 Å in diameter, corresponding to DNA
molecules complexed with proteins (see also Fig. IV-15). These fibrils
are parts of the chromosomes. During cell division each chromosome
coils into a compact configuration and is readily identifiable as an individual unit (Section 4.2.5). In cells not in division, chromosomes exist in
a relatively uncoiled state. In euchromatin, this uncoiling is maximal; in
heterochromatin, the chromosomes or parts of them are in a more compact coiled array. Some nuclei show a stable clearly nonrandom pattern
of heterochromatin. For example, in many cells of some female mammals, a small specific heterochromatic body is present. This is an X
chromosome, one of the chromosomes that determines the sex of the
animal (Fig. II-12). This heterochromatic region is absent in males,

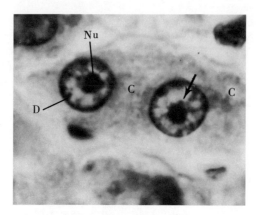

Fig. II-11 *Two hepatocytes in a section of rat liver prepared by widely used routine procedures: fixation in formaldehyde, embedding in paraffin, preparation of sections 5–10 μm thick, and staining with hematoxylin (colors chromatin blue) and eosin [gives the cytoplasm (C) a contrasting pink color]. Prominent nucleoli (Nu) are present within the nuclei. Much of the chromatin is dispersed and difficult to see in a black and white photograph; the rest is in accumulations close to the nucleolus (arrow) or the nuclear envelope (D). While fixation can clump dispersed chromatin, producing "artifacts" (Section 1.2.2), dense accumulations such as these are often seen in living cells. The term* heterochromatin *is used for condensed form of chromatin (see also Fig. II-12). × 1500.*

which have only one X chromosome as opposed to the two present in the cells of the females. Cells from males and females can be distinguished on the basis of this cytological characteristic. Evidence outlined later (Section 4.4.6) supports the tentative conclusion that a given region of chromatin is active when in the diffuse euchromatic state and inactive when converted to the condensed heterochromatic state.

Nucleoli are especially rich in RNA and proteins. Although the pattern varies from organism to organism, often there are from one to four nucleoli per nucleus; generally they take the form of large spherical or ovoid bodies. Nucleoli are prominent in RNA metabolism (Section 2.3.4).

It is believed that intranuclear structures are surrounded by a more or less fluid component generally called *nuclear sap*. Among the substances dissolved or suspended in it are probably those in transit from nucleus to cytoplasm. Structures visible in the nuclear sap include fine fibrils of unknown nature, and small granules with diameters of 200–500 Å or more (Fig. II-14). Some of the granules disappear in preparations digested with RNase and thus are believed to contain RNA, perhaps in transit to the cytoplasm. Of these, some approximately 400 Å in diameter are sometimes referred to as *perichromatin* granules. Other, smaller granules sometimes are called *interchromatin* granules.

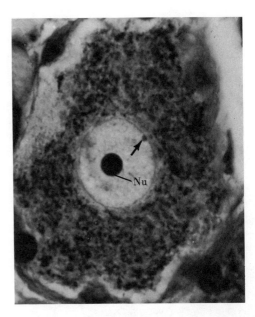

Fig. II-12 *Within the nucleus of the nerve cell of a cat, a prominent nucleolus is visible* (Nu). *Most of the chromatin is in the form of fine threads dispersed throughout the nucleus. The arrow indicates the* sex heterochromatin *or* Barr body, *a characteristic accumulation of dense chromatin found in several cell types in females of many mammals.* × 1000. (Courtesy of M. L. Barr.)

2.2.6   **THE NUCLEAR ENVELOPE**   The structure of the "nuclear membrane" is of great interest for nuclear–cytoplasmic interactions. The electron microscope has revealed it to be a flattened and expanded part of the endoplasmic reticulum that surrounds the nuclear material. The term *nuclear envelope* is used to convey the fact that the nucleus is surrounded by a flattened sac, rather than a simple membrane. The outer surface of the sac, the one exposed to the cytoplasm, has ribonucleoprotein granules attached to it. These are probably ribosomes similar to those of the cytoplasm. The inner surface of the envelope, exposed to the nuclear contents, lacks ribosomes. Often much chromatin is aggregated along this surface. Specific attachments of chromosomes to the envelope is known to occur in a number of cases (e.g., in various species the paired chromosomes of the special *meiotic* cell divisions of gamete formation (Section 4.3.1) attach to the envelope by their ends, at certain stages); some investigators are convinced that special associations between chromatin and the nuclear envelope, perhaps involving the nuclear pores discussed below, are present in many cell types. In addition, in a number of cell types the inner envelope surface is coated with a layer of fibrillar, electron-dense material, which may be a few hundred Å thick. Cytochemical studies and work on isolated nuclear envelopes have shown a general biochemical and enzymatic similarity to the rest of the endoplasmic reticulum. Thus far no special striking biochemical features have been found that might be significant for the functioning of the envelope, but such investigations are still at an early stage.

Nuclear pores occur at intervals along the envelope. These appear as roughly circular or polygonal areas, where inner and outer surfaces of the sac are fused, and the sac is thus interrupted (Figs. II-13 and II-14). The pores measure 500–800 Å across, and there may be as many as thousands of them, scattered across the nuclear surface. Commonly they occupy 10–30 percent or more of the nuclear surface area. In a very few cell types, such as some mature sperm or the micronuclei of protozoa (Section 3.3.1), pores are much sparser than this, or even absent. This may reflect the relatively inactive states of such nuclei.

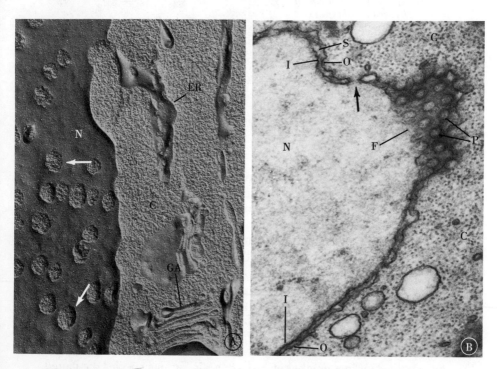

Fig. II-13   *The nuclear envelope as seen with different preparative procedures for electron microscopy. **A.** Portion of an onion root tip cell prepared by freeze-etching (Fig. II-4). At (N) a portion of the nuclear surface is seen; the arrows indicate two of the pores that appear as circular depressions or interruptions of the surface. In the cytoplasm (C) portions of the Golgi apparatus (GA) and endoplasmic reticulum (ER) are visible. × 30,000. (Courtesy of D. Branton.) **B.** Portion of an oöcyte of the toad, Xenopus, prepared by conventional methods of fixation, embedding, and sectioning. (N) indicates the nucleus and (C) the cytoplasm. In cross section, the nuclear envelope appears as a membrane-enclosed sac (S). The membrane at the surface of the sac facing the cytoplasm (O) and that at the surface facing the nucleus (I) are continuous at the edge of the pores; thus in cross section the pores appear (arrow) as gaps in the envelope. At (F) the envelope is twisted and the section provides a face view showing the pores (P) as circular openings. × 40,000. (Courtesy of J. Wiener, D. Spiro, and W. R. Loewenstein.)*

The pores serve as routes of nuclear–cytoplasmic interchange. Granules, or other bodies, are sometimes seen adjacent to both nuclear and cytoplasmic surfaces of the nuclear envelope and even inside the pores (Fig. II-14); some of the structures seen in these configurations are thought to contain RNA, presumably on its way to the cytoplasm. Experiments with marker substances indicate that material can also move in the other direction through the pores. When particles of gold up to 100 Å in diameter are injected into the cytoplasm of amebae, some of the particles gain access to the nucleus. The electron opacity of the gold permits direct visualization of the particles in the electron microscope, and one can see some of the particles within the pores as if caught in the act of traversing the envelope.

However, the pores are not simply holes in the membrane. Much of the aperture is occupied by a cylindrical or ring-like arrangement (annulus) of moderately electron-dense granular or fibrillar material (Fig. II-14). This plus the membrane bordering the pore constitutes a *pore complex*. Investigators have yet to agree upon the finer details of the organization of the complex, but it is generally thought that the complex shows an overall octagonal structure when seen in face view, and that the central region of each complex is different in organization from the more peripheral portions. The movement of microscopically

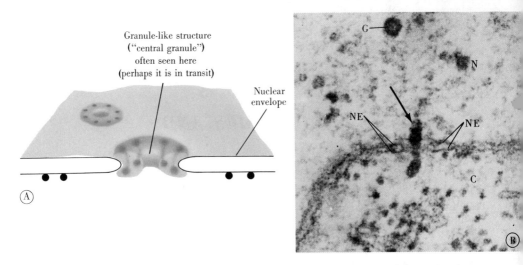

Fig. II-14 *The nuclear envelope.* **A.** *One interpretation of the major features of the pore complex. Several other models, differing in details, have also been proposed. (After Franke, Gall, Kessel, Maul and others.)* **B.** *Small portion of a cell from the salivary gland of the insect, Chironomus. The arrow indicates a structure that seems to be passing through a pore in the nuclear envelope (NE). The structure is thought to be a granule, similar to the one seen at (G) and containing RNA in passage from the nucleus (N) to the cytoplasm (C)* × *100,000. (Courtesy of B.J. Stevens and H. Swift.)*

visible materials through the pores appears to be confined to the central region of the complex.

The mechanisms by which molecules pass through the pores are not known. Some nuclei are large enough that one can insert electrodes within them and thus evaluate their permeability to inorganic ions by electrophysiological techniques. Other studies of nuclear permeability have utilized a variety of tracers, introduced in the cytoplasm or in the medium surrounding isolated nuclei. The findings indicate that even in the absence of dramatic differences in pore frequency, different nuclei may differ greatly in permeability, and that the nucleus of a given cell type may vary in its permeability under altered conditions. In some but not all cases, small molecules and ions pass through the envelope at rates much lower than expected if the pores were substantially open passages through which diffusion occurs freely. Whether this means that enzyme-mediated or other special transport mechanisms function at the pores has yet to be adequately studied.

In some metabolically active cells, the nuclear envelope shows many infoldings which increase the surface available for nucleo–cytoplasmic exchange. In other cells, "annulate lamellae" are seen near the envelope. These are stacks of parallel flattened sacs with "pores" similar to the nuclear pores. One proposal is that these lamellae derive from the nuclear envelope of the cells and serve some function in the cytoplasm. Some investigators maintain that they give rise to endoplasmic reticulum, while others believe that their chief role is in the transport of RNA from nucleus to cytoplasm. There are also reports, yet to be fully verified, that annulate lamellae can form from the ER.

c h a p t e r **2.3**

NUCLEOLI, RIBOSOMES, AND POLYSOMES

Ribosomes, the intracellular sites of protein synthesis, are present in virtually all cells. The few exceptions (for example, mature mammalian red blood cells and mature sperm cells have few, if any, ribosomes) show no protein synthesis. These exceptional cells have very restricted functions and relatively short life spans.

Ribosomes possess RNA of distinctive size and base composition known as *ribosomal RNA* (rRNA). Production of this RNA is a major function of nuclei. RNA-DNA hybridization experiments on a few organisms indicate that well over a hundred (sometimes many hundreds) of apparently identical rRNA molecules can bind simultaneously to the DNA from one nucleus. This indicates that many duplicate copies of DNA sequences responsible for rRNA production are probably present in the chromosomes. A distinctive intranuclear organelle, the *nucleolus*, is prominent in most cells of eucaryotes; it is the site of rRNA synthesis.

2.3.1   ***PROTEIN***   As this process is currently pictured (Fig. ***SYNTHESIS***   II-15), specific "activated" amino acids are attached in the cytoplasm to specific *transfer RNAs* (tRNA), relatively small molecules approximately 75–90 nucleotides long and with folded compact configurations. Some feature of tRNAs, perhaps particular sequences of bases, permits a specific enzyme to recognize a tRNA molecule and attach the proper amino acid to it. The amino acid–tRNA complexes become aligned on a molecule of *messenger RNA* (mRNA) which has derived its base sequence from a specific region of DNA, the gene reponsible for the synthesis of a particular protein. This mRNA base sequence can be pictured as a series of sets of three nucleotide bases, each of which is complementary to a three-nucleotide sequence of a tRNA. By base pairing, the nucleotide sequence of the mRNA specifies the alignment of the specific tRNA sequence. Since each tRNA carries a specific amino acid, amino acids are thus aligned according to the mRNA base sequence. The amino acids are linked together sequentially to form a growing polypeptide chain which, when complete, is released from the RNA (Fig. II-15). A protein molecule consists of one or several such chains folded in a specific three-dimensional form. The amino acid sequence in a polypeptide chain results from an mRNA base sequence and thus, ultimately, from a particular base sequence in DNA. It is the sequence of amino acids that determines the specificity of the protein. As will be outlined later (Section 4.1.2), the folding of polypeptide chains is based upon their amino acid sequences. Different enzymes differ in amino acid sequences. Thus they differ in their *active sites*, the region of the molecule that binds to and acts upon the substrate. The active site of a particular enzyme consists of specific amino acids arranged in a specific geometrical array.

The interaction of tRNA and mRNA, and protein synthesis, take place on a ribosome–mRNA complex. A given mRNA molecule is simultaneously complexed at different points with a number of ribosomes to form a polyribosome or *polysome* (Figs. II-16 and II-30). Present theory suggests that the ribosomes and mRNA molecule move with respect to one another as successive portions of the mRNA molecule are "read" or "translated" into amino acid sequence.

Each ribosome of a polysome apparently synthesizes a polypeptide chain so that several chains are probably being synthesized simultaneously along a given polysome. The number of ribosomes per polysome varies for different cells and proteins; on occasion polysomes are reported with as many as 10–20 ribosomes or, rarely, many more.

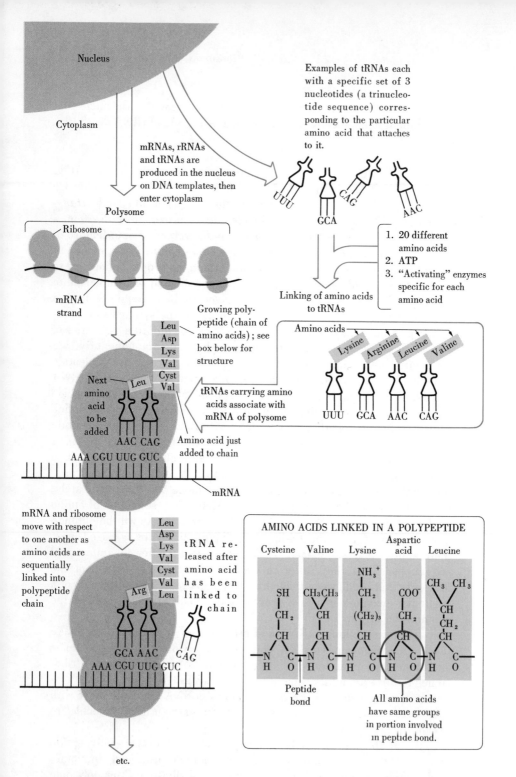

Fig. II-15 *The current concept of protein synthesis. The diagram shows only a few of the many different tRNAs and of the 20 amino acids. The trinucleotide sequences shown for the tRNAs are hypothetical, but they are plausible from knowledge of the genetic code and tRNA structure. The actual sequences may include modified bases in addition to the usual four.*

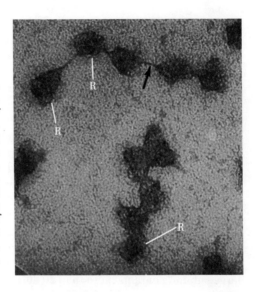

Fig. II-16 Two polysomes *isolated from human reticulocytes (progenitors of red blood cells). Each appears to be composed of 5 ribosomes (R). The polysome at the top of the micrograph was stretched during preparation, and the thin strand (arrow) running between its ribosomes may be a strand of mRNA.* × *350,000. See also Figure II-30. (Courtesy of A. Rich, J. R. Warner, and C. E. Hall.)*

Many functioning proteins are composed of several polypeptide chains, which may be coded for by separate genes and translated from separate mRNAs. As Section 4.1.2 will outline, the chains often can associate with one another spontaneously in the test tube, but how they assemble in the cell is not certain. Proteins may also undergo several types of modifications subsequent to their synthesis, such as addition of carbohydrates (Section 2.5.2) and limited enzymatic cleavage (Sections 2.4.1 and 3.1.2).

Even a relatively small bacterium may contain 5000–10,000 ribosomes; a larger cell may contain from 100,000 to many million. Thus many proteins can be made simultaneously by one cell. On the average, snythesis of a complete polypeptide chain may take from 10 to 20 seconds to a minute or two, depending on the cell type and on the length of the chain. For example, in mammalian cells producing the blood protein *hemoglobin*, about one minute is required for completion of a polypeptide containing 150 amino acids. Bacterial cells may synthesize their proteins at a somewhat faster rate than this.

2.3.2   ***RIBOSOMES***   Ribosomes have been well characterized by a variety of techniques. They appear in the electron microscope as spherical to ellipsoidal bodies, roughly 150-250 Å in diameter, and consisting of two distinct subunits (Fig. II-25). The subunits can be reversibly separated by various treatments, for example, by lowering the Mg^{2+} concentration of the medium. The two subunits differ in size. Generally, ribosome sizes are spoken of in terms of the speed with which they sediment in a centrifugal field. The *Svedberg unit*, designated as S, is the unit used in measuring this speed; it is a rough measure of particle or molecule size, but it is influenced also by the density and shape of the particles and by the medium in which they are suspended. Intact ribosomes of eucaryotic cells sediment at 80 S and have a molecular weight of about 5 million. These figures are derived from mammalian cells; other eucaryotes may differ slightly, and bacterial cell ribosomes are distinctly smaller (Section 3.2.3). The two ribosomal subunits sediment respectively at about 60 S for the larger and 40 S for the smaller; the larger subunit accounts for 60–70 percent of the weight of the ribosome. The larger subunit contains an RNA molecule (28 S) the molecular weight of which is about 1.5 million, twice that of the RNA (18 S) in the smaller subunit; the number of nucleotides can be estimated from the fact that each nucleotide contributes a molecular weight of approximately 330. A small RNA molecule (5 S; molecular weight 40,000) is also found in the large subunit. Each subunit is a ribonucleoprotein particle with roughly equal amounts of RNA and protein. In bacteria which have been most adequately studied the large subunit contains more than 30 different protein molecules, and the small one approximately 20. In eucaryotes the corresponding numbers may be somewhat larger. Each protein is presumed to occupy a specific site in the ribosome's structure. In the functioning polysome, the mRNA is associated with the small subunits of the ribosomes, and the newly synthesized polypeptide, as it forms, is transiently associated with the large subunit.

Ribosomal RNA molecules are large, and they may account for a substantial proportion of the total cellular RNA (often 75 percent; in some bacteria 90 percent). But as yet the details of their roles are unknown. They are not translated into corresponding proteins. In comparison to ribosomal RNAs, the tRNAs are small; they have molecular weights of 25,000. Messenger RNAs vary considerably, as would be expected. Each contains at least the three nucleotides per amino acid required to synthesize a polypeptide chain, and the lengths of these chains vary greatly in different proteins. Insulin contains about 50 amino acids, the pancreatic digestive enzyme precursor *chymotrypsinogen*, about 250; other single polypeptide chains may have many more amino acids (500–1000 or more have been reported for some bacterial and viral proteins). Most known mammalian cell mRNAs fall in the range of 300–5000 nucleotides. Much longer ones are known in bacteria or viruses, where a single RNA may code for several proteins (Section 3.2.5). The length of the particular mRNA molecule present is presumably a chief determinant of the size of a polysome. (For example, each of the polysomes that synthesize hemoglobin polypeptides seems to include five ribosomes; Fig. II-16).

Experiments may prove that the RNA in ribosomes holds the components of the complex system of protein synthesis together in some specific manner. Very likely, tRNAs have special base sequences for binding to the ribosomes and perhaps for binding to one another as well as to mRNA. It may be speculated that rRNAs stabilize the interaction of mRNA and tRNA by binding to both, thus holding them in fixed relation to one another; or perhaps that the rRNAs hold the ribosomal proteins in some specific configuration. The functions of these proteins are being studied through reconstitution experiments. The ribosomal proteins and RNAs can be separated and then, by mixing them under suitable conditions, functional subunits can be reassembled. By leaving specific proteins out of the reconstitution mixtures, one can determine the effect of the lack of particular ones on the functioning of the particle. Most of the proteins, but apparently not all, appear to be essential for protein synthesis. Some probably play structural roles, while others may have enzymatic functions (for example, *peptidyl transferase*, the enzyme that brings about the actual formation of a peptide bond, may be an integral part of the large ribosomal subunit).

Because ribosomes play an essential role in protein synthesis, the mechanism by which they themselves are made is of great interest. Recent experimental analysis points toward the nucleolus as playing a central role in the formation of rRNA.

2.3.3   NUCLEOLI   Autoradiographic evidence shows nucleoli to be the sites of extensive RNA synthesis (see Fig. I-22). Cytochemical and cell fractionation studies indicate that at least 5–10 percent of the nucleolus is RNA; the rest is mainly protein. Preparations of isolated nucleoli also contain some DNA, mainly in the *nucleolus associated chromatin* that is attached to the nucleoli and sends strands deep into their substance. When discrete chromosomes become visible in cell division, nucleoli are often seen associated with specific *nucleolar organizer* regions of specific chromosomes (Fig. II-18). Presumably these are the same chromosome regions that contribute to the nucleolus associated chromatin when the cells are not in division.

As seen in the electron microscope, nucleoli are divided into zones of distinct structure. In addition to chromatin strands, there are areas rich in granules, roughly 150 Å in diameter, and regions that are primarily fibrillar or amorphous (Fig. II-17). The three-dimensional arrangement of these components varies among different cell types. They may be intermingled in apparently complex fashion (Fig. II-17), or, for example, as in hepatocytes and other cells in amphibia, the granules may be concentrated at the nucleolar periphery, surrounding a fibrillar core. Observations of the changes induced in nucleolar structure by digestion with RNase and proteolytic enzymes indicate that both the

granular and the fibrillar zones contain RNA and proteins. A number of
investigators have also suggested that DNA from the chromatin may
extend into some of the fibrillar zones, but further study is needed to
clarify the details and frequency of this. For one special case, that of
amphibian oöcyte nucleoli, there has been a dramatic microscopic demonstration of the interrelations of DNA and RNA in the fibrillar core
(Fig. IV-18).

No membrane delimits the nucleolus. However, contacts between
nucleoli and nuclear envelopes are frequently seen; these might facilitate passage of nucleolar products into the cytoplasm.

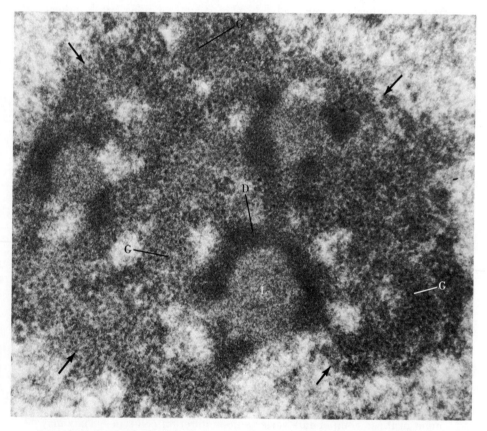

Fig. II-17   Nucleolus *from a cell in the vaginal lining (human). Arrows
indicate the approximate edges of the nucleolus. The organelle contains
numerous granule-like structures* (G), *dense fibrillar regions* (D), *and
amorphous somewhat less dense regions.* (L). *The pale irregular areas
probably contain strands of chromatin.* × 60,000. *(Courtesy of J. Terzakis.)*

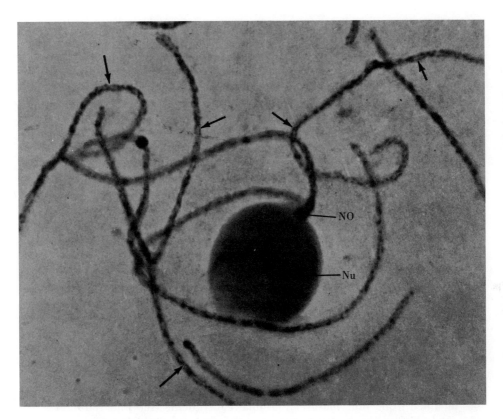

Fig. II-18   *Preparation of a cell in corn during male gamete formation. The cell has been squashed prior to fixation, thus separating organelles and permitting observation of some details. The thread-like structures are pairs (arrows) of closely associated chromosomes (pachytene stage of meiosis; see Chapter 4.3 and Figs. IV-26 and IV-27.) Note the darker regions (chromomeres) present at intervals along the length of the threads. The nucleolus is prominent (Nu). It is associated with a specific chromosomal site, the nucleolar organizer (NO). Approx. × 2000. (Courtesy of M. M. Rhoades and D. T. Morgan, Jr.)*

2.3.4   NUCLEOLI AND RIBOSOME SYNTHESIS   Cytologists noted long ago that nucleoli are particularly prominent in cells that have high rates of protein synthesis. Information outlined in the preceding paragraph suggests a role in RNA metabolism. A much more detailed analysis of nucleolar function is possible from several additional lines of evidence.

One important series of observations has been made on a mutant strain of the clawed toad *Xenopus*. It was initially noted that matings of certain *Xenopus* individuals produce offspring of which one quarter die. (The explanation for this percentage of inviable offspring will be outlined in Section 4.3.2.) For the present purposes, the key fact is that cytological and genetic studies show that death can be attributed to an inheritable chromosomal defect affecting nucleolus formation. The nuclei of embryos carrying defective nucleolar chromosomes contain either no visible nucleoli or grossly abnormal ones. The embryos develop normally until just after hatching and then die. Death occurs shortly after the stage when normal embryos show a great acceleration in ribosome synthesis; before this, both normal and abnormal embryos rely on ribosomes stored in the egg during its maturation preceding fertilization. The embryos that die cannot make ribosomes, and the lack of nucleoli is a visible manifestation of this inability. Ribosomal RNA purified from normal *Xenopus* cannot hybridize with DNA from abnormal embryos, indicating that the DNA base sequences for rRNA formation are missing in the mutant chromosome. These observations demonstrate a central role for nucleoli in ribosome formation.

Initially it was believed that the small granules normally present in the nucleolus represented newly formed ribosomes that had accumulated there. However, the evidence now available does not support this idea. Properly isolated nucleoli contain few complete ribosomes. They do contain much RNA metabolically related to rRNA, but it seems to be in the form of extremely large molecules (up to 45 S; molecular weight about 4 million) that must be enzymatically fragmented and otherwise modified to become mature ribosomal constituents (see Fig. II-19). A single 45 S molecule gives rise to one 18 S and one 28 S RNA. The sum of the molecular weights of the two rRNAs is less than 2.5 million. This means that much of the precursor molecule is left over. Evidently each precursor contains a segment corresponding to an 18 S rRNA, one corresponding to a 28 S RNA, plus "spacer" segments (Section 4.2.6) that may be important in the processing of the precursor into the rRNAs but are destined for eventual degradation within the nucleus.

The complexing of ribosomal proteins with the rRNA also begins in the nucleolus. The proteins apparently are synthesized in the cytoplasm but migrate into the nucleus to participate in the initial steps of the assembly of ribosomes. The detailed sequence of events leading to functional particles is being actively investigated. Present concepts suggest that the large and small ribosomal subunits pass into the cytoplasm separately. At least in some cells, a ribonucleoprotein particle containing the 18 S RNA from a given precursor molecule appears in the cytoplasm a few minutes earlier than does one with the 28 S RNA made from the same precursor. (Additional features of ribosome assembly are discussed in Section 4.1.6)

The most obvious places to search for DNA template sequences

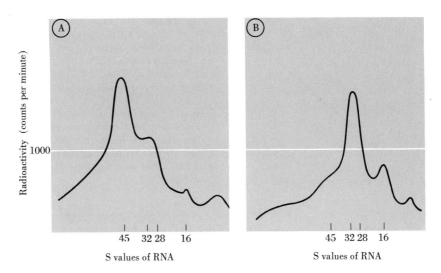

Fig. II-19 *RNA synthesis in HeLa cells. The cells were exposed for 25 minutes to radioactive RNA precursor (^{3}H-uridine). A. At this time a sample was taken from which RNA was extracted and centrifuged on a density gradient (see Chapter 1.2B) to separate RNA molecules of different size classes (measured here in terms of S; see Section 2.3.2). B. RNA extracted from cells exposed to radioactive precursor then grown for an additional 20 minutes in nonradioactive medium in which the drug* **Actinomycin D** *was included. This drug prevents the transcription of new RNA molecules on DNA templates. Thus changes in distribution of radioactivity during the 20-minute exposure reflect modifications of molecules made before the cells were placed in the drug-containing medium rather than synthesis of new RNA molecules. Comparison of* **B** *with* **A** *indicates that there are fewer radioactive 45 S molecules and more 16–18 S and 32 S molecules. This is interpreted as meaning that 45 S molecules give rise to 16–18 S and 32 S molecules; probably a given 45 S RNA molecule is fragmented to produce one 16–18 S and one 32 S molecule. The 16–18 S and 32 S RNAs are precursors of the RNAs in the small and large ribosomal subunits. (From J. E. Darnell and J. Warner.)*

responsible for transcription of rRNA (the *rDNA* sequences) would be in the nucleolar chromatin. Autoradiographic evidence for some cell types indicates that the sites of initial incorporation of radioactive RNA precursors in nucleoli are in portions of the nucleolus-associated chromatin and in fibrillar zones of the nucleoli. Subsequently the labeled molecules appear in the granular zones. The tentative explanation is that the fibrillar zones contain large rRNA precursor molecules, newly transcribed from rDNA, and that as these are modified, they pass into the granular zones (see Fig. IV-18 for an electron micrograph of the precursor RNAs). Another type of evidence for the presence of rDNA sequences in the expected locales is summarized in Figure II-20. This

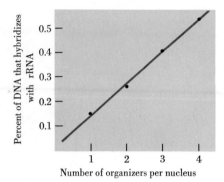

Percent of DNA that hybridizes with rRNA

0.5
0.4
0.3
0.2
0.1

1   2   3   4

Number of organizers per nucleus

Fig. II-20   *An experiment by F. M. Ritossa and S. Spiegelman on the relations between nucleolar organizers and rRNA in the fly* Drosophila. *The DNA purified from nuclei of individuals with different numbers of nucleolar organizer regions in the chromosomes is tested for the number of sites that can hybridize with purified rRNA (graphed as the percent of the DNA per nucleus that hybridizes).*

indicates that different strains of the fruit fly *Drosophila*, carrying different numbers of nucleolar organizer regions, show proportional differences in the number of DNA sites in their nuclei that can hybridize with purified rRNA. The complementarity in base sequence implied by the hybridization indicates that these DNA sites could serve as rRNA templates. There is some dispute about details of such experiments. For example, some cells of flies with one organizer may undergo an interesting "compensatory" increase in rDNA and thus possess more rDNA than expected. But the basic conclusions seem sound. Careful bookkeeping of the sizes and amounts of hybridizing molecules indicates that a normal *Drosophila* nucleus (two nucleolar organizers) contains sufficient DNA sites to bind about 250 28 S rRNA molecules and the same number of 18 S rRNA molecules. In other words, the nucleolar organizer involves a cluster of multiple copies, apparently identical, of the rDNA. Section 4.2.6 will discuss some additional features of the arrangement of DNA in the nucleolar organizer.

The chromosomal sites of repetitive DNAs can be detected by "*in situ*" RNA-DNA hybridization techniques. In these procedures radioactive RNAs are hybridized with preparations of fixed cells rather than with purified DNAs. Figure II-21 shows the result of such an experiment on the localization of rDNA.

Some significant features of nucleolar behavior remain to be clarified. For example, during cell division the nucleolus usually disappears as a discrete entity, reappearing only in the terminal stages of division. Studies on plant cells indicate that the first sign of nucleolar reappearance is the occurrence of many small nucleolus-like bodies that eventually fuse to form the one or two nucleoli of the usual interphase cell. Although the origin of the small bodies is not yet known, such observations have led to the hypothesis that the nucleolar organizing region or the forming nucleolus collects either material synthesized at many points along the chromosomes or perhaps material that is dispersed

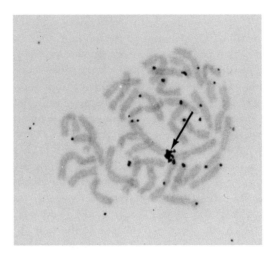

Fig. II-21 *Autoradiogram of chromosomes from* Xenopus *(mitotic metaphase; see Fig. IV-8) prepared by an* in situ *hybridization technique (also referred to as* cytological *hybridization). The chromosomes were exposed to alkaline conditions to "denature" the DNA, separating the two strands of the double helix. Then they were incubated with ribosomal RNA which was highly radioactive (it was synthesized in the test tube using highly radioactive precursors). The radioactive RNA has hybridized (Fig. II-10) with the DNA at the nucleolar organizer sites present on two chromosomes. These two sites are closely associated with one another in this preparation so that the autoradiography shows only one cluster of grains (arrow). Individual grains are scattered through the remainder of the chromosomes but this probably is due to nonspecific binding; such effects presently limit the usefulness of this technique to cases in which repetitive DNA sequences are studied and much RNA binds to a small region, resulting in a definitive cluster of grains (see Fig. IV-17 for another example). ×1500. (From M. L. Pardue,* Cold Spring Harbor Symposium *38: 475, 1974.)*

among the chromosomes when the nucleolus disappears early in cell division (Section 4.2.10).

Finally there are some organisms whose cells form many small nucleoli either normally or under experimental conditions; presumably the cells have many separate nucleolar organizer regions in their chromosomes rather than only a few.

2.3.5   ***mRNA AND POLYSOMES***   Studies on bacteria support the following picture of ribosome functioning. Messenger RNA molecules containing the information for proteins are transcribed from the corresponding DNA by sequential addition of nucleotides to a growing mRNA chain. As the mRNA molecule grows, it associates with a ribosomal subunit of the small type plus a specific tRNA linked to a modified amino acid (*N*-formyl methionine) and also protein molecules known as "initiation" factors. This complex, in turn, associates with a large ribosomal subunit, forming a functional polysome that begins the synthesis of a polypeptide chain. The "initiation" amino acid is the amino acid methionine with a formyl (CHO) group attached. Each amino acid in a polypeptide chain is linked to its two neighbors through an NH and a CO group (Fig. II-15): with *N*-formyl methionine the NH group is blocked, hence this amino acid can form only one link, and thus can occur only at one end of the chain. (Subsequent to completion of a polypeptide chain, the *N*-formyl methionine, or its formyl group, may be removed so that not all completed chains start with the same amino acid.) The base sequence AUG codes for the tRNA that carries *N*-formyl methionine, but there is still some uncertainty as to how the mRNA–ribosome complex "identifies" the proper point to start translation.

The mRNA base sequences UAA, UAG, and UGA, which do not code for amino acids, seem to be used as termination "punctuation." When the ribosome reaches these sequences growth of the polypeptide stops, and with the involvement of protein molecules referred to as "release" factors, the polypeptide is released from the polysomes. As this occurs, the ribosomal subunits separate from one another and rejoin the pool of subunits available for formation of new polysomes. Some investigators suspect that there are few if any actual ribosomes that are not parts of polysomes. When not complexed with mRNA, the large and small subunits may remain separate.

Less is known of the details of ribosome and mRNA functioning in eucaryotes. It is well established that the genetic code is essentially universal; in all organisms the same set of trinucleotide sequences can code for the same amino acids. The fundamentals of protein synthesis probably are pretty much the same in all organisms. But there are some interesting differences in detail between eucaryotes and procaryotes. Procaryotes' ribosomes are somewhat smaller than those of eucaryotes, and they differ slightly in the relative proportions of RNA and proteins. The known mRNAs of eucaryotes usually each carry information for only one polypeptide chain, whereas some mRNAs of bacteria code for several. Procaryotes lack nucleoli. Eucaryotes seem not to utilize *N*-formyl methionine for initiation, although they may use the AUG base sequence, and methionine without a formyl group. Perhaps most striking is the fact that in procaryotes, translation of a message into protein may start before synthesis of the mRNA is complete; ribosome subunits and the other protein synthesis components can be attached to one end of the mRNA before transcription of the other end has been completed. In eucaryotes, RNA synthesis occurs in the nucleus, but protein synthesis takes place in the cytoplasm. Thus messenger RNAs must traverse the nuclear envelope before they are used. They may also undergo some special intranuclear "processing" as will now be described.

Considerations of experimental simplicity (Section 3.1.3) have led to the use of viruses and bacteria for most studies of mRNA. Until very recently little was known of eucaryotes' mRNAs. Advances have now come from several directions. First, to minimize the problem of mRNA heterogeneity associated with the synthesis of many proteins by a given cell, attention has been focused on cell types that specialize in the production of large quantities of particular proteins. Favored ones include reticulocytes (p. 87) and cells secreting egg albumin (ovalbumin) in hens or silk in silkworms. In such cells, polysome-associated RNA molecules have been identified, which possess the size and other properties expected of appropriate messengers and can support protein synthesis in the test-tube. Identification may also be based on the presence of suspects at the proper time—usually just before synthesis of the protein begins. Thus, mRNAs for histones are made in a particular period preceding cell division (p. 311).

A second advance was the recognition that many messenger RNA molecules in the cytoplasm (those for histones are exceptions) have attached to one end a chain of nucleotides, all of which carry the base, adenine—a *poly A* (polyadenylic acid) chain, which may include 200 nucleotides. This has made possible identification of the probable mRNAs in complex RNA mixtures by determining which of the molecules can hybridize (Fig. II-10) with synthetic polynucleotides that carry only thymine or uracil bases. Such hybridization can also be adapted as a preparative technique for mRNA purification.

Much attention currently is focused on the intranuclear RNAs that give rise to cytoplasmic messages. They seem to be included in a complex fraction, aptly called heterogeneous nuclear RNA (hnRNA). The relations of hnRNAs and their possible mRNA offspring are difficult to evaluate. This is due, in part, to the fact that 80 percent or more of the hnRNA made in nuclei never reaches the cytoplasm; instead, it breaks down in the nucleus. In addition, many hnRNA molecules are much larger than mRNAs; they range from 1000 to 50,000 nucleotides while in a few cases in which intranuclear messenger precursors have been identified with fair certainty (for example, silk gland, hen oviduct, slime molds) the precursors seem not to be much larger than the corresponding cytoplasmic messengers. The working hypotheses presently adopted by many investigators are (1) some of the hnRNA molecules give rise to mRNAs; (2) this may involve enzymatic processing in which portions (at least small ones) of the hnRNAs are degraded; and (3) polyA chains are added to mRNAs chiefly in the nucleus, and primarily by enzymatic mechanisms that do not require the presence of long stretches of T in DNA. These hypotheses stem in part from direct studies on mRNAs, such as some indicating that the hnRNA fraction includes polyA-containing molecules, some of which seem to give rise to cytoplasmic RNAs. They fit well with present conceptions of rRNA production (Section 2.3.4) and with what is known of the organization of pertinent regions of DNA (Section 4.2.6). Work in progress should soon settle many of the remaining questions: Does the hnRNA fraction contain several distinct categories of molecules—some serving as mRNA precursors and others functioning entirely in the nucleus? What is the role of polyA? In what form does an mRNA leave the nucleus? Is the intranuclear breakdown of RNA a degradation of the fragments of molecules left over after processing or are certain RNAs somehow selected for use while others, not used, are degraded *in toto*? Are some mRNAs produced directly, not from a larger precursor?

The turnover rate of mRNA is variable in different cell types and probably varies for different mRNAs in a given cell. An mRNA may last for a few minutes (bacteria), several hours, or many days (reticulocytes and other cells specializing in one product). In bacteria the rapid turnover of mRNA is associated with metabolic flexibility. For example, bacteria can rapidly adjust synthesis of some proteins to changing environmental conditions (Section 3.2.5). Presumably this is facilitated by the fact that newly synthesized mRNAs are continually replacing old molecules in directing protein synthesis. If the synthesis of one particular mRNA is greatly increased, this mRNA will rapidly provide an increase in the synthesis of the protein for which it carries the information. In reticulocytes or hen oviduct cells, on the other hand, the relatively long-lived messenger RNA molecules are each used repeatedly for the massive production of particular proteins. The mechanisms controlling mRNA breakdown are only beginning to be studied. Those controlling synthesis are better understood, as will be seen in Sections 3.2.5 and 4.4.6.

2.3.6   ***FREE AND*** Two distinct populations of ribosomes are
           BOUND recognized: those *bound* to membranes
        RIBOSOMES (endoplasmic reticulum); and those that are
                    free, that is, not visibly bound to membranes. Both types form polysomes with mRNA and both play similar roles in protein synthesis, but the free ribosomes leave the newly synthesized proteins in the hyaloplasm (the soluble fraction; see Chapter 1.2B). Bound ribosomes transfer the protein into the interior (or membrane) of the endoplasmic reticulum. Evidence for this difference in function rests on comparative studies of different cell types.

Animal cells that secrete proteins such as digestive enzymes or hormones have a high proportion of bound ribosomes. The proteins enter the highly developed system of rough endoplasmic reticulum and, by mechanisms considered in Chapters 2.4 and 2.5, are "packaged" into membrane-delimited granules. Eventually the granules are released at the cell surface.

Similar synthesis on bound ribosomes and packaging in membrane-delimited granules probably occurs with lysosomal and peroxisomal enzymes. Under usual circumstances, most cells of multicellular animals do not release the contents of lysosomes and peroxisomes from the cell.

Free ribosomes are especially abundant in cells synthesizing much protein (enzymes and so forth) used internally in rapid growth but not specially packaged inside membrane-delimited structures (for example, cancer cells and most cells of embryos). In *reticulocytes*, the cells that mature into red blood cells, most of the ribosomes are of the free variety. These cells synthesize the *hemoglobin* that ultimately fills the cytoplasm. The hemoglobin is not included within special membranes and is not exported from the cell.

Hepatocytes synthesize much protein (such as serum albumin) that is secreted into the blood and other proteins for internal use (for example, in replacing proteins lost in turnover). These cells contain many free and many bound ribosomes. Presumably not only the proteins in the hyaloplasm but also some of those in nonmembrane-delimited organelles are synthesized on free ribosomes.

As with most generalizations, the ones in this section must be treated cautiously. For example, it may be premature to rule out the synthesis of a few proteins of the hyaloplasm on bound ribosomes.

We have little insight into how a polysome "knows" whether or not to attach to the ER. One might imagine that this information is somehow encoded in the nucleic acids in such a way that the mRNA for pertinent proteins binds to some component in the ER. But many now feel that the distinction depends on such features of bound polysomes as a special short sequence of amino acids that is made during the initial period of polysome functioning. The pertinent features of the ER membranes also remain to be described. Some investigators feel there may be specific receptor-like sites in the membrane that bind ribosomes.

c h a p t e r **2.4**
ENDOPLASMIC RETICULUM (ER)

A great contribution of electron microscopy was the demonstration that in many cells of eucaryotes an extensive membranous system, the endoplasmic reticulum (ER), traverses the cytoplasm. (The resolving power of the light microscope is too low for identification or analysis of this system.) The lipoprotein membranes delimit interconnecting channels that take the form of flattened sacs (known as *cisternae*) and tubules. Rough ER is studded with ribosomes most of which are probably in the form of polysomes. Rough ER and smooth (ribosome-free) ER are part of one interconnected system. The proportions of these two ER types vary in different cell types. Proteins (including certain enzymes), lipids, and probably other materials are transported and distributed to various parts of the cell through the ER. In some cases these substances may accumulate and may be stored within the ER for considerable periods. In striated muscle the ER takes a special form, the *sarcoplasmic reticulum*, involved in coupling nerve excitation to muscle contraction (Section 3.9.1).

The endoplasmic reticulum is much more than a passive channel for intracellular transport. It contains a variety of enzymes playing important roles in metabolic sequences, for example, in synthesis of steroids.

As mentioned earlier (Chapter 1.2B) the ER, as such, cannot be isolated intact from cells. Fragments of ER are the major component of the *microsome* fraction. The fragments are in the form of membrane-delimited vesicles (Fig. I-23). The vesicles appear to form as pinched-off portions of the ER during homogenization. The interior of the vesicles corresponds to the interior of the ER cisternae and tubules. The microsomes from rough ER retain their ribosomes which thus stud the external surface of the vesicles.

2.4.1   **PROTEIN-SECRETING CELLS**   In cells actively engaged in synthesizing and secreting proteins, *rough* ER is particularly highly developed.

The most intensively studied secretory cell is the type responsible for production of digestive enzymes in the guinea pig pancreas (Figs. II-22 and II-23); other secretory cells will be discussed in Chapter 3.6. The fate of specific proteins (precursors of digestive enzymes made by the pancreas) has been followed by combining electron microscopy with biochemical analysis of isolated organelles and with autoradiography. By autoradiography, labeled amino acids used in the synthesis of these enzymes can be shown first over the rough ER, later in the region of the Golgi apparatus, still later in secretion (*zymogen*) granules, and finally in the extracellular space at one pole of the cell where secretions are released. (Upon release the proteins enter ducts leading to the intestine where they function after some modification.) From the polysomes where the proteins are manufactured (Sections 2.3.1 and 2.3.6) they

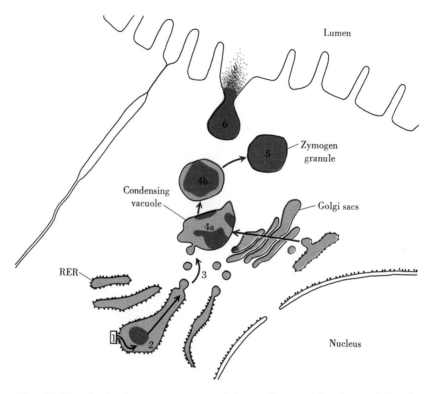

Lumen

Zymogen granule

Condensing vacuole

Golgi sacs

RER

Nucleus

Fig. II-22   *Path of secretory material from polysome (1) to lumen (6) in the exocrine gland cells (Section 3.6.1) of the guinea pig pancreas. RER is rough endoplasmic reticulum. (After G. E. Palade, J. Jamieson, and others.)*

enter the cisternae of the ER and move toward the Golgi apparatus. Then, apparently by transport in small vesicles, they pass from the reticulum either directly into "condensing" vacuoles, or into the sacs of the Golgi apparatus and from these into vacuoles derived from the sacs (Fig. II-22). "Packaged" into vacuoles, the secretory material undergoes concentration (condensation) which results in the formation of dense secretion granules. The vacuoles containing the secretion granules move to the cell surface where the granules eventually are discharged by a kind of reverse pinocytosis (exocytosis) in which the vacuole membrane fuses with the plasma membrane (see Fig. II-33).

The passage of proteins synthesized on bound ribosomes into the cisternae of the ER is one of the key devices by which eucaryotic cells achieve compartmentalization—the segregation of different macromolecules into different structures. Important progress has been made in understanding how proteins, which in their usual state are too

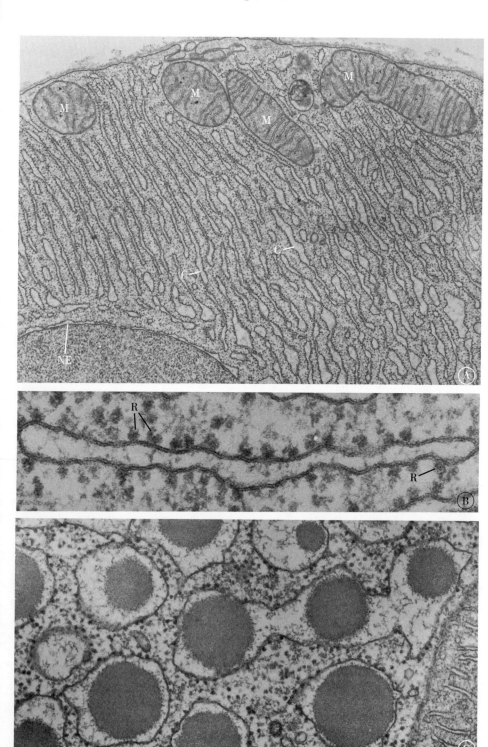

large to cross membranes, nonetheless are able to enter the cisternae. Many of the key experiments have been done with microsome fractions, and thin slices of pancreas. This permits one conveniently to manipulate the biochemistry; one can control levels of energy sources such as ATP, or add inhibitors affecting different phases of the process. Figure II-24 illustrates one such experiment. The data are most simply interpreted as reflecting direct passage of newly synthesized proteins from the ribosomes into the cavities of the ER cisternae. Radioactivity accumulates first at the ribosomes, then ribosomal labeling levels off as the synthesis of new proteins is balanced by the passage of completed ones into the interior of the microsomes. Experiments of these types have provided no support for an alternative possibility, the release of proteins from the ribosomes to the hyaloplasm outside the ER and their subsequent accumulation by the ER. Furthermore, as far as can be told from fragmentary information, the movement from ribosomes into the ER does not require a special input of energy or the participation of transport enzymes.

Rough-surfaced microsomes have been treated with EDTA, an agent that complexes with ions such as Mg^{++}, to dissociate the attached ribosomes into their two subunits. Under carefully controlled conditions of dissociation, most of the small ribosomal subunits can be freed from the microsomes, while most of the large subunits remain attached. Since electron microscopy indicates that ribosomes are attached to the endoplasmic reticulum by only one of their subunits (Fig. II-25), it appears that the larger subunit is the one responsible for this anchoring. The messenger RNA of a polysome is associated with the small ribosome subunit but, as a forming polypeptide chain elongates, it is associated with the large subunit. A widely accepted hypothesis asserts that the growing polypeptide chain passes directly from this subunit into the membrane as a relatively thin strand of linked amino acids. From the dimensions and flexibility of the amino acids one would guess that such a strand might be on the order of 5–10 Å thick over much of its length. As it enters the ER cavity, the chain can begin to fold into its mature compact three dimensional structure. At this point, passage back through the membrane becomes very unlikely, since the overall diameter is now much greater than that of the unfolded polypeptide.

◀ *Fig. II-23* *Features of the* rough *ER. (Courtesy of G. E. Palade.) The three micrographs are from pancreas exocrine gland cells. A. Portion of a cell (rat pancreas) showing numerous flattened cisternae (C) of rough ER. Mitochondria arc seen at (M) and the nuclear envelope at (NE). × 15,000. B. Higher magnification micrograph of a single cisterna (guinea pig pancreas). The membrane surface is studded with numerous ribosomes (R). × 120,000. C. Portion of a cell from a fasted guinea pig, in which large granules, probably of protein, have accumulated inside the cisternae. × 50,000.*

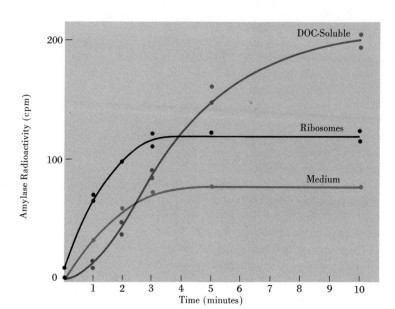

Fig. II-24 *A microsome fraction derived chiefly from rough ER (pigeon pancreas) was incubated in a medium containing a radioactive amino acid (^{3}H-leucine). At successive intervals samples were taken and treated with a detergent, deoxycholate (DOC), which frees the ribosomes and solubilizes the membranes. The deoxycholate-soluble subfraction represents the contents of the microsomes; it is the experimental equivalent of the contents of the ER cisternae. The ribosomes can be separated from this by centrifugation. The* medium *is the solution in which the microsomes were suspended and presumably is the experimental equivalent of the cell's hyaloplasm. The medium, DOC-soluble material, and ribosomes were studied for the extent of incorporation of radioactivity into the protein, amylase, an enzyme secreted by the pancreas. CPM is counts per minute.*

The fact that some radioactive amylase does appear in the medium may reflect leakage from microsomes damaged during preparation or incubation.

Additional more detailed studies strengthen the conclusion that transport is directly from ribosomes to DOC-soluble fraction. (For example, the specific activity, *the radioactivity per unit weight, is found always to be higher for amylase in the ribosomes and DOC-soluble fractions than in the medium.) (From C. M. Redman, P. Siekevitz, and G. E. Palade.)*

Isolated microsomes are relatively permeable to inorganic ions and small molecules, but even so, some special local organization of the membrane might be needed to permit rapid enough passage of a polypeptide chain. It has been suggested that there is some kind of channel through the membrane, at the point of attachment of the ribosome.

Subsequent to its entry in the ER, a polypeptide chain may undergo considerable modification before reaching its final destination. As

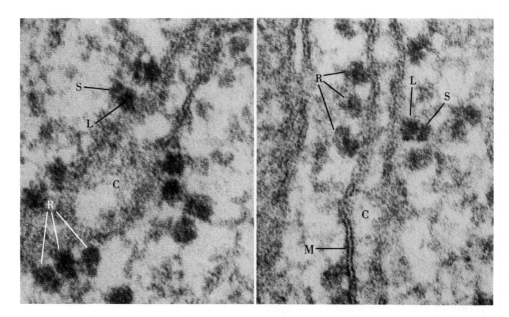

Fig. II-25   *Portions of two cisternae (C) of endoplasmic reticulum in a guinea pig hepatocyte. The delimiting membrane shows a three-layered membrane appearance (M). The ribosomes (R) consist of large (L) and small (S) subunits; the large ones apparently attach the ribosomes to the membrane. × 275,000. (Courtesy of D. D. Sabatini, Y. Tashiro and G.E. Palade.)*

will be discussed in Section 2.5.2, carbohydrates are often added. Several polypeptide chains may associate to form a multichain protein. Where and when this occurs is not known. A surprising number of proteins appear to undergo proteolytic processing steps—portions are nicked off by enzymes. For example, as finally released from the cell, the hormone insulin is composed of two polypeptide chains. As initially synthesized, these two are part of a single "proinsulin" molecule. This is modified into insulin by action of specific proteases which remove part of the molecule, yielding the two insulin chains. Precisely where the proteases and proinsulin are brought together is still being studied.

2.4.2   **PROTEINS IN ER**   The accumulation of protein within the rough ER can be demonstrated microscopically in a few cell types. In these cells, proteins condense to form readily visible accumulations while inside the rough ER cisternae. Granules resembling secretion granules are found in the ER of secretory cells in dog pancreas and in the pancreas of fasting guinea pigs (Fig. II-23). This is direct visual confirmation of the indirect evidence from autoradiography that secretory proteins are transported in the ER. (See also Fig. III-33.)

Cytochemical methods for demonstration of enzyme activities (Section 1.2.3) can also be used to detect proteins within the ER. For example, cells of the thyroid gland and some cells of salivary glands *synthesize* and *secrete* peroxidase enzymes. When preparations of these tissues are incubated in the proper cytochemical medium for demonstration of peroxidase activity (Section 2.1.4), the enzyme-produced reaction product is distributed as in Figure II-26A. As with several other microscopically demonstrable enzymes, the nuclear envelope shows the reaction product like the rest of the ER.

An immunohistochemical method (p. 22) has been used to determine the sites of accumulation of newly synthesized antibodies. When foreign proteins (*antigens*) are injected into a rabbit, *plasma cells* synthesize antibody proteins (Section 3.6.4) that can bind specifically to the injected antigen. In the experiment with the outcome shown in Figure II-26B, the antigenic protein injected was *peroxidase*, purified from horseradish. Two to four days after the last injection, the rabbit's spleen was removed and fixed. Sections were cut and then soaked in a solution of peroxidase (as were control spleen sections from animals that had not been injected with the antigen). Wherever antibody to peroxidase was present in the plasma cells, the specific antigen–antibody reaction caused binding of the peroxidase to these regions. When the spleen sections were then incubated in a medium producing an electron-opaque enzyme reaction product at peroxidase sites, the results were like those in Figure II-26B. Similar treatment of the control sections resulted in no reaction product since no antibody and therefore no peroxidase was present. Thus it may be concluded that the dilated ER cisternae contain the antibody protein synthesized in response to the antigen. (In interpreting this experiment, it should be noted that peroxidase is used twice in different ways. Initially it is an antigen injected into the animal so that the plasma cells are induced to make specific antibodies. Later when the cells, which contain newly made antibodies but no peroxidase, are soaked in a solution of peroxidase, the enzyme is being used as a device to detect the antibodies. The techniques of enzyme cytochemistry are employed to locate peroxidase activity in soaked cells. Where such activity is present, peroxidase has been picked up from the solution and bound. The agents responsible for this binding are the specific antibody molecules whose locations are thus demonstrated by an indirect method.)

In the study just described it was noted that a few of the plasma cells containing detectable antibodies showed reaction product only in some ER cisternae and not in others. This raises, as an important mat-

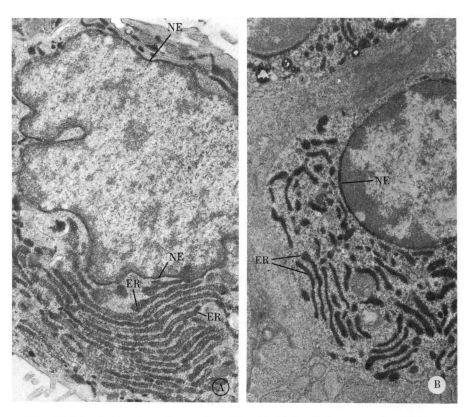

Fig. II-26   **A.** *Portion of a rat thyroid epithelial cell from a section incu-bated to demonstrate sites of peroxidase enzymes synthesized by these cells. Deposits of electron-dense reaction product are present in the nuclear en-velope (NE) and in the many cisternae of rough ER. The enzyme responsible for staining is thought to be one that ultimately is released into the gland's lumen where it may participate in adding iodine to secreted protein (see Section 3.6.1). × 15,000.* **B.** *Portion of a developing plasma cell treated with peroxidase as described in the text (Section 2.4.2). The ER and nuclear envelope (NE) show electron-dense reaction product formed by the enzy-matic action of the peroxidase bound by the specific antibodies synthesized by the cell. × 7500. (Courtesy of E. Leduc.)*

ter for further study, the possibility that different large regions of the rough ER in a given cell may synthesize different proteins. If there are regions of rough ER specialized for the synthesis of particular proteins, it will be interesting to discover how the appropriate polysome–ER as-sociations are controlled. Related to this is the question, how are differ-ent molecules transported in the ER kept separate, brought together, or carried to the proper destination in the cell? For example, hepatocyte

ER is thought to transport lipoproteins, lysosomal hydrolases, peroxisomal enzymes, probably albumin and other molecules. The lipoproteins eventually are released from the cell after inclusion in vacuoles derived from ER or Golgi apparatus (Figs. II-30 and II-34). The hydrolases are packaged into lysosomes, many of which form near the Golgi apparatus (Section 2.8.1). Peroxisomes arise as enzyme-containing bodies that appear to form directly from the ER (Fig. II-57). At present it is not even certain whether transport in the ER is primarily a matter of flow within the membrane-delimited channel, or whether it reflects movement of the entire cisterna, delimiting membrane plus contents, or both.

2.4.3   LIPID SYNTHESIS IN ER   The ER is a major site of lipid synthesis, perhaps the major site in most cells. Enzymes involved in synthesis of triglycerides, phospholipids, and steroids (Fig. II-2) are present in isolated microsome fractions. Autoradiography shows that radioactive lipid precursors are incorporated rapidly into material in the ER. Both rough and smooth ER seem able to synthesize lipids.

Under some normal conditions, and some abnormal ones, lipids may accumulate as visible deposits within the endoplasmic reticulum. Thus when fats are fed to a rat, triglycerides are broken down in the intestine into smaller molecules, chiefly monoglycerides, glycerol, and fatty acids. These then cross the plasma membrane of the absorptive cells lining the intestinal cavity. Within the absorptive cells the smaller molecules enter the ER and enzymes in the ER membranes convert them into triglyceride again. The triglycerides synthesized by the ER are visible as small droplets within the meshwork of smooth endoplasmic reticulum and vesicles that form from this ER (Fig. II-27). Triglyceride is also seen in the cavities of the rough ER, including the nuclear envelope.

Normally much of the lipid synthesized in the ER of hepatocytes complexes with proteins, also made in the ER, to form lipoprotein droplets that are secreted by the cell via vacuoles most or all of which are derived from the Golgi apparatus (Fig. II-30). These lipoproteins are of great importance for the transport of cholesterol and other lipids in the blood. When the compound orotic acid is added to the diet of rats, the animals develop an abnormal "fatty liver". Apparently lipoprotein transport via the Golgi apparatus is blocked and numerous lipid droplets accumulate in the ER, both rough and smooth. The ER breaks into vesicles and the hepatocytes become full of ER vesicles, each containing one or more lipid droplets. The entire process can be reversed by adding adenine (see Fig. II-9) to the diet.

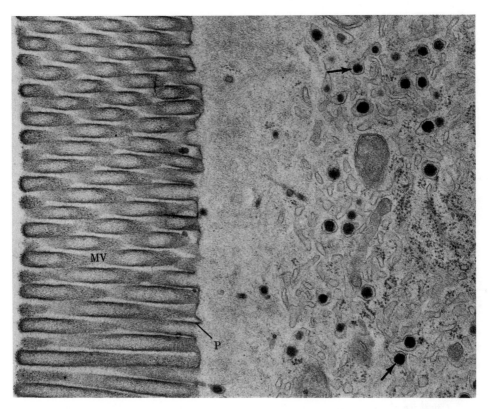

Fig. II-27 *Portion of an absorptive cell in the intestine (see Fig. III-19) of a rat fed a fat-rich meal. Many lipid droplets (arrows) are present inside the endoplasmic reticulum. At the lumen of the intestine, the plasma membrane (P) is seen to be extensively folded into closely packed microvilli (MV).* × *30,000. (Courtesy of S. Palay and J. P. Revel.)*

2.4.4   **SMOOTH ER;**   Whereas the rough ER is extensively devel-
ER FORMATION   oped in protein secreting cells, it is the
smooth ER that is extensive in cells secret-
ing steroids, such as the hormone-secreting cells of the cortex of the
adrenal gland and others (Fig. II-28). Key enzymes of steroid synthesis
are found in microsome fractions which, in these cells, derive mainly
from fragmented smooth ER.

In hepatocytes, extensive deposits of glycogen are found inti-
mately associated with the smooth ER; the polysaccharide is in the form
of small granules that lie in the cytoplasm close to the tubules of the
smooth ER meshwork (Fig. I-7). The significance of this relationship is
not known.

In hepatocytes, both rough and smooth ER can break down or

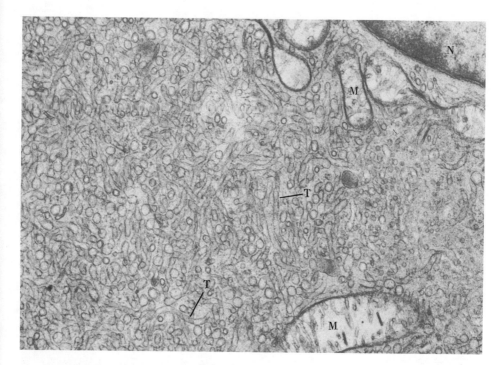

Fig. II-28   *Portion of an* interstitial cell *in opposum testis; these cells produce* steroid hormones. *Tubules of smooth ER* (T) *are abundant in the cytoplasm. An edge of the nucleus is seen at* N *and mitochondria are indicated by* M. × 23,000. *(Courtesy of D. W. Fawcett.)*

otherwise inactivate (detoxify) drugs such as phenobarbital; the enzymes responsible include a set that oxidize the coenzyme NADPH. Following administration of phenobarbital to the animal, dramatic increases occur in the amounts of hepatocyte smooth ER and of the drug detoxifying enzymes. The changes reverse when drug administration ceases. Marked ER changes also occur in rat hepatocytes at about the time of birth. Just before birth, much rough ER is formed. Just after, much smooth ER appears. These and similar situations provide favorable experimental material for study of the formation, maintenance, and change of ER membranes. For example, there have been investigations of the relative amounts of proteins, such as glucose-6-phosphatase, cytochrome b_5, and drug detoxification enzymes, and of different lipids present in hepatocyte ER membranes under various conditions of increase and decrease of the reticulum. From these studies, and from work on the turnover of ER constituents in the steady state (see Section 2.8.3), it has been tentatively concluded that at least some of their proteins and lipids can enter or leave the membranes without noticeably affecting the integrity of the structure. In other words, the membranes seem to be dynamic systems, rather than immutable ones whose original constituents all remain together from the time the structure forms to the time it is destroyed. Apparently the ER in a given cell can change both by addition (or removal) of membrane and by modification of existing membrane.

Increases in the amounts of ER apparently occur by expansion of the preexisting reticulum. Precisely how this takes place is not known. Many of the proteins and lipids participating in the expansion probably are synthesized by the ER itself. Some investigators are convinced that the new molecules are inserted all throughout the preexisting reticulum, so that one does not have detectable "old" and "new" regions. However, by itself this would not account for processes such as the growth specifically of smooth ER, following phenobarbital administration. To explain the latter it has been suggested that smooth ER can be produced by synthesis of membrane in rough ER, which then becomes smooth by detachment of ribosomes. Or perhaps the proteins are made in the RER but the SER contributes to its own growth by synthesis of lipids—the ability of macromolecules to move in the plane of the membrane (Section 2.1.2) raises several possibilities for processes by which this could come about. Perhaps there are several mechanisms for ER increase, with one or another being most important under given circumstances.

chapter **2.5**
THE GOLGI APPARATUS

The history of our knowledge of the Golgi apparatus illustrates well the dependence of understanding of cell organelles upon the techniques available at a given time. The structure was discovered in 1898 by Camillo Golgi, who named it the "internal reticular apparatus." His "metallic impregnation" method, involving long-term soaking in osmium tetroxide (later modified by using silver salt solutions) was difficult to control; despite several improvements by Golgi and others, it failed to give consistent results. Furthermore, the structure had not been seen in living cells. Thus a lively controversy developed between those who thought it to be a real cell structure and those who insisted it was simply an artifact of the fixing and staining procedure. A flood of publications failed to clarify the situation. With the development of electron microscopy, however, the controversy was settled unequivocally. All eucaryotic cells studied, with rare exceptions such as red blood cells, possess the Golgi apparatus. Wherever the classical cytologists had shown the Golgi apparatus in their drawings, electron microscopists found stacks of flattened sacs, each bounded by smooth-surfaced membrane. Located nearby, and sometimes connected to the flattened sacs, were vesicles of various sizes. It was this membrane system in which the osmium or silver was deposited during the classical impregnation procedures.

A major recent advance in understanding Golgi apparatus functions is the development of methods for isolating cell fractions rich in fragments of Golgi apparatus; these procedures have been applied to a growing number of animal and plant cells. The fractions are found to be highly enriched in enzymes known as *glycosyl transferases*. These enzymes catalyze polymerization of sugars into polysaccharides that cells release to extracellular spaces, for example, certain components of plant cell walls (Section 3.4.1) and several secretions of animal cells. Glycosyl transferases also are responsible for the attachment of sugars to glycoproteins (Section 2.1.5), molecules that are prominent in many secretions and in plasma membranes. Microscopists have long associated the Golgi apparatus with roles in secretion. The biochemical findings are beginning to provide essential clues to the molecular events underlying such roles.

2.5.1   ***MORPHOLOGY*** Terminological and conceptual problems abound in the literature on the Golgi apparatus. In part these reflect difficulties that sometimes arise in making unambiguous distinctions between the Golgi sacs and vesicles, and other vesicles or smooth ER that may be present nearby. Moreover, with the usual thin sections used for electron microscopy, it is difficult to evaluate the three-dimensional arrangement of the apparatus. Thus in many animal cells, the apparatus has the form of a continuous network, but a given electron-microscope section will include only scattered portions of the network and thus may give the impression that the cell contains many separate stacks of Golgi sacs. The overall three-dimensional arrangement can better be appreciated by light microscopy of sections incubated in a cytochemical medium for demonstration of the enzyme *thiamine pyrophosphatase* (TPPase), which is present in high levels in the apparatus (Figs. I-16 and II-29). Additional information has begun to come from use of relatively thick sections (0.5–5 μm) in the electron microscope; more of the apparatus is included in a given section than is true in the usual thin sections.

There is great diversity in the size and shape of the Golgi apparatus in different cell types. For instance, in neurons an extremely elaborate network surrounds the nucleus (Fig. I-16), whereas in animal cells that secrete proteins or carbohydrate-rich materials, and in absorptive cells, the apparatus is more compact and is usually located between the nucleus and the cell surface where secretion or absorption takes place (Fig. II-29). In many cells of higher plants, and in a very few animal cells, the Golgi apparatus appears to consist of many unconnected units, called *dictyosomes*. Plant cells may contain dozens to hundreds of these dictyosomes, each a stack of Golgi sacs.

The number of sacs in the Golgi stack varies from three to seven in most higher animal and plant cell types (Figs. II-30 and II-31) to ten to twenty in some cells such as the unicellular organism *Euglena*. Some investigators believe that certain lower plants utilize a single sac for their Golgi apparatus. In the stacks the sacs are separated from one another by a relatively constant distance of 200–300 Å. The nature of the material or forces holding them together as a stack is unknown, but in a few cells a thin layer of electron-opaque, sometimes fibrillar, material is seen in the middle of the 200–300 Å space. The sacs themselves are of variable width, sometimes quite flat and sometimes dilated by materials that accumulate within them.

In most cell types, plant and animal, the Golgi stack is morphologically polarized—one face looks different from the other (Fig. II-31). We will use the terms *outer* and *inner* to refer to the two faces. Others employ "convex and concave," "forming and maturing" or "cis and trans." The sacs at the two faces may differ in a variety of features, few of which are understood in functional terms. For example, in many cell types it is only the sac at the outer face that shows osmium deposits with metallic impregnation methods, while only the one at the inner face shows reaction product in tissues incubated to demonstrate TPPase activity.

An interesting feature of the TPPase-rich Golgi sac in certain neurons and in hepatocytes is the presence of numerous interruptions or *fenestrations* (see frontispiece); indeed, the sac looks like a meshwork of

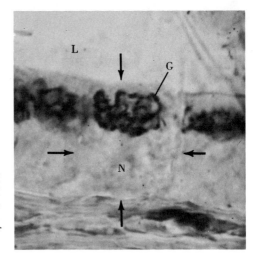

Fig.   II-29   Thiamine pyrophosphatase preparation (see Fig. I-16.) showing the Golgi apparatus (G) of a cell of the rat epididymis, a male reproductive gland. As is usually the case for secretory and absorptive cells, the Golgi apparatus is situated between the nucleus (located at N but unstained) and the lumen (L). Arrows indicate the approximate locations of the borders of the cell. × 900.

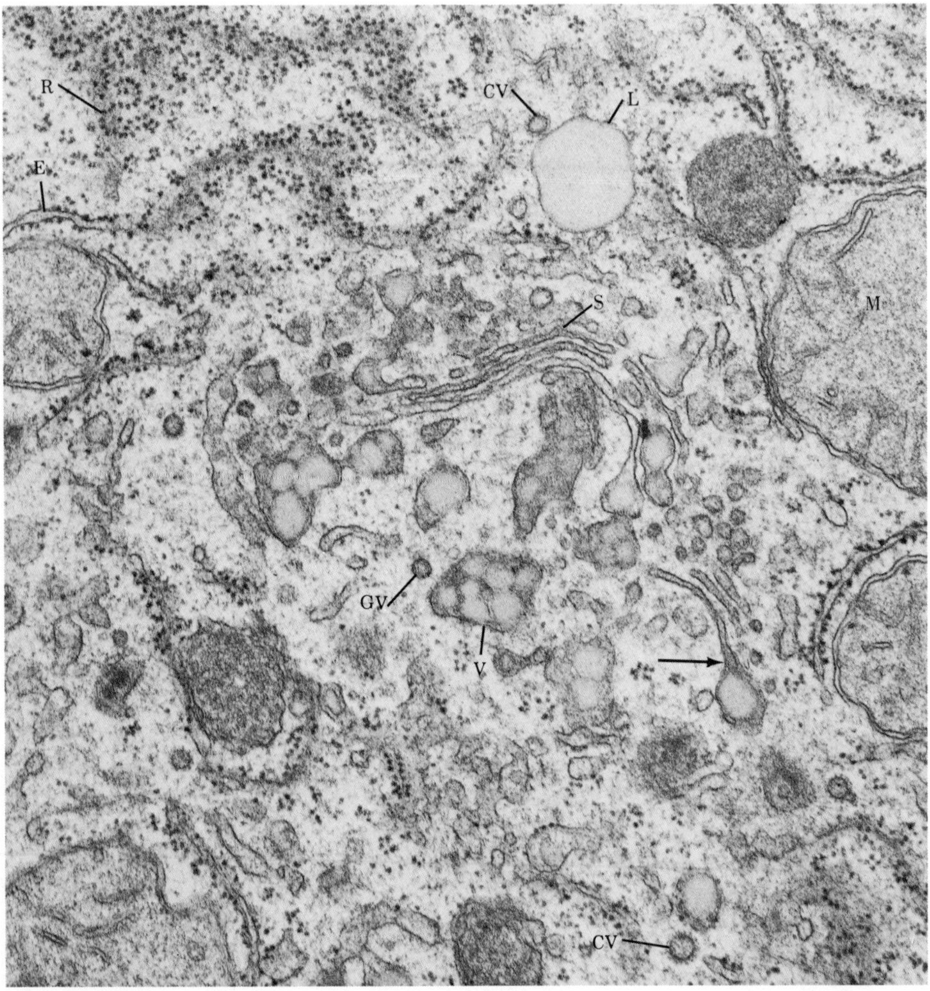

Fig. II-30 *Portion of a rat hepatocyte. M indicates a mitochondrion; E, endoplasmic reticulum; L, a lipid droplet; S, Golgi sacs, GV, small vesicles of the Golgi apparatus; and CV, "coated vesicles" (vesicles with a fuzzy coat on their membrane surface). The structure at V is a large Golgi vesicle (vacuole) filled with lipoprotein droplets; the arrow indicates one such vesicle, probably in formation as a dilatation (swollen region) of a Golgi saccule. At R the endoplasmic reticulum has been sectioned so that the cisterna surface is seen in face view. The ribosomes are arranged in spiral, hairpin, and other patterns that probably reflect ribosome grouping in polysomes.* × *30,000. (Courtesy of L. Biempica.)*

tubules, interconnected in polygonal arrays. Fenestrations, generally less extensive, may be present in the other sacs of the stack and also in the endoplasmic reticulum near the Golgi apparatus. Their functional significance is uncertain. They could provide an increased surface area

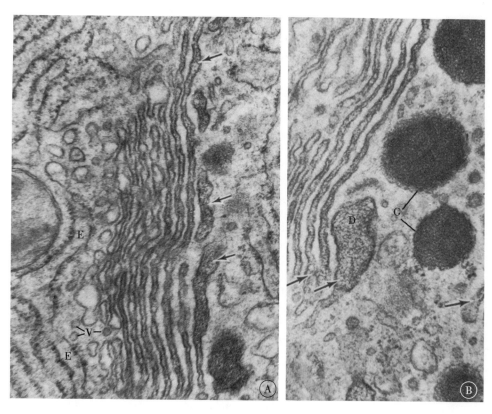

Fig. II-31 *Small portions of cells in glands (Brunner's glands) of the intestine (mouse) showing the Golgi apparatus. Dense material is present within the Golgi sacs (arrows). This material appears more concentrated in the sacs toward one surface of the stack (the inner surface—toward the right) than in the sacs near the other surface (the outer one). As these figures suggest, secretory granules appear to form at least in part by accumulation of the dense material within dilated portions (D) of the Golgi sacs at the inner surface of the stack. Eventually the dilated regions pinch off and give rise to membrane-enclosed granules (G). E indicates ER, and V, small Golgi vesicles. A. × 54,000. B. × 55,000. (Courtesy of D. Friend.)*

for interactions of the elements of the stack with adjacent structures. Or they might somehow relate to the fact that the functioning of the Golgi apparatus involves extensive rearrangements of membrane-delimited compartments, such as the fusion of vesicles with the sacs or the budding of vesicles from the sacs.

An interesting instance of polarity in the Golgi stack occurs in the fungus *Pythium ultimum*, where the membranes of different sacs do not have the same thickness. At the outer face the sac membrane is relatively thin, similar to that of the ER. The closer the sac is to the inner face, the thicker its membrane. In many secretory cells there seems to be a similarly oriented gradient in the concentration of materials visible within the sacs (Fig. II-31). Such observations have been taken to indicate that the sacs move down the stack, from outer to inner, and undergo changes as they move. Often secretory granules form at the inner face, apparently by the "packaging" of materials within vacuoles or vesicles that bud from the inner sac (Fig. II-31). Movements of these types do seem likely, but techniques are not available to establish unequivocally that they occur, and there is no basis for believing that passage from one surface of the stack to the other is an invariable feature of Golgi apparatus functioning. For example, while the prominent secretion granules of secretory cells appear to separate from the inner sacs, membrane-delimited vesicles probably also separate from the lateral margins of the sacs further up the stack. In the white blood cells known as polymorphonuclear leukocytes (Section 2.8.2) packaging of enzymes into lysosomes occurs at the inner face of the Golgi apparatus, while later in the cells' maturation, a different set of enzymes is included in the "specific granules" that form at the outer Golgi face. Such observations hint at controls and complexities in the movement through the Golgi apparatus that are still to be analyzed.

2.5.2   ***ROLE IN***   For a large variety of gland cells, morpho-
                   SECRETION   logical studies indicate that secretory ma-
                                      terials are packaged within membrane-
delimited bodies by the Golgi apparatus or related structures. Autoradiography (ARG) has clarified important facets of the packaging. (In what follows, bear in mind that the processes of tissue preparation for microscopy usually wash out small molecules so that ARG shows only the labeled macromolecular propulation.)

When cells secreting proteins are administered amino acids labeled with tritium (p. 23), the first organelles to show radioactivity are the ribosomes and ER. In the mammalian pancreas (Fig. II-22) this is observed at 3–5 minutes after label administration. Only later (20–40 minutes) is radioactive protein present in or near the Golgi apparatus (Fig. II-32). The likely explanation is that proteins are synthesized by the ribosomes of the rough ER and that some time is required for them to be transferred to the Golgi apparatus. (The mechanism of such transfer will be discussed in the next section.)

With suitable tritium-labeled sugars, the ARG results may be dramatically different. Most plant cells secrete polysaccharide components of their surrounding extracellular walls (Figs. III-17 and IV-23 and Section 3.4.1). Some animal cells also secrete polysaccharides; for example, cartilage cells release chondroitin sulfate, a major component of the stiff extracellular cartilage matrix (Section 3.5.1) and goblet cells

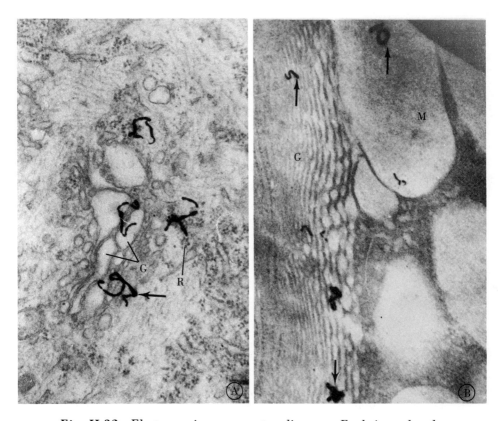

Fig. II-32 **Electron microscope autoradiograms.** *Each irregular dense structure (arrows) would be seen as a single grain in the* **light** *microscope (Fig. I-21).* **A.** *Part of a neuron in a rat spinal ganglion fixed 10 minutes after exposure to radioactive amino acid (³H-leucine). Two of the grains lie close to clusters of membrane bound ribosomes (R) while the other two lie over Golgi sacs (G). Approx. × 40,000. (Courtesy of B. Droz.)* **B.** *Portion of a mucus-secreting cell (a goblet cell, see Fig. III-19) in rat intestine fixed 20 minutes after exposure to radioactive sugar (³H-glucose). Grains are seen over the Golgi saccules (G) and over a large Golgi vesicle (vacuole) containing mucus (M). × 45,000. (Courtesy of M. Neutra and C. P. Leblond.)*

in the intestinal lining (Fig. II-32) secrete *mucin* which contributes to the protective layer of mucus that lines the intestine. With these cell types radioactive sugars are seen to accumulate first in the Golgi apparatus, not in the ER. This holds also for radioactive sulfate in cells, such as those of cartilage, secreting sulfated polysaccharides. Apparently the Golgi apparatus is the site of synthesis of secretory polysaccharides.

With cells secreting glycoproteins rather than polysaccharides, ARG indicates that the addition of carbohydrates to the protein molecules begins in the ER and is completed in or near the Golgi apparatus. This is true, for example of the *thyroglobulin* molecules synthesized and then stored extracellularly by the thyroid gland (Section 3.6.1 and Fig. II-52). Sugars are added sequentially to the first ones, which are attached to amino acids of the proteins. Different radioactive sugars will give different labeling patterns, depending on whether they occur at the beginning or at the end of the carbohydrate chain. Galactose or fucose often occur near the end, and these are incorporated initially near the Golgi apparatus whereas mannose is incorporated in the ER since it often occurs near the beginnings of the carbohydrate chains.

The discovery of glycosyl transferase activities in Golgi-enriched fractions, alluded to near the beginning of this chapter, provides direct biochemical confirmation for the synthetic capacities suspected for the Golgi apparatus from the ARG data.

The membrane-delimited bodies that leave the Golgi region often contain mixtures of several different proteins and polysaccharides, or sometimes complexes of lipids and proteins (Fig II-34) and their contents may be highly concentrated. How the proper mixtures are put together in the Golgi region is not known. As for the concentration or "condensation" process, some speculation suggests that there is an active extrusion of water from forming secretory bodies. However, this may not be necessary. In mixtures of proteins and polysaccharides, molecules can associate to form large multimolecular aggregates. If this occurs in a membrane-delimited structure such as a vacuole containing secretory materials, the result could be a spontaneous osmotic efflux of water; the osmotic properties of a solution depend upon the total number of independent molecules in that solution, and aggregation reduces this number.

The secretions packaged by the Golgi apparatus and related structures are released from the cell by exocytosis (Figs. II-33 and II-34). Upon appropriate stimulation (Section 3.6.3) the membranes delimiting the secretory bodies fuse with the plasma membrane, releasing the contents to the extracellular space. In the fasted and refed guinea pig pancreas, radioactive secretions begin to be released into the secretory ducts within one to two hours after initial administration of labeled amino acids. We will discuss secretory processes further in Sections 3.6.1 and 3.6.3.

2.5.3 **ER–GOLGI APPARATUS INTERRELATIONS** In all cell types that have been studied in this regard, electron microscope observations indicate that macromolecules transferred from the ER to the Golgi region remain within membrane-delimited structures. But different transfer

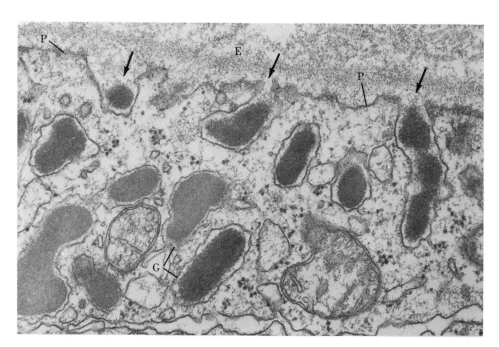

Fig. II-33   *Portion of the surface of a rat pituitary gland cell that produces the hormone* prolactin. *Secretion granules are seen at G. The arrows indicate granules in process of release from the cell by exocytosis, the fusion of the membranes enclosing the granules and the plasma membrane (P). E indicates the extracellular space.* × *23,000. (Courtesy of M. Farquhar.)*

mechanisms appear to be involved in different cells and circumstances. Occasionally continuities between ER cisternae and Golgi sacs are found; whether they are long lasting or intermittent is not known. Often the ER and Golgi apparatus seem structurally separate, with transport dependent upon a kind of "membrane flow." Vesicles (Figs. II-22 and III-24B), or sacs and tubules, with their contents separate from the ER and fuse with or transform into Golgi components. Sometimes this "flow" seems to occur at the outer Golgi face, with the sacs then passing down the stack, ultimately to give rise to vesicles or larger secretory structures. In other cases other routes may apply (Fig. II-22). Sometimes, closely apposed to Golgi sacs, one finds ER cisternae with ribosomes only on the surface facing away from the Golgi apparatus (frontispiece), as if the ER had been caught in the act of contributing to the apparatus. Apparently the net passage of materials from the ER to the Golgi region is balanced in the steady state by passage of materials out of the region in the form of secretory bodies and other structures formed there. (Some investigators believe that vesicles can shuttle back to the ER.) But when cells are stimulated to enhance their rate of production

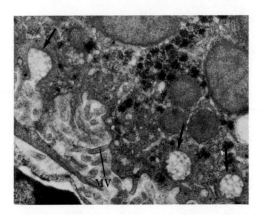

Fig. II-34 *Region of a rat hepatocyte near its sinusoidal surface (see Figs. I-1 and I-2). Arrows indicate vacuoles filled with lipoprotein droplets (see Fig. II-30). The vacuole at the upper left shows fusion of its delimiting membrane with the plasma membrane and presumably release of its contents into the extracellular space (compare with Fig. II-33). One of the surface microvilli is seen at MV. × 20,000. (Courtesy of N. Quintana.)*

of secretions, they may show increases in the size of the Golgi stacks and in the prominence of the other structures of the region.

An interesting observation, pertinent to our discussion, has been made on amebae. When an ameba is deprived of its nucleus by microsurgical techniques, the Golgi apparatus is no longer detectable. If a new nucleus is implanted, the apparatus reappears, apparently by transformation of ER sacs.

In the mammalian pancreas, when energy metabolism is interrupted, by exposing tissue slices to respiratory inhibitors such as cyanide, transport of proteins from the ER to the secretion granules forming in the Golgi region is blocked. The simplest explanation is that the vesicle-mediated transport from the ER that occurs in these cells (Fig. II-22) is an energy-dependent process.

In many cell types, lysosomes, which are membrane-delimited "packages" of hydrolytic enzymes (acid hydrolases), form in the vicinity of the Golgi apparatus. Sometimes they may originate from Golgi sacs. This is true for the distinctive granules of phagocytic white blood cells (Section 2.8.2). In various cells some of the small vesicles that appear to bud from Golgi sacs contain cytochemically demonstrable acid phosphatase. These vesicles probably contain other hydrolases and function as lysosomes. But in several cell types, such as neurons and hepatocytes, there seems to be a specialized route for transfer of hydrolases directly from the ER to lysosomes forming near the Golgi apparatus. Near the inner face of the Golgi apparatus in these cells there is a region

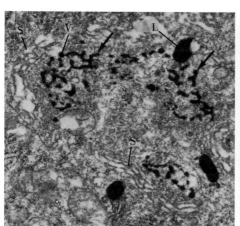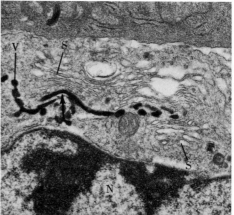

Figs. II-35 and II-36 *Enzyme localizations in GERL, a special membrane system that appears to be a part of the ER associated closely with the Golgi apparatus. Figure II-35 shows the reaction product for acid phosphatase in GERL (arrows) in a small vesicles (V) and in larger lysosomes (L) in a rat neuron. Golgi sacs are indicated by S. × 20,000. In Figure II-36 (showing a similar region from a cell of mouse* melanoma, *a pigmented cancer)* GERL *shows reaction product for* tyrosinase. N *indicates part of the nucleus.* × 20,000.

of smooth ER that is unusual in that it contains cytochemically demonstrable acid hydrolases (Fig. II-35). This ER is continuous with rough ER, and several types of lysosomes appear to form from it. The name GERL (*G*olgi associated *ER* from which *L*ysosomes form) has been proposed for this structure. GERL may also provide a direct route for transport of proteins other than the hydrolases. For example, in cells of melanomas, a type of tumor, the enzyme tyrosinase seems to pass into forming pigment granules through GERL (Fig. II-36). Once in the granules, the enzyme catalyzes the formation of the pigment melanin, which comes to fill the granules.

The precise structural and functional interrelations of GERL and the Golgi apparatus have yet to be determined.

2.5.4   **MEMBRANE**   In *Ameba* filamentous material like that
    TRANSFORMATION   bordering the cell (Fig. II-7) is seen lining
                       Golgi vacuoles. Earlier in this chapter (Section 2.5.1) we mentioned that in the fungus *Pythium* there is a gradient of membrane thickness in the Golgi stack; interestingly, the vacuoles that separate from the inner Golgi face are delimited by relatively thick

membranes, comparable to the plasma membrane, with which these vacuoles appear to fuse. Such observations are part of a body of circumstantial evidence suggesting that important components of plasma membranes and carbohydrate-rich surface coats enter the cell surface by fusion (exocytosis-like) of Golgi-derived vacuoles or vesicles.

This raises again interesting questions about membrane formation, initially discussed in our consideration of the ER (Section 2.4.4). The evidence we have been considering in this chapter suggests that, at least in some cells, membrane from the ER moves into the Golgi sacs, that membranes surrounding secretory granules bud from the sacs, and that membrane from the Golgi apparatus can be incorporated in the cell surface. Yet in their biochemical and cytochemical properties, the membranes of these various structures often differ considerably from one another. Does this mean that the ER really does not contribute membrane to the Golgi apparatus or the apparatus to secretory structures—that we have somehow been fooled by the microscopic images? Or are our still fairly crude cell fractionating procedures at fault? Does, for example, the ER make a special Golgi apparatus type of membrane that passes to the apparatus but is not evident in microsome preparations since it is only a small subfraction, not readily separated from the remainder of the ER? An attractive alternative explanation for part of the diversity is that one type of membrane might transform into another. Perhaps ER membrane can modulate into Golgi apparatus or cell surface membrane. The gradient in the *Pythium* Golgi apparatus might reflect a progressive change of this type. If membranes actually can gain and lose macromolecules without extensive disruption (Section 2.4.4), the modifications being postulated might have a simple basic mechanism, but how a particular set of molecules could be "directed" to add to membranes at a particular locale needs to be explained.

Sections 4.1.1 and 4.1.6 will present a brief general discussion of membrane assembly, tying together some points already raised with some additional ones to be considered in the next few chapters.

c h a p t e r 2.6
MITOCHONDRIA

Mitochondria are present in virtually all eucaryotic cells, both plant and animal. Their number may range from a few per cell to 1000 in a rat hepatocyte, up to 10,000 in one of the giant amebae. Although their sizes vary considerably in different cell types, diameters of 0.5–1.0 μm and lengths of 5–10 μm or more are common. Mitochondrial structure is striking and readily recognizable in the electron microscope.

Mitochondria are easily isolated from homogenates in fairly pure fractions. Dry weight proportions are typically 25–35 percent lipid and 60–70 percent protein. It was learned relatively early that mitochondria are the major sites of ATP production linked to oxygen consumption, and especially by virtue of their possession of the Krebs cycle enzymes, they play important roles in many metabolic pathways, including the breakdown and synthesis of carbohydrates, fats, and amino acids (Chapter 1.3). They illustrate well the grouping of a number of sequentially acting enzymes within a membrane-limited organelle. This promotes efficiency, in part since the products of one reaction do not have far to travel before they are likely to encounter the enzymes catalyzing the next. Such effects and others may be especially striking when the enzymes are held in close relationship with one another by actual incorporation in a membrane. The profound functional implications of the arrangement of enzymes in mitochondrial membranes are becoming increasingly evident, largely through reconstitution experiments in which the membrane macromolecules are isolated and recombined.

In regard to terminology important for this chapter, several equivalent definitions of *oxidation* and *reduction* are often used. In chemical terms, oxidation–reduction reactions involve the transfer of electrons from one molecule to another. The molecule losing electrons is *oxidized*, the one gaining electrons is *reduced*. In biological systems, many such electron-transfer reactions include the exchange of hydrogen atoms; this is the case in the reactions discussed earlier. (Section 1.3.1), where oxidation of a substrate takes place by the transfer of hydrogen atoms to a coenzyme which thus becomes reduced. The enzymes catalyzing such reactions are referred to as *oxidative enzymes*; included in the category is a large class known as *dehydrogenases*.

2.6.1   COMPARTMENTS WITHIN THE MITOCHONDRIA   Figure II-37 illustrates the basic structure of the mitochondrion. The organelle is delimited by an outer membrane, separated from the inner membrane, by a space 60–100 Å wide. The inner membrane is thrown into many folds, or *cristae* (Fig. II-38). Enclosed by the inner membrane is the *matrix.* Techniques are available for obtaining separate outer and inner membranes and for isolating some of the matrix substances. With these techniques, the segregation and organization of different enzymatic functions in the different membranes and compartments are being studied. As more is learned, the description presented here will undoubtedly required modification.

As Chapter 1.3 indicated, some mitochondrial enzymes are readily

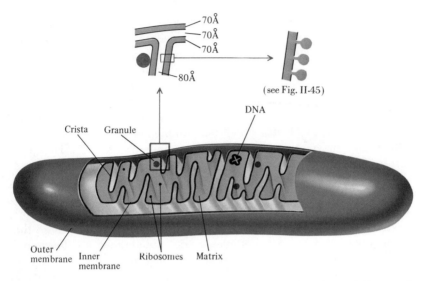

(see Fig. II-45)

Fig. II-37   *Diagrammatic representation of a mitochondrion. (After work of L. Ornstein, G. E. Palade, F. Sjostrand, H. Fernandez-Moran, and others.)*

solubilized when the organelles are disrupted. Presumably these could include enzymes in solution within the matrix or in the compartment between the inner and outer membranes. Or perhaps some enzymes are loosely attached to one or the other membrane and these detach during the disruption. Many Krebs cycle enzymes are in the readily solubilized category, and the usual interpretation is that they come from the matrix in the inner compartment. Enzymes for fatty acid breakdown also are present there. But certain Krebs cycle enzymes (such as succinic dehydrogenase) are linked to the inner membrane.

The electron transport components responsible for respiratory metabolism, and the systems that carry out oxidative phosphorylation, are both tightly bound to the inner membrane. This joint localization might be expected since, in the intact mitochondrion, electron transport and oxidative phosphorylation are *coupled*—as electron transport occurs, leading ultimately to oxygen consumption, phosphate groups are

Fig. II-38   *Mitochondria as seen by usual electron microscopic techniques. A. In a pancreas cell (bat). The outer (O) and inner (I) mitochondrial membranes and the flat cristae (C) are readily visible. Many intramitochondrial granules (G) are present; most cell types show fewer such granules. × 50,000. (Courtesy of D. W. Fawcett.) B. In the protozoan Epistylis. As in many protozoa and algae, the cristae are tubular. × 60,000. (Courtesy of P. Favard.) C. In a muscle (bat). The cristae are numerous and closely packed, correlating with the high state of activity of the muscle. At the arrows it is evident that the cristae are infoldings of the inner mitochondrial membrane. × 45,000. (Courtesy of S. Ito.)* ▶

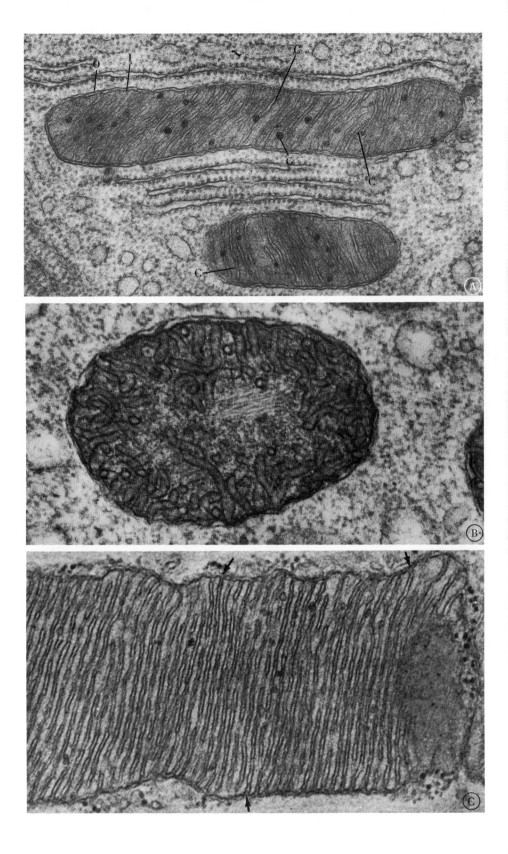

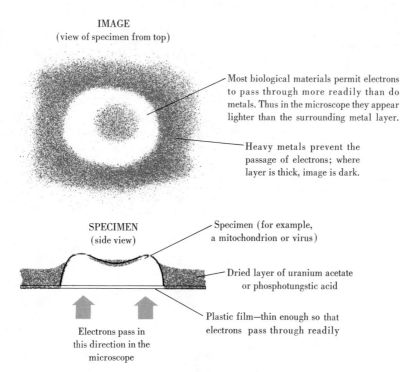

IMAGE
(view of specimen from top)

Most biological materials permit electrons to pass through more readily than do metals. Thus in the microscope they appear lighter than the surrounding metal layer.

Heavy metals prevent the passage of electrons; where layer is thick, image is dark.

SPECIMEN
(side view)

Specimen (for example, a mitochondrion or virus)

Dried layer of uranium acetate or phosphotungstic acid

Plastic film—thin enough so that electrons pass through readily

Electrons pass in this direction in the microscope

Fig. II-39 *Negative staining, one electron microscope technique for examining three-dimensional and surface aspects of cell structures. The specimen is not sectioned. It is placed on a thin plastic film and covered with a drop of solution containing heavy metal atoms (usually* uranium or tungsten *compounds such as uranyl acetate or phosphotungstic acid). The solution is allowed to dry leaving the specimen in a thin layer of electron-dense material.*

added to ADP, to produce ATP. Usually the coupling is tight, meaning that neither ATP formation nor electron transport can occur alone; both must take place simultaneously. (The next section will take up these processes in detail.)

In electron micrographs of sectioned cells, the inner membrane often appears to be a three-layered membrane (Sections 2.1.1 and 2.1.2). It is thinner (50–70 Å) than most plasma membranes (70–110 Å). "Negative staining" (Fig. II-39) of isolated mitochondria (Fig. II-40) shows a great many small spheres covering the surface of the inner membrane that faces the matrix. Each sphere is attached by a stalk to the membrane. When these were first discovered, it was thought that each sphere might be a *respiratory assembly* containing all the enzymes responsible for electron transport and associated phosphorylation. This proved not to be the case; the spheres are too small to hold all the needed enzymes and, when the spheres are removed experimentally,

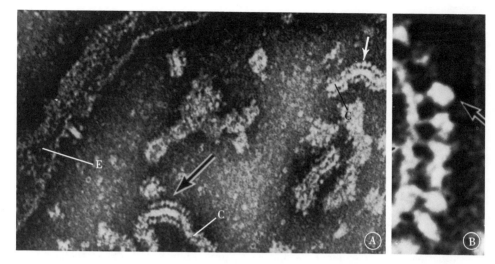

Fig. II-40 *Portion of mitochondria from heart muscle (cow). The organelles have been isolated, broken open, and then negatively stained (see Fig. II-39). A is a view (× 80,000) showing part of the edge of a mitochondrion (E) and several cristae (C). Along the cristae surfaces facing the matrix (Fig. II-37) are arrays of spheres attached to the cristae by stalks (arrows). B is at higher magnification (× 600,000) and shows several spheres. (Courtesy of H. Fernandez-Moran, T. Oda, P. B. Blair, and D. E. Green.)*

the remaining membranes still possess most essential components of the respiratory chain. The spheres contain an enzyme, called F_1. When extracted from the mitochondrion in soluble form, F_1 acts as an ATPase, that is, it splits ATP into ADP and phosphate. However, in the intact mitochondrion it is considered that F_1 acts in the opposite direction and participates in coupling phosphorylation of ADP with electron transport.

Thus far, structures corresponding to respiratory assemblies have not been visualized by electron microscopy. However, there are at least two reasons for concluding that the enzymes of the inner membrane exist as repetitive arrays, in which enzymes of sequential metabolic steps are held in close relation with one another. First F_1 has an intimate functional relation to the enzymes of the respiratory chain, and it is visible in the form of a repetitive unit. Second, when mitochondrial membranes are fragmented into small pieces (by ultrasonic vibration, for example, or by agitation with glass beads), the fragments can carry out electron transport and oxidative phosphorylation (Fig. II-45); the various relevant enzyme molecules are present in the same ratios in the fragments as in the intact mitochondrion. This would not be expected if there were large specialized patches of membranes, each devoted to just one step in the sequence. From the membrane fragments, multienzyme complexes can be isolated. These are discrete assemblies of relatively few molecules that contain several sequentially acting enzymes (see Fig. III-6 and Section 3.2.4). It has been estimated that a square micrometer of membrane surface has many hundreds of respiratory assemblies, each providing a highly organized arrangement of the enzymes responsible for electron transport and oxidative phosphorylation; individual mitochondria of mammalian cells probably contain 5000–20,000 or more assemblies.

Much current work is directed toward interrelating electron microscope observations with biochemical studies in order to obtain a more detailed picture of molecular arrangements. For example, it remains to be determined whether or not the exact form of the F_1 units of living cells is accurately reflected in the small spheres seen with negative staining. Eventually it should be possible to specify the position and function of each molecule within an F_1 unit. Figure II-42 illustrates a current proposal for the arrangement of mitochondrial electron transport enzymes.

In addition to the enzymes just discussed, the major constituents of the inner mitochondrial membrane include lipids, particularly phospholipids, and a possibly heterogeneous group of "structural" proteins. (Overall, the inner membrane is quite rich in proteins. For hepatocytes, some estimates suggest it is 75–80 percent protein and 20–25 percent lipid, by weight; see p. 45.) The "structural" proteins and lipids may constitute a framework into which the enzymes of the respiratory assemblies are anchored. However, thus far "structural" proteins are defined by the negative fact that no enzymatic role is known for them at present; the details of their possible structural roles are not well understood.

The number of cristae per mitochondrion tends to be greater in cells with particularly intense respiratory activity (Fig. II-38); this finding is consistent with the inner membrane's structure and role in producing ATP. The mitochondria themselves are often concentrated in intracellular regions of greatest metabolic activity, for example, close to the contractile fibrils in muscle cells, at the base of kidney tubule cells where molecular exchange with the blood is rapid, or wrapped around the flagella that move sperm cells.

The physiological significance of the various forms of cristae in different cell types is less clear. The cristae may appear (Fig. II-38) as simple plates (most cells of higher organisms) or as tubules (protozoa, algae, cortex of the mammalian adrenal gland), in some cases they may be organized into more complex three-dimensional structures.

The outer mitochondrial membrane has several oxidative enzymes different from those of the inner membrane, such as *monoamine oxidase*. It is composed of lipids and proteins in roughly equal amounts, by weight. Very little is known about the space between the two membranes.

The mechanisms by which mitochondria exchange molecules with their surroundings may be varied and complex. The outer membrane appears fairly permeable to many small molecules and ions. The inner membrane is not and it serves as a barrier regulating the entry and exit of many molecules. Some substances, like water or certain small fatty acids, enter the inner mitochondrial compartment readily while others, such as reduced NAD, do not (hydrogens from NAD can enter mitochondria, but this involves a special, circuitous, metabolic route). For ADP, ATP, and some intermediates of the Krebs cycle, carrier systems seem to mediate passage across the inner membrane.

Isolated mitochondria can accumulate ions such as Ca^{++} from the medium; this is an energy-dependent process which stops if mitochondrial respiration is inhibited (for example, by adding cyanide to the medium). Phosphate is taken up with the Ca^{++}, and there may be a simultaneous extrusion of H^+ ions. When large amounts of calcium (or Ba^{++} or Sr^{++}) and phosphate have been taken up, prominent granular deposits of calcium phosphate are seen in the mitochondria. Somewhat similar though usually less electron-dense granules, 200–500 Å in diameter, are present normally in mitochondria (Fig. II-38). Some workers infer that in the intact cell these serve to bind calcium and phosphate and perhaps act as biologically significant storage depots for such materials, but this is still unsettled.

2.6.2   **OXIDATIVE**   Chapter 1.3 outlined the major chemical
              PHOSPHO-   steps in the Krebs cycle and indicated those
              RYLATION   steps at which reduction of coenzymes takes
                                place. The electrons from the coenzymes are
then transported, in orderly manner, from one carrier to another in the electron transport system. This system is outlined in Figure II-41; prominent among the participants are different iron-containing proteins called cytochromes. One can think of the chain as starting at a position of high energy and ending at one of low energy. Each step in the chain involves loss of electrons (oxidation) by one molecule and electron gain (reduction) by the next, and each oxidation–reduction event results in a release of energy. Some of the energy made available in this way is used for the formation of ATP from ADP plus inorganic phosphate. Precisely how this *oxidative phosphorylation* occurs is not known.

For many years the prevailing theory was that at each of three steps in electron transport, energy was transferred from electron transport components to other molecules and that these latter molecules then engaged in a series of reactions leading to ATP formation. This "chemi-

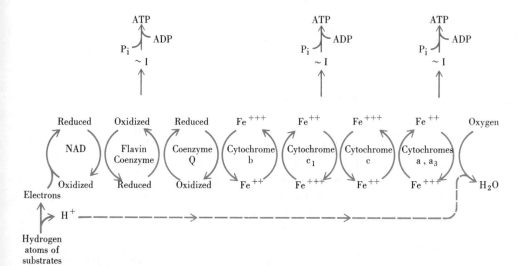

Fig. II-41 *Major steps in mitochondrial electron transport. Each compo-*
nent of the chain receives electrons from the previous component and passes
them to the next. In the case of the cytochromes the receipt of electrons
reduces iron ions in the molecule to Fe^{++}. *Subsequent loss of the electrons to*
the next component regenerates the oxidized, Fe^{+++} *form. Together, cyto-*
chromes a and a_3 *constitute "cytochrome oxidase."*
     Protons (H^+) *derived from Krebs-cycle substrates may take part in a*
few reactions in the sequence (perhaps they are transported across the
mitochondrial membrane; Fig. II-42) and ultimately participate in forming
H_2O *from* O_2.
     The red arrows illustrate the "chemical intermediate" hypothesis of
oxidative phosphorylation (see also Fig. II-42). Energy from electron trans-
port is used to generate a hypothetical high energy intermediate compound
($\sim$I) which then participates in reactions forming ATP from ADP and
phosphate (Pi). Phosphorylation is considered to be coupled to 3 portions of
the electron transport chain since for the entire chain the observed P:O
ratio (molecules ATP formed per atom of oxygen consumed) is 3. (After A.
L. Lehninger and others.)

cal intermediate" hypothesis has been challenged in recent years, particularly since the participating molecules have never been identified.
The so-called "chemiosmotic hypothesis" is the major alternative now
under investigation. According to this scheme, electron transport produces a concentration gradient of protons (H^+) or some other ions between the inner mitochondrial compartment and the spaces outside.
The gradient results from the suspected asymmetric arrangement of the
enzymes in the inner membrane (Fig. II-42) which could permit certain
electron transport components to acquire hydrogen atoms or H^+ ions at
one membrane surface, and other components to expel the H^+'s at the

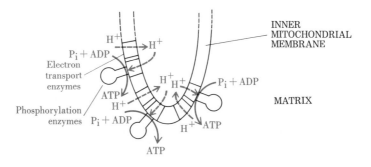

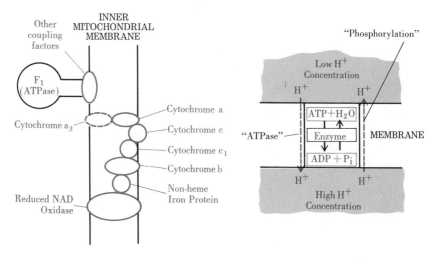

Fig. II-42   A. *The chemiosmotic hypothesis of oxidative phosphorylation. Electron transport enzymes are thought to carry protons (H^+) from the matrix across the inner mitochondrial membrane. The protons then interact with phosphorylation enzymes leading to the formation of ATP.*

*   *   **B.** *Asymmetric arrangement of respiratory chain and phosphorylation enzymes as currently postulated. The dotted line indicates the current uncertainty regarding the location of cytochrome a_3. The roles and locations of nonheme-iron proteins are also unsettled.*

*   *   **C.** *One way in which an H^+ concentration gradient might be used to generate ATP. An enzyme system built into the membrane, through which H^+ cannot pass by passive diffusion, might be capable of coupling H^+ transport either to the hydrolysis of ATP ("ATPase" activity) or to the formation of ATP from ADP and Pi ("phosphorylation" activity). The ATPase activity can occur if the enzyme transports H^+ from a region of low concentration to one of high concentration—an active transport process for which energy is required (Section 2.1.3). But when a concentration gradient has been established, for example, through action of electron transport enzymes, the enzyme can carry H^+ down the gradient and in so doing produce ATP. This can be thought of as reversing active transport and using energy represented by the concentration gradient for phosphorylation.*

*   *   *(Based on work by P. Mitchell, E. Racker, and others.)*

opposite surface. Since the inner membrane is impermeable to ions, the H^+ ions cannot simply diffuse back. The energy from electron transport in a sense is "stored" in the concentration gradient and then is used to generate ATP through action of the phosphorylation enzymes (F_1 and other factors) of the inner membrane. This could occur in several ways. The one illustrated in Figure II-42 is a relatively simple possibility. More sophisticated proposals involve subtle interactions of ions with membrane-associated enzymes. The net distribution of H^+ across the inner membrane would represent a balance between the electron transport system that tends to produce a gradient and the phosphorylation system whose activities reduce the gradient. Proponents of the chemiosmotic view cite several observations as tentative support. (1) Actively respiring mitochondria are able to transport ions across their inner membrane; (2) when the ion permeability of the membrane is increased, for example, by adding valinomycin (p. 55), phosphorylation is inhibited; (3) there have been experiments with chloroplast membranes and some other systems in which membrane-associated "ATPases" have been made to run in reverse, producing ATP, by subjecting them to artificial concentration gradients. It may turn out that both "chemical intermediate" and "chemiosmotic" events are involved in oxidative phosphorylation. (See also p. 200, bottom).

The rate at which the respiratory chain functions is controlled by various factors, including the level of available ADP. The more ADP available for phosphorylation, the more rapid is electron transport, since the coupling of transport and phosphorylation requires that ADP be converted to ATP as electrons move down the chain. The higher the rate at which a cell utilizes ATP, the higher the ADP level, leading to higher rates of electron transport, oxygen consumption, and ATP production. Thus an effective mechanism is provided for balancing ATP production with its utilization.

Certain chemicals like *dinitrophenol* "uncouple" ATP production and permit the respiratory chain to function without simultaneous phosphorylation. The extent to which similar effects occur naturally in cells is unknown, although such uncoupling has been proposed as a source of cell abnormality in one or two diseases.

2.6.3   ***DNA, RNA, AND***   The power of molecular biology and its rapid
         PROTEIN   progress are evident in the success of recent
         SYNTHESIS;   experiments on mitochondrial nucleic acids.
MITOCHONDRIAL   Morphological observations had shown the
         DUPLICATION   presence of 20–30-Å wide fibrils and ribo-
                                somelike particles, and cytochemical inves-
tigations had studied the removal of the fibrils by DNase (which hyd-

rolyzes DNA specifically) and of the ribosome-like granules by RNase (an RNA-hydrolyzing enzyme). The results of these efforts indicated that mitochondria might have traces of both DNA and RNA. When molecular biologists, armed with the techniques applied so successfully in the study of nucleic acids and protein synthesis in microorganisms, turned to this problem, it took only a year or two to establish the presence in purified mitochondria of DNA and RNA and of machinery for synthesizing protein. Mitochondrial nucleic acids and ribosomes were shown to be of different sizes and to have properties unlike comparable components of the rest of the cell. Only small amounts are present in mitochondrial fractions isolated from cells, but the distinctive properties permitted the firm conclusion that the nucleic acids or ribosomes found in isolated mitochondria are not contaminants derived from other organelles during isolation. Interestingly, mitochondrial DNA and ribosomes share some properties with DNA and ribosomes of bacteria; some possible evolutionary implications of this fact will be discussed later (Section 4.5.2).

The DNA of mitochondria (mtDNA) is a double helix, as it is in the cell nucleus. In many of the organisms studied, however, it has been shown to exist as a circular molecule. This has been most thoroughly documented for higher animals (Fig. II-43). Many investigators believe that circularity is a widespread, perhaps nearly ubiquitous feature of mitochondrial DNA's, but perhaps due to problems in DNA isolation (such as breakage of the circles) some plants and other organisms show

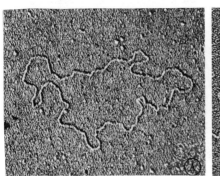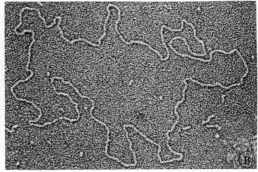

Fig. II-43 Electron micrographs of circular DNA molecules from mitochondria of leucocytes from a leukemic human. As is true of mitochondrial DNA molecules from many animal cells, the molecule in A is 5 μm long. (This is the length of the thin fiber, a DNA double helix, that delineates the perimeter of the distorted circle in the micrograph.) The molecule in B is 10μm long and probably is a dimer made of two 5-μm molecules. Dimers and interlocked pairs of 5μm circles have been found (in small numbers) in mitochondrial DNA preparations from several sources. × 35,000. (Courtesy of D. A. Clayton and J. Vinograd.)

linear mtDNA. Mitochondria possess machinery for replicating their DNA. Presumably such replication is an integral part of mitochondrial duplication (see below) and transmission of genetic information (see p. 275 and Section 4.3.6;).

Mitochondrial RNAs hybridize with mtDNAs suggesting that mitochondria synthesize their own RNAs. Among these are some tRNAs and the RNAs of mitochondrial ribosomes. Often mitochondrial ribosomes and ribosomal RNAs are markedly smaller than the corresponding components in the cytoplasm outside the mitochondrion, but their sizes and other properties seem to vary considerably among different organisms. Mitochondria also can synthesize proteins; one interesting feature of this process is the probable use of *N*-formyl methionine for initiation, as in bacteria.

Despite these various capacities, mitochondria are not capable of producing all of their own proteins. In fact, available evidence suggests they make only a few. The total amounts of DNA in mitochondria are small. In an L cell (a type of cultured mouse fibroblast) the 250 mitochondria contain a total amount of DNA equal to 0.1–0.2 percent of the DNA in the nucleus. Each mitochondrion contains up to six 5-μm circular DNA molecules, but there is little basis to believe that the six molecules (or the DNAs of different mitochondria in a given cell) differ significantly from one another. Thus when one considers that the DNAs must also code for rRNAs and other components, a single mitochondrion probably contains enough DNA to code for only a few proteins (for estimates of the information content of nucleic acids, see Section 3.1.3). In unicellular organisms and some higher plants, mitochondrial DNAs are longer than in cells of higher animals (up to 20 μm or more). Perhaps there is correspondingly more information encoded within mitochondrial DNAs in such cells. In yeast, under some circumstances, the total DNA content of mitochondria may approach 10 percent of the nuclear content.

Genetic data (Section 4.3.6) suggest that mitochondria are under the control of two sets of genes, those in the mitochondria and genes in the nucleus. There is complementary biochemical evidence suggesting a dual origin of mitochondrial proteins. Cells of yeast and of the mold *Neurospora* have proved especially useful for these experiments since they can be grown as large fairly homogeneous populations into which drugs and radioactive precursors can conveniently be introduced and from which mitochondria can be isolated. Moreover, with yeast extensive formation of mitochondria can be evoked by simple changes in the growth medium, and the cells can also survive without functional mitochondria (see below). The key results are obtained by exposing cells either to the drug *cycloheximide* (CHI), which stops protein synthesis on ordinary cytoplasmic ribosomes, but does not affect the mitochondrial system, or to the drug *chloramphenicol* (CAP), which has the opposite effects (mitochondria share with bacteria their sensitivity to this drug). The synthesis of many mitochondrial proteins, including such characteristic ones as cytochrome c (Fig. II-41) is not inhibited by CAP, but is by CHI, indicating origin on the nonmitochondrial ribosomes. The synthesis of a few proteins of the inner mitochondrial membrane is not inhibited by CHI. Included are elements of the ATPase system and certain molecules participating in electron transport. The polypeptides made in mitochondria appear to be quite hydrophobic (Section 2.1.2) and probably pass directly from the ribosomes into the membranes. It is not known how the proteins made outside the organelles enter and associate with those made inside. Although such findings are unusual, in some preparations of yeast cells ribosomes are seen apparently adhering to the outer surfaces of mitochondria. Do these send polypeptides through the outer membrane (as in the ER, Section 2.4.1)? Occasionally continuities are noticed between ER and mitochondrial membranes, and the outer membranes of mitochondria bear some resemblence in composition and enzyme activities to ER membranes. But it is not yet justified to conclude from this that the ER contributes proteins to mitochondria. Related to these considerations is the fact that most cellular lipids including mitochondrial molecules apparently are synthesized by the ER.

The question of the origin of mitochondrial proteins is part of the more general question: How are mitochondria duplicated? Early cytologists observed the division of certain cells containing only two or three mitochondria (for example, developing sperm cells of invertebrates). They showed that new mitochondria could arise by division of preexisting ones. Similar observations have been made recently in electron microscope studies of algae; in these cases mitochondrial division is coordinated with the division of the cells. Electron microscope images of nondividing cells, such as hepatocytes, also suggest that mitochondria can divide.

When the protozoan *Tetrahymena* is grown in tritiated thymidine, the DNA of its mitochondria becomes labeled. If the cells are then transferred to a nonradioactive growth medium, the fate of this labeled DNA during cell growth can be followed by autoradiography. As the cells grow and divide, the mitochondria increase proportionally, so the number *per cell* remains fairly constant. By counting the number of autoradiographic grains present over mitochondria, one can demonstrate that the DNA synthesized during the labeling period is randomly distributed throughout the entire mitochondrial population. In other words, the average label per mitochondrion is halved each time the total mitochondrial population doubles. This is as expected if mitochondria duplicate by some sort of division process in which a parental organelle contributes its molecules, including DNA, more or less equally to "daughter" mitochondria. Similar results have been obtained using labeled lipid precursors with the mold *Neurospora*. Experiments of these types are not completely unambiguous since, especially with lipids, there are mechanisms other than growth and division through which radioactive macromolecules can spread from labeled organelles to initially unlabeled ones. Transient organelle fusions may occur, and also "exchanges" in which individual macromolecules enter and leave otherwise intact structures. However, the results obtained have helped convince most investigators that growth and division is the predominant mode of mitochondrial duplication under most circumstances.

The most important other known pathway for mitochondrial origin is found in yeast when they shift from anaerobic to aerobic metabolism (Section 1.3.1). Thus when yeast are maintained in the absence of oxygen, they lack almost all cytochromes, and the electron microscope shows few if any typical mitochondria. They possess, instead, a few small membrane-bounded bodies with little internal structure. Upon addition of oxygen to the medium, there is both a marked increase in the number of such bodies, and an extensive synthesis of cytochromes; eventually the bodies become typical mitochondria. The precise manner of this transformation is variable, depending upon concentrations of glucose, lipid precursors, and other components of the growth medium. Apparently the premitochondrial bodies can divide and perpetuate themselves in the absence of oxygen. Occasionally it has been claimed that mitochondria arise in yeast possessing no recognizable premitochondrial bodies. Presumably simple primordia containing DNA would be present, but not detectable because they lack a distinctive structure. However, these claims are disputed, and there is not yet a fully documented and universally accepted case of mitochondrial origin in the absence of membrane-delimited precursor structures.

2.6.4   ***PLASTICITY IN***   The study of living cells by phase contrast
       THE LIVING CELL   microscopy and by microcinematography
                                (Section 1.2.2) reveals that in many cells mitochondria are continually in motion. They show dramatic changes in shape and volume, and they can probably fuse with one another or separate into several parts. The static images of the electron microscope cannot reveal such movement. In some muscles of the rat, for example, highly branched mitochondria are seen in electron micrographs; these may actually be snapshots of mitochondria in the process

of fusing or fragmenting. Recently it has been reported that yeast grown in media rich in glucose contain one or a few large multiply branched mitochondria, while in other media the cells seem to possess many small spherical mitochondria. The thyroid hormone *thyroxine*, calcium ions, and other physiological substances can induce mitochondrial swelling in the test tube. Form alterations in isolated mitochondria are also observed upon addition of *dinitrophenol* and other nonphysiological agents that uncouple phosphorylation from oxidation.

Figure II-44 illustrates morphological changes that may be importantly related to mitochondrial function. Isolated mitochondria assume the *orthodox* configuration when the medium is depleted of ADP so that phosphorylation and the coupled respiration are stopped. If ADP is added, phosphorylation and respiration resume and the mitochondria show the *condensed* configuration. Similar changes can be induced in cells (Fig. II-44) by treatment with 2-deoxyglucose, which produces an increased level of intracellular ADP. Some investigators propose that the normal state of mitochondria in cells falls between these extremes and that mitochondria often appear in the orthodox configuration in electron micrographs due to inactivation of metabolism during fixation. Others suggest that the configurational changes reflect or parallel struc-

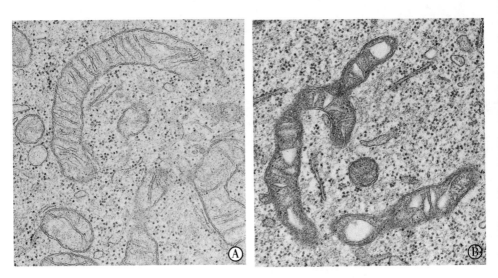

Fig. II-44 *Mitochondria in Ehrlich ascites tumor cells (Chapter 3.12). A shows the "orthodox" mitochondrial configuration in an untreated cell. B shows a deoxyglucose-treated cell. The mitochondria have assumed the "condensed" configuration. The matrix is more electron dense than in A and there are enlarged spaces within many cristae.* × *27,000. (From C. R. Hackenbrock, J. Cell Biology, 51:123. Courtesy of the author and publisher.)*

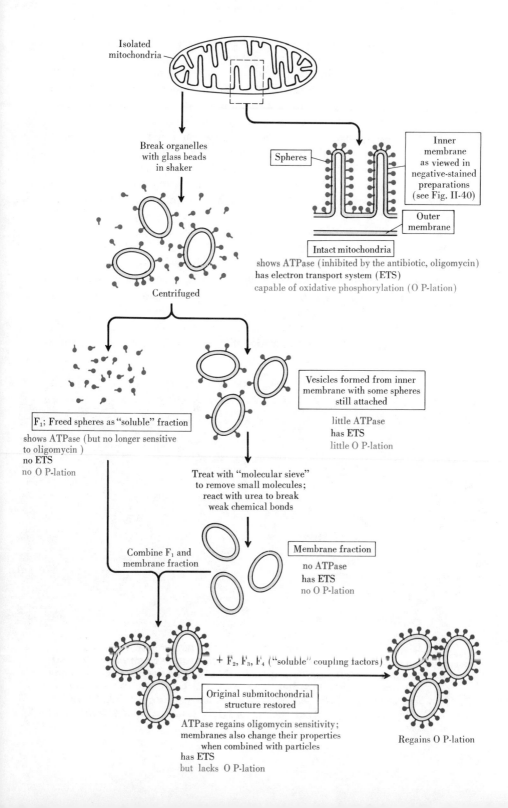

Isolated mitochondria

Break organelles with glass beads in shaker

Spheres

Inner membrane as viewed in negative-stained preparations (see Fig. II-40)

Outer membrane

Intact mitochondria

shows ATPase (inhibited by the antibiotic, oligomycin)
has electron transport system (ETS)
capable of oxidative phosphorylation (O P-lation)

Centrifuged

F_1; Freed spheres as "soluble" fraction

shows ATPase (but no longer sensitive to oligomycin)
no ETS
no O P-lation

Vesicles formed from inner membrane with some spheres still attached

little ATPase
has ETS
little O P-lation

Treat with "molecular sieve" to remove small molecules; react with urea to break weak chemical bonds

Combine F_1 and membrane fraction

Membrane fraction

no ATPase
has ETS
no O P-lation

+ F_2, F_3, F_4 ("soluble" coupling factors)

Original submitochondrial structure restored

ATPase regains oligomycin sensitivity; membranes also change their properties when combined with particles
has ETS
but lacks O P-lation

Regains O P-lation

tural rearrangements (conformational changes) in inner membrane proteins that are directly important for ATP formation.

2.6.5   **RECONSTI-**   Initially reconstitution experiments involved
            TUTION   only a few enzymes, but as successes in-
   EXPERIMENTS   creased the attempts became more ambi-
                    tious. The reconstitution of mitochondria
begins with fragmentation; from the fragments separate extracts are
made containing a few enzymes linked in complexes, single enzymes,
and proteins or lipids. Attempts are being made to reconstitute entire
sequences, such as the respiratory assembly or oxidative phosphorylation, by mixing extracted components together under carefully controlled conditions. Results of one impressive series of experiments are
diagrammed in Figure II-45. It may not be surprising that certain properties of the enzyme F_1 (such as its sensitivity to inhibition by oligomycin) differ in isolated enzymes from the intact organelle. However, it is
particularly interesting that the original pre-isolation properties of F_1
return when other mitochondrial components are added to it; portions of
the initial structure are then restored. This is one of many examples of
the intimate relations of mitochondrial structure and function.

In addition to aiding in the analysis of functional architecture,
reconstitution experiments may also provide information about the way
in which organelles form or grow. In revealing the manner in which
complex structures can be disassembled and reassembled in the test
tube, they may furnish clues to the types of bonds that hold structures
together and the modes of assembly in the cell (Section 4.1.6).

c h a p t e r **2.7**
CHLOROPLASTS

The plastids of plant cells are of various types, and they contain
different proportions of several pigments and of components such as

◀ **Fig. II-45** *Disruption of isolated beef heart mitochondria and separation
of the resulting fragments into membranous and "soluble" components followed by reconstitution of some mitochondrial structure and function. The
drug* oligomycin *is known to inhibit* F_1-*related functions in intact
mitochondria. Thus sensitivity to this drug is one measure of the extent to
which reconstitution succeeds in reproducing conditions similar to those in
mitochondria. (Based on experiments of E. Racker and colleagues.)*

starch. The most familiar and abundant plastids are *chloroplasts*, which contain large quantities of the green pigment *chlorophyll*. They are responsible for the photosynthetic use of the energy of sunlight to effect the transformation of carbon dioxide and water into carbohydrates with the simultaneous release of oxygen.

Chloroplasts are relatively large organelles that vary in size and shape from species to species. In some algae the one or two chloroplasts are in the form of cups or elongate spirals that fill much of the cytoplasm. In higher plants a great many chloroplasts may be present in each cell in the form of ovoid or disklike bodies. Leaf cells contain several dozen chloroplasts, each measuring 2–4 by 5–10 μm. Typical dry weight figures for chloroplast composition are 40–60 percent protein, 25–35 percent lipid, 5–10 percent chlorophyll, 1 percent pigments other than chlorophyll, and small amounts of DNA and RNA.

2.7.1   **STRUCTURE**   Chloroplasts are bounded by two membranes. Within they contain additional membranes surrounded by a matrix, the *stroma*. Granules containing starch may be scattered in the stroma. In many algae, however, starch may accumulate near a special region known as the *pyrenoid* (see Fig. II-46 and Chapter 3.4.A), in which the starch is apparently synthesized from glucose.

The internal membrane systems of the chloroplast are usually in the form of flattened sacs called *lamellae* or *thylakoids*. In many algae the sacs are arranged in parallel arrays and run the length of the plastid (Figs. II-46 and III-12). In higher plants the structure varies somewhat but usually resembles the arrangement shown in Figure II-47. The *grana* consist of stacked sacs resembling a pile of coins, and they are connected to each other by membranes running in the stroma.

2.7.2   **FUNCTION**   Photosynthesis involves two major sets of reactions, each set consisting of many steps. One set depends on light and cannot occur in the dark. The other set, referred to as the *dark reactions,* is not directly dependent on light. The light-dependent reactions result in the production of ATP from ADP (photophosphorylation), the formation of reduced NADP from NADP (a coenzyme related to NAD), and the release of oxygen derived from water. Two electron transport chains cooperate in these reactions; "photosystem I" reduces NADP, "photosystem II" generates O_2 from H_2O, and the two systems interact to produce ATP.

In the dark reactions carbon dioxide is built up into carbohydrates using the ATP and reduced NADP from the light-dependent reactions.

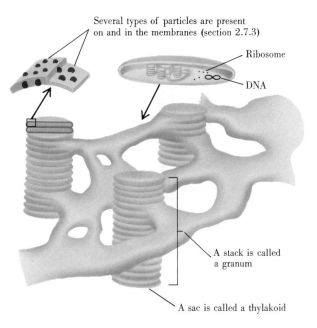

Several types of particles are present on and in the membranes (section 2.7.3)

Ribosome

DNA

A stack is called a granum

A sac is called a thylakoid

(A) Structure of a chloroplast of a higher plant. Details of arrangement of thylakoids, grana etc. vary from species to species. (After Weier, Arntzen, Staehelin and many others)

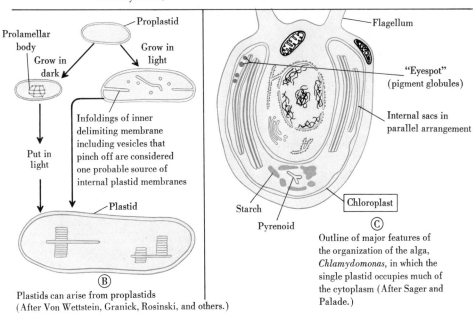

Proplastid

Prolamellar body

Grow in dark

Grow in light

Infoldings of inner delimiting membrane including vesicles that pinch off are considered one probable source of internal plastid membranes

Put in light

Plastid

Plastids can arise from proplastids (After Von Wettstein, Granick, Rosinski, and others.)

(B)

Flagellum

"Eyespot" (pigment globules)

Internal sacs in parallel arrangement

Chloroplast

Starch

Pyrenoid

(C)

Outline of major features of the organization of the alga, *Chlamydomonas*, in which the single plastid occupies much of the cytoplasm (After Sager and Palade.)

Fig. II-46 *Schematic representation of some structural and developmental features of chloroplasts.*

The best known pathway for this is the "carbon reduction cycle" (Calvin cycle), a sequence of reactions in which CO_2 is "fixed" through its reaction with a 5-carbon sugar yielding two 3-carbon molecules which are then modified and combined to build up larger molecules.

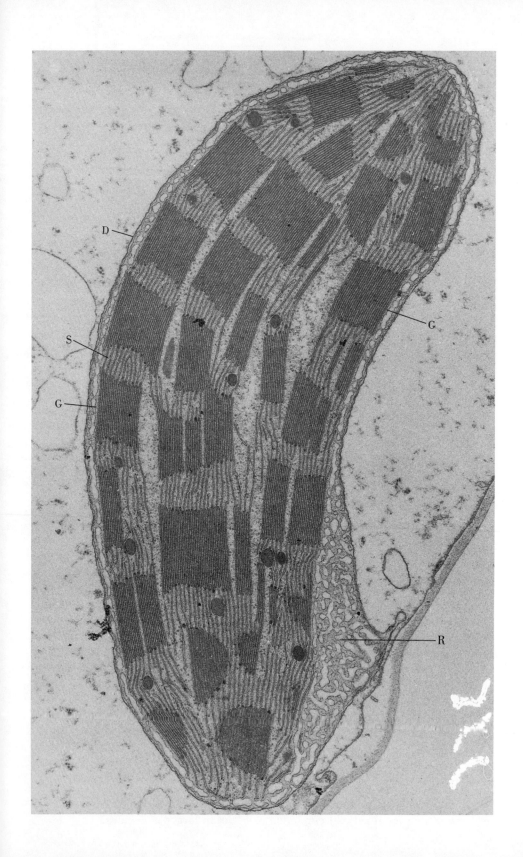

2.7.3   **RELATION OF STRUCTURE FUNCTION**   As is true of pigments in general, the color of chlorophyll results from its selective absorption of light with certain wavelengths (see p. 14). The transformation of light energy into chemical energy in chloroplasts depends upon the ability of chlorophyll to enter an "excited" state when it absorbs light energy. The energy in light exists in discrete packets or *photons*. The absorption of photons by chlorophyll molecules can raise the "energy levels" of electrons in the molecules; under some circumstances it results in ejection of an electron from a chlorophyll (photoionization). The generation of ATP and reduced NADP depends on electron transport, beginning with light-excited chlorophyll and proceeding through a chain that involves iron-containing proteins (cytochromes, ferredoxin), a copper-containing protein (plastocyanin), and compounds related to lipids (plastiquinone).

This sequence of events implies a high degree of molecular organization. Like the mitochondria, plastids may be isolated and fragmented and portions of the structure separated (by centrifugation) into distinct fractions. When chloroplasts are broken open, many of the enzymes of the dark reactions are released, but most components involved in the light-dependent reactions are not. Thus it appears that, while much of carbohydrate synthesis occurs in the stroma, the formation of ATP, reduced NADP, and O_2 takes place in association with the sacs. Further subfractionation indicates that the sac membranes contain chlorophyll and the components involved in electron transport. Chlorophyll molecules are similar in size and solubilities to lipid molecules and could participate in the formation of membranes of the types discussed in Chapter 2.1.

Beyond this lies a realm of uncertainty. How are the pigments and enzymes put together on the membranes providing a large surface for light absorption and the molecular arrangements that result in efficient function? In dry-weight the membranes are about one half protein, one eighth chlorophyll and other pigments (including the orange *carotenoids*), and three eighths lipids similar to those discussed earlier (Section 2.1.2). Extremely small fragments of chloroplast membranes can engage in multistep reactions requiring the sequential operation of several enzymes and other components. At least some of the plastid macromolecules are in repeating multienzyme arrays. Analysis of the membranes indicates a great excess of chlorophyll. The number of

◀ **Fig. II-47** *A chloroplast from a leaf cell of corn. The two delimiting membranes are seen at* D. *The grana* (G) *consist of stacks of sac-like thylakoids. Stroma membranes* (S) *run between the grana. The reticulum seen at* R *is present only in some plants (Chapter 3.4B).* × 75,000. *See Fig. II-46A for a diagrammatic interpretation. (Courtesy of L. K. Shumway and T. E. Weier.)*

chlorophyll molecules is 10–500 times greater than that of other key participants (notably, electron transport enzymes) in the light-dependent reaction. This raises the question of how the membranes are arranged so that energy absorbed by large numbers of chlorophyll molecules is efficiently transferred to a relatively small number of electron transport systems.

One theory proposes that the membranes contain units each containing up to several hundred closely packed chlorophyll molecules and other pigments associated with one or a few reaction centers. A reaction center (sometimes called an *active center*) contains special chlorophyll molecules linked to an electron transport system. Light may be absorbed by any one of the pigment molecules. The energy from light absorbed by a given pigment molecule is carried to the reaction center by passage from one molecule to another. The physical processes underlying this transfer are based on the fact that one chlorophyll (or other pigment) molecule in an excited state can transfer energy to another nearby chlorophyll molecule; it raises the second molecule to the excited state and returns to its normal unexcited state. It is possible that energy is passed essentially at random from molecule to molecule in the unit until, by chance, it reaches the reaction center. It is also possible that some organization exists that guides the passage. In either event, when the energy reaches the reaction center, electron transport commences with the ejection of an electron from a special chlorophyll molecule. In the unit, then, most of the pigments provide a large "light-harvesting" system or "antenna" which funnels energy to chlorophylls at the reaction center; here electron transport releases the energy for conversion into usable form, that is, into ATP and reduced NADP. Such hypothetical units have been called "quantasomes."

This theory is based on a substantial body of evidence. For example, among the chlorophyll molecules in a plastid, a few percent have properties expected of photosystem I reaction center molecules that would accept the energy from other "light-harvesting" pigment molecules and pass it to the electron transport system. Due perhaps to unique orientation in the membrane these chlorophylls are found to differ slightly in light absorption characteristics; their maximal absorption is of light with a wavelength of about 700 nm, in contrast to most other chlorophylls that absorb maximally at 675–685 nm. The wavelength of light is related to the energy it carries—in quantum physics a lower wavelength indicates a higher energy. For our considerations this means that the "excited" state of the special chlorophyll molecules is at a relatively slightly lower energy level. This would imply that once energy is transferred to the special chlorophylls, it is "trapped" there and cannot be transferred to other chlorophylls. It can, however, be transferred further "downhill" via the electron transport enzymes. (The carotenoids and other "accessory" pigments of chloroplasts absorb light at wavelengths much lower than 675 nm; the energy from this light can be passed to the chlorophylls. Thus the pigments permit chloroplasts to use a wider range of wavelengths for photosynthesis than would be provided by the chlorophylls alone).

However, there is no strong evidence that the pigments and enzymes of the membrane are simply bundled into units, each capable of all the steps of the light reactions. Although it seems likely that reaction centers do occur as discrete assemblies of the requisite electron transport components, the bulk of the chlorophyll may not be separated into well-defined units. Instead, chlorophyll might be spread more or less uniformly in the membrane, and a given molecule might transfer the energy of absorbed light to any one of a number of nearby reaction centers. A key point requiring clarification in structural terms is the relations between the two photosystems. They interact "cooperatively" in function, but the electron transport components of the two can be separated physically into different subfractions made from disrupted chloroplasts.

When viewed in freeze-etch preparations, chloroplast thylakoid membranes show an abundance of particles of the sizes (100–200 Å) expected for clusters of enzymes (Fig. II-46). Some of these occur at the membrane surface, while others penetrate deeper within the membrane interior. A major effort is currently underway to establish the roles of these particles. Many investigators believe that the 90–150 Å particles at the membrane surface include coupling factors similar to the mitochondrial F_1. Like the mitochondrial particles, these can be removed from the thylakoid membranes. This disrupts phosphorylation but leaves the electron transport components still present in the membrane. Others of the surface particles may represent an enzyme involved in the initial steps of the dark reactions, *ribulose diphosphate carboxylase*. The particles deeper in the membrane are larger than the coupling factor. They are prime suspects as probable sites of the electron transport components. Still to be determined is the number of types of these particles (is there one for photosystem I and another for II?) and their precise relations to membrane architecture (some investigators believe there are two classes of intramembrane particles differing slightly in size and location).

The mechanism of photosynthetic phosphorylation is being debated in much the same terms as is oxidative phosphorylation (Section 2.6.2 —chemiosmotic hypotheses versus chemical intermediate hypotheses. Movements of ions across chloroplast membranes and ATP formation by chloroplast fragments induced by artificial pH gradients are among the evidence supporting chemiosmotic proposals.

Volume changes and alterations in the arrangements of internal membranes have been induced by subjecting isolated chloroplasts to various conditions of illumination or otherwise altering their metabolism. The significance of these effects for chloroplast function is still to be determined. In some mutant algae stacking of the thylakoids is abnormal and the frequency and distribution of certain membrane particles is altered. Perhaps there is some causal relation between membrane stacking and particle presence and arrangement.

Section 2.9.2 will consider metabolic interactions of chloroplasts with peroxisomes.

2.7.4   *Reproduction*   Plastids contain their own DNA and RNA, distinct from both nuclear and other cytoplasmic nucleic acids. They are capable of protein and RNA synthesis. Plastid ribosomes are somewhat smaller than other cytoplasmic ribosomes, often resembling ribosomes of procaryotes in size and in RNA molecular weights. Plastid protein synthesis is sensitive to chloramphenicol (and insensitive to cycloheximide) as is true also with mitochondria (Section 2.6.3) and bacteria. The DNA isolated from chloroplasts of higher plants such as spinach or peas is in the form of 40-μm circles. These appear capable of coding for a fair number of proteins, in addition to the chloroplast RNAs. However, as with mitochondria, chloroplasts obtain some of their proteins from outside. For the key dark reaction enzyme, ribulose diphosphate carboxylase, there is the intriguing possibility that one of the enzyme's subunits is coded for by a nuclear gene and the other subunit by a chloroplast gene.

In algae, plastids are seen to divide regularly as part of the cell division cycle, and thus maintain a constant number. In higher plants also, plastids can divide. In addition, plastids can develop from much simpler organelles, the proplastids. These appear to be self-duplicating, but do not contain the elaborate patterns of internal membranes found in mature plastids. They are present in gametes and embryos. Higher plants grown in the dark provide an interesting system for studying plastid formation from proplastids. Their tissues are usually colorless, containing no fully developed plastids and little chlorophyll. Modified plastids are present (Fig. II-46, and II-48); they contain arrays of membraneous tubules, often in highly regular arrangements, called prolamellar bodies. The latter seem to contribute to the formation of normal internal membranes when the plants are illuminated. On exposure to light, the small amounts of protochlorophyll present are converted to chlorophyll, and additional chlorophyll synthesis begins; this is paralleled by the development of the usual internal chloroplast structure.

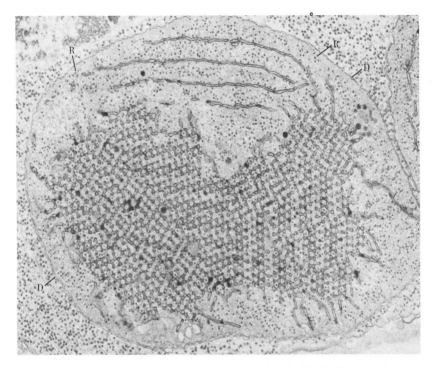

Fig. II-48 *Plastid from an oat plant grown in the dark. Plants grown in the dark are called etiolated, and the plastids like that shown here are etioplasts. Most of the organelle is occupied by a prolamellar body, a network of regularly arranged tubules. D indicates the pair of membranes that delimit the plastid, and R, a few of the many plastid ribosomes. Ribosomes are also numerous in the cytoplasm outside the plastid. × 35,000. (From B. E. S. Gunning, in* Ultrastructure and Biology of Plant Cells, *Gunning and Stern, Eds. London: Arnold Co. 1975.)*

Mutants of the unicellular alga *Chlamydomonas* (Fig. II-46) show various alterations in details of chloroplast structure and development. One such strain is proving useful in attacking the question: How does the enzymatically complex membrane of chloroplasts form? *Chlamydomonas* normally can synthesize chlorophyll in the dark, but the mutant cannot, and thus when grown in the dark its plastids do not mature into chloroplasts. When dark-grown mutant cells are illuminated, they begin synthesis of chlorophylls and other chloroplast molecules; now chloroplast membranes form. This provides an excellent experimental system. When the cells are illuminated in the presence of chloramphenicol, chloroplast membranes appear, but they lack some key proteins. Thus they are nonfunctional; they also do not form their normal stacked arrays. If the cells are now shifted to cycloheximide, the chloroplasts develop into normal functioning organelles. Apparently the abnormal membranes formed initially are "cured" by the addition of proteins made in the chloroplast. This demonstrates that the membranes can be made in a multistep process—a framework can be built from some of the components and completed by the insertion of others. Perhaps some of the subassemblies present in and on the mature membrane form as multimolecular arrays before adding to the membrane (Sections 4.1.1 and 4.1.6).

Invaginations of the inner delimiting membrane (Fig. II-46) are frequently seen during maturation of proplastids into plastids. It is often claimed that thylakoid membranes form from these invaginations, and some investigators also assert that extensive continuities persist between the inner membrane and the internal membrane systems of mature chloroplasts. However, definitive evidence is still being sought. In *Chlamydomonas* there are few if any invaginations of the delimiting membranes. For the mutants discussed above it has been proposed (1) that thylakoids form by expansion of small sacs and other membranes that persist in the plastids of dark-grown cells; (2) that this expansion involves continual addition of new macromolecules throughout the expanse of the growing structure (there may be no "new" and "old" membrane regions; Section 2.4.4); and (3) that chloroplast ribosomes bound to the growing membranes may be important in the insertion of new proteins.

The fact that chloroplast membrane formation, and synthesis of some pertinent macromolecules, is tied closely to the synthesis of chlorophyll, represents an important cellular "control." Photosynthesis cannot proceed without chlorophyll and evolution apparently has led to mechanisms that avoid formation of an elaborate photosynthetic apparatus lacking this pigment.

c h a p t e r **2.8**
LYSOSOMES

Historically, the study of organelles usually begins with the accumulation of morphological observations and then passes to the isolation of the organelle in relatively pure fraction and biochemical study. For lysosomes, however, this pattern was reversed. Their discovery began with an investigation of *hydrolytic enzymes*, enzymes catalyzing reactions of the type A_1—A_2 + H_2O → A_1—H + A_2—OH. Biochemical analyses of cell fractions separated from rat liver homogenates revealed that five such enzymes, all acting optimally in *acid* media, sedimented together in centrifugation. Further, it appeared that these enzymes were inactive toward their potential substrates if the fraction was carefully prepared to avoid disruption of the fragile organelles. This led to the hypothesis that the *acid hydrolases* were packaged together in an organelle previously undescribed, since the enzymes it contained were hydro*lytic*, the organelle was called a *lysosome*. From their sedimentation characteristics it was predicted that hepatocyte lysosomes had a certain size, about 0.4 μm in diameter. From the observation of *latency*, that is, the fact that substrates added to isolated lysosomes were not split unless disruptive procedures were used, it was predicted that the organelle was surrounded by a membrane. When the acid hydrolase-containing fractions were examined by electron microscopy, morpholorgically distinctive cytoplasmic particles were indeed found, and the predictions concerning both size and outer delimiting membranes proved correct.

In subsequent years many acid hydrolases were shown to be present in lysosomes (Fig. II-49). Because the activities of one of the lysosomal enzymes, *acid phosphatase*, could be localized cytochemically in microscopic preparations, evidence accumulated for the existence of a wide variety of forms of lysosomes. It soon became evident that lysosomes are versatile organelles involved in a variety of cell functions and showing different morphologies.

2.8.1 **FORMS AND** Biochemical and cytochemical studies have
FUNCTIONS already demonstrated lysosomes in protozoa, insects, amphibia, mammals, and a variety of other vertebrates. It is probable that with a few exceptions such as mammalian red blood cells lysosomes are ubiquitous in animal cells. This is probably true of plant cells as well, but this matter is still unsettled. Some plant lysosomes will be described in Section 3.4.2.

A tentative determination that a cytoplasmic particle is a lysosome is possible (1) if electron microscopy shows it to be membrane delimited, and (2) if cytochemistry shows it to have one or more of the hydrolase activities found in lysosomes as studied biochemically. Identification from cytochemistry is tentative since reliable methods are available for only a few of the lysosomal hydrolases, so the presence of other hydrolases can only be *assumed*. Acid phosphatase activity is the most widely used "marker" enzyme demonstrable by staining procedures, although fairly reliable methods are also available for two or three other lysosomal enzymes. By such cytochemical methods, and in a growing number of instances by parallel biochemical studies, a variety of morphologically different structures may be identified as lysosomes.

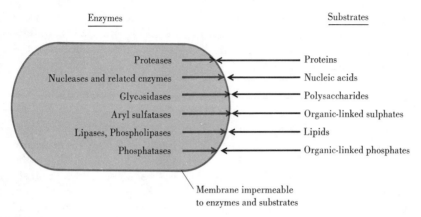

Enzymes   Substrates

Proteases ⟶ ⟵ Proteins
Nucleases and related enzymes ⟶ ⟵ Nucleic acids
Glycosidases ⟶ ⟵ Polysaccharides
Aryl sulfatases ⟶ ⟵ Organic-linked sulphates
Lipases, Phospholipases ⟶ ⟵ Lipids
Phosphatases ⟶ ⟵ Organic-linked phosphates

Membrane impermeable
to enzymes and substrates

Fig. 11-49   *The biochemical concept of* lysosomes *according to the work of C. deDuve, his collaborators, and others. There are now more than 40 known hydrolases in lysosomes. Several different enzymes hydrolyze each class of substrates shown to the right; the several enzymes within each group differ in specificity and details of their action.*

It would be of obvious advantage to be able to study all suspected lysosomes by biochemical methods and thus confirm their status. However, for one reason or another, many tissues are poorly suited for organelle isolation procedures. For example, some organs have large numbers of intermingled cell types, so that one cannot easily identify the source of isolated organelles as cells of a given type. In other cases it is difficult to obtain amounts of tissue large enough or to achieve adequate purity of fractions. Cytochemistry with its direct access to single cells can avoid many such problems. Ideally, cytochemistry and biochemical studies on isolated organelles are used together.

Many lysosomes are in the same size range as the first liver lysosomes studied, about 0.5 μm in diameter. The largest, however, such as the "protein droplets" in cells of mammalian kidney (Section 3.5.4), are several micrometers in diameter. The smallest are the *Golgi vesicles;* these are only 25–50 nm in diameter. The identification of some of the vesicles of the Golgi apparatus as probable lysosomes rests upon the presence of a delimiting membrane and upon cytochemical evidence that they possess acid phosphatase and aryl sulfatase activities. In addition, some autoradiographic observations suggest that the vesicles carry proteins, presumably acid hydrolases, from the Golgi apparatus to vacuoles responsible for enzymatic digestion of material phagocytosed by certain white blood cells and other phagocytes.

The size heterogeneity of lysosomes is paralleled by heterogeneity in form, origin, and function. All lysosomes are related, directly or indirectly, to *intracellular digestion*. The material to be digested may be of exogenous (extracellular) or endogenous (intracellular) origin. Collectively, the lysosomal enzymes are capable of hydrolyzing all the classes of macromolecules in cells (Fig. II-49), and presumably would do so if they were not confined in structures delimited by membranes. There is considerable speculation about the possibility that in abnormal situations, lysosomal enzymes may escape into the cytoplasm and either kill the cell or bring about dramatic metabolic changes. However, in most of the known functions of lysosomes, the material on which the acid hydrolases act must gain access to the interior of the lysosome, and the enzymes remain confined within the organelle.

There is much variation in the mechanisms by which material to be digested becomes enclosed inside a lysosome with the acid hydrolases. Figure II-50 is a diagrammatic representation of the major types of lysosomes and their possible modes of origin. The diagram illustrates the proposal that lysosomal enzymes, manufactured at the ribosomes, are transported by the endoplasmic reticulum to lysosomes, either directly or via packaging in the Golgi apparatus. In some cells,

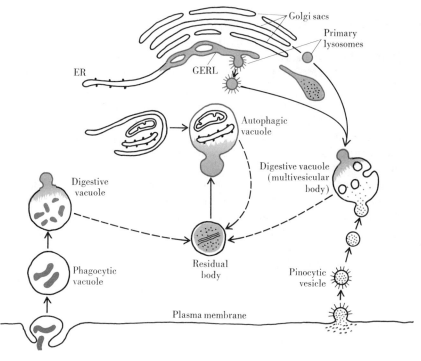

Fig. II-50   *Diagram suggesting possible origins and functions of lysosomes of different types. Illustrated are: (1) primary lysosomes arising from Golgi sacs and GERL; (2) fusion of lysosomes (primary or secondary) with phagocytic, pinocytic, and autophagic vacuoles; (3) formation of residual bodies from digestive vacuoles, autophagic vacuoles, and GERL.*

primary lysosomes appear to be produced by the Golgi apparatus or the Golgi-associated ER (GERL, Section 2.5.3). These lysosomes are thought of as packages that transport hydrolases to other membrane-delimited bodies with which the primary lysosomes fuse; the Golgi vesicles referred to above probably fall into this category. When both the enzymes and the material to be digested, being digested, or already digested are present within a lysosome, the lysosome is referred to as a *secondary lysosome*. Secondary lysosomes may accumulate large quantities of undigested or indigestible molecules; the resulting structures are known as *residual bodies*. The classification of lysosomes in functional categories such as "primary," "secondary," "residual body," and so forth is useful in establishing relationships among structures that may differ considerably in morphology. This morphological variation reflects the variety of materials digested and the fact that lysosomes may form in several ways.

2.8.2   HYDROLYSIS OF EXOGENOUS MACROMOLECULES (HETEROPHAGY)   Many small molecules enter cells individually by passing through the plasma membrane. But macromolecules and larger particles generally enter by the endocytic bulk transport processes, phagocytosis or pinocytosis (Section 2.1.5). The endocytic vacuoles formed by the plasma membrane usually fuse with lysosomes, contributing their content to the lysosome interior where digestion takes place. The lysosomal membrane is permeable to amino acids, small sugars, and other molecules released through the hydrolysis of macromolecules. Thus the products of digestion can diffuse out of the lysosomes into the adjacent cytoplasm. There are only a few special situations, such as viral penetration into cells (Section 3.1.1) in which macromolecules or larger structures can enter the cytoplasm substantially intact. In protozoa, hydrolysis of endocytosed material is a central mechanism in feeding. Most multicellular organisms do not rely heavily on endocytosis for nutrition, but many of their cell types are capable of pinocytosis and a few, such as white blood cells, are phagocytes.

The phagocytic white blood cells of mammals are very important in the body's defense against bacteria and other invading organisms or toxic materials. They are characterized by distinctive cytoplasmic granules. These have been extensively studied by microscopy and cytochemistry and through biochemical work on isolated fractions. There are two major classes of granules, called specific granules and azurophilic granules (Fig. II-51). Both arise from the Golgi apparatus (Section 2.5.2). The azurophilic granules are primary lysosomes. When the phagocytes engulf bacteria or other materials, there is a rapid movement of granules to the phagocytic vacuoles. The membranes of the granules and vacuoles fuse, and the granule contents are emptied into the vacuoles. Within a few minutes the phagocytic vacuoles have acquired acid hydrolases from the azurophilic granules and thus are converted into (secondary) lysosomes. Through the actions of the hydrolases and of other enzymes and nonenzymatic material contributed by the two granule populations, most bacteria are destroyed. (A very few microorganisms, such as bacteria responsible for leprosy and tuberculosis, survive phagocytosis and multiply inside mammalian phagocytes; some have coats that resist the lysosomal hydrolases, others can somehow inhibit fusion of lysosomes with the vacuoles in which they enter the cell.)

Phagocytes known as *macrophages* are numerous in the liver, spleen, lymph nodes, and other tissues of higher animals. In addition, macrophage precursor cells (monocytes) are present in the circulation; these accumulate at sites of injury or infection and develop into macrophages. Monocytes and macrophages produce primary lysosomes in their Golgi apparatus. Along with the white blood cells macrophages participate in defense mechanisms (see Section 3.6.4 for some addi-

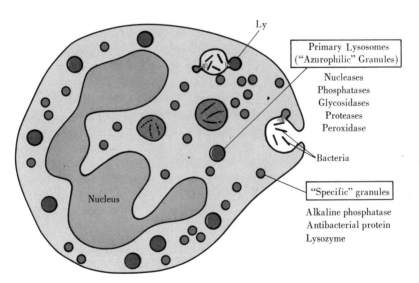

Fig. II-51 *Diagram of a rabbit white blood cell (polymorphonuclear leukocyte; neutrophil) ingesting bacteria. Two classes of granules fuse with the phagocytosis vacuoles and contribute digestive enzymes and other components; the granules colored red are lysosomes. (After the work of M. Baggiolini, D. F. Bainton, Z. A. Cohn, C. de Duve, M. G. Farquhar, and J. G. Hirsch.)*

tional discussion). In addition, they act as scavengers of cellular debris released in the course of tissue destruction following injury or infection, and in normal ("programmed") cell death. Extensive tissue breakdown occurs, for example, in the development of digits in the hand and foot and the resorption of the frog's tail during metamorphosis.

Endocytosis and subsequent lysosomal degradation is important for turnover of the cells of blood. Thus in humans red blood cells are produced and released by the bone marrow, circulate for 120 days, and then are destroyed. Normal degradation of "aged" red blood cells involves macrophages of the spleen and, to a lesser degree, of other organs. How the phagocytes discriminate among aged and younger cells is not known. This may depend in part upon changes that occur in red blood cell surface properties and upon alterations in mechanical properties, such as flexibility, that retard the movement of cells through special narrow spaces in the circulatory system of the spleen. Turnover of blood plasma proteins also is thought to involve endocytosis, occuring largely in the liver.

In the thyroid gland, endocytosis and partial hydrolysis in lysosomes has considerable physiological importance (Fig. II-52). The glycoprotein *thyroglobulin* is secreted into the extracellular lumen of the gland and stored there (Fig. II-52 Section 3.6.1). When the gland is stimulated to release thyroid hormone (for example, by hormones from other glands) the stored thyroglobulin is engulfed by the secretory cells in large pinocytosis vacuoles. Electron microscopic and cytochemical observations show that lysosomes fuse with those pinocytic vacuoles. Apparently the thyroglobulin is partially hydrolyzed within the vacuole, and the thyroid hormone *thyroxine* is one of the digestion products. Thyroxine, an iodine-containing derivative of the amino acid *tyrosine*, is somewhat larger than an amino acid. Precisely how it leaves the lysosomes and enters the blood capillaries at the base of the cell (the opposite pole of the cell from where thyroglobulin is released) is still to be determined.

A useful example of the uptake of exogenous material is the endocytosis of protein (from calf serum included in the growth medium) by the macrophages obtained from the abdominal cavity of the mouse, and grown in tissue culture. This is an excellent system for eventual elucidation of the control mechanisms that regulate acid hydrolase synthesis and lysosome production. When exposed to high concentrations of serum, the cells pinocytose very actively and accumulate large numbers of both primary and secondary lysosomes. Accompanying this change in lysosome number, dramatic rises can be measured in the levels of lysosomal enzyme activities (Fig. II-53). If the cells are then returned to a low-serum medium, most lysosomes disappear and levels of lysosomal enzymes return to normal. Interestingly, if in place of proteins the cells

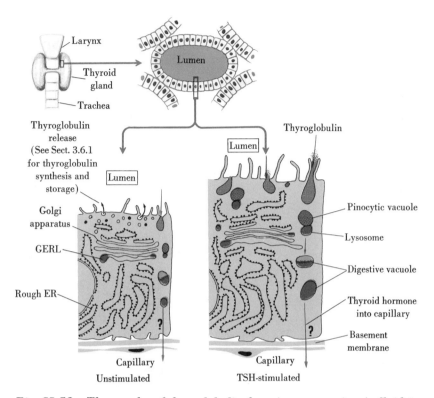

Larynx

Thyroid gland

Trachea

Lumen

Thyroglobulin release (See Sect. 3.6.1 for thyroglobulin synthesis and storage)

Golgi apparatus

GERL

Rough ER

Lumen

Lumen

Thyroglobulin

Pinocytic vacuole

Lysosome

Digestive vacuole

Thyroid hormone into capillary

Basement membrane

Capillary

Capillary

Unstimulated

TSH-stimulated

Fig. II-52   The uptake of thyroglobulin from its storage sites (colloid in the lumen) has been studied under normal conditions but is more readily observed when the thyroid gland cells are stimulated by horomonal treatment (thyroid-stimulating hormone TSH). TSH increases the uptake of thyroglobulin from the lumen by pinocytosis and the release of thyroid hormone to the blood capillaries at the base of the cell. The diagram illustrates the key role of lysosomal enzymes (proteases) in the partial hydrolysis of thyroglobulin and consequent liberation of thyroid hormones.

are induced to take up nondigestible particles, such as "latex" spheres made of synthetic polymers, they do not show an increase in hydrolase levels. The pertinent metabolic controls apparently depend in some way upon products of digestion.

The incorporation of nondigestible materials in lysosomes has proved useful in preparing lysosome fractions that are largely free of the extensive contamination with mitochondria and peroxisomes present with conventional techniques. Thus large amounts of the detergent Triton WR 1339 accumulate in the lysosomes of liver cells in rats administered this compound—Triton-loaded lysosomes are much less dense than normal, and can be purified by centrifugation in a density gradient (Fig. I-23).

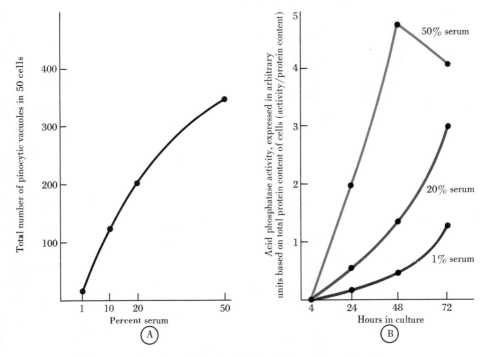

Fig. II-53 *Results of studies by Z. A. Cohn and colleagues on macrophages from the body cavity of a mouse. The cells were cultured in media containing varying proportions of newborn calf blood serum (a complex mixture of proteins and other components). Similar, though less dramatic, effects are seen with sera from three other mammals. The higher the concentration of serum, the greater the number of pinocytosis vesicles (counted by phase microscopy, **A**), the greater the number and size of lysosomes (many of the lysosomes accumulate near the Golgi apparatus), and the higher the levels of activities of acid phosphatase, **B**, and of the two other acid hydrolases studied (β-glucuronidase and cathepsin). The specific mechanisms and factors responsible for the induction of pinocytosis and for the effects on enzyme synthesis and transport are currently being sought. Also of interest is the fact that when the cultured cells are transferred from a medium with 50% serum to one with 1% serum, the number of lysosomes and the concentrations of acid hydrolases decrease markedly, although neither lysosomes nor acid hydrolases have been found to be released from the cells to the culture medium.*

Primary lysosomes are not the only source of the acid hydrolases that enter vacuoles containing exogenous material. Fusion of secondary lysosomes with newly formed pinocytosis or phagocytosis vacuoles apparently is common in many cells. A striking experiment demonstrates such fusions. Tissue culture cells (mouse fibroblasts) were permitted to engulf an electron-opaque material, finely dispersed iron particles. Like other engulfed material, these iron particles accumulate in secondary lysosomes. Having "marked" these lysosomes with iron, the cells were then permitted to engulf a second "marker," finely dispersed gold particles in a mixture of DNA and protein. Since the gold particles are readily distinguishable from the iron particles used in the first "feeding," the results were unequivocal (Fig. II-54). The presence of both iron and gold in the same lysosome demonstrates that fusions occur between old secondary lysosomes and newly formed digestive vacuoles. The same lysosomal enzymes apparently may be used for more than one round of digestion.

Some interesting problems relating to the hydrolysis of exogenous

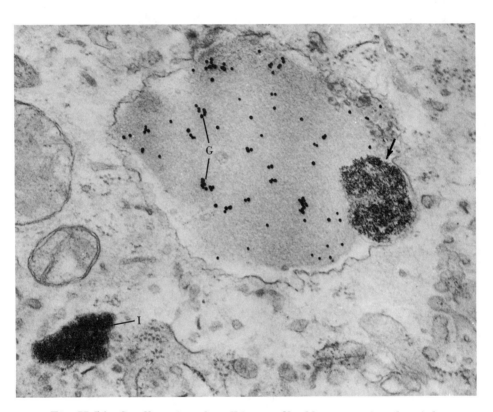

Fig. II-54 *Small portion of a cell (mouse fibroblast grown in culture) that was exposed first to colloidal iron and several hours later to colloidal gold (see text). A residual body (I) has accumulated a mass of electron-dense iron particles. The larger gold particles are seen within a digestive vacuole (G). The presence of iron particles (arrow) in the vacuole containing gold strongly suggests that a lysosome formed as a digestive vacuole during the exposure to iron has fused with the more recently formed gold-containing vacuole. × 50,000. (Courtesy of G. B. Gordon, L. R. Miller, and K. Bensch.)*

molecules are actively being studied. For example, if phagocytes are permitted to ingest yeast stained with pH-dependent indicator dyes, it is possible to follow pH changes in the phagocytic vacuoles by observing changes in the dye color. The interior of digestion vacuoles rapidly becomes quite acid, which presumably favors the action of the acid hydrolases. How the pH changes are brought about is still not understood.

What mechanisms control lysosome movement within the cell? What features of their membranes permit lysosomes to fuse with endocytic vacuoles but not with mitochondria or other organelles? We mentioned earlier that certain microorganisms inhibit lysosome fusions with endocytic structures. This apparently also occurs when the plasma membrane is coated with a specific plant protein, *concanavalin A* (p. 280) prior to endocytosis. Such cases are being studied for clues to the control of lysosome fusions. Also still to be determined are the features of the lysosome surface that protect it from digestion by the hydrolases within and whether, for example, in primary lysosomes hydrolases are somehow stored in inactive form.

2.8.3 HYDROLYSIS OF ENDOGENOUS MACROMOLECULES: INTRACELLULAR TURNOVER

Most or all eucaryotic cells have a mechanism through which bits of their own cytoplasm are surrounded by a membrane and subsequently degraded. This process has been named *autophagy* to suggest self-(*auto*)phagocytosis. "Autophagic" vacuoles, a type of secondary lysosome, have been described in protozoa, cells of vertebrates and invertebrates, and plant cells. Autophagic vacuoles are identified by their content of one or more recognizable cytoplasmic structures such as mitochondria, plastids, bits of ER, ribosomes, glycogen, or peroxisomes (Fig. II-55).

At first glance, autophagy often appears to be an unselective process. The vacuoles seen in a given cell frequently appear almost to be random gulps of cytoplasm whose contents depend largely on the relative abundance of different structures in the cell. However, there are cases in which autophagy seems quite selective indeed. For example, during development of the larvae of the butterfly *Calpodes* the cells of the "fat body" undergo waves of sequential selective organelle destruction in which first many peroxisomes, later mitochondria, and then ER are degraded through autophagy and replaced by newly formed organelles. Less dramatic autophagic selectivity may be a more widespread phenomenon than presently recognized.

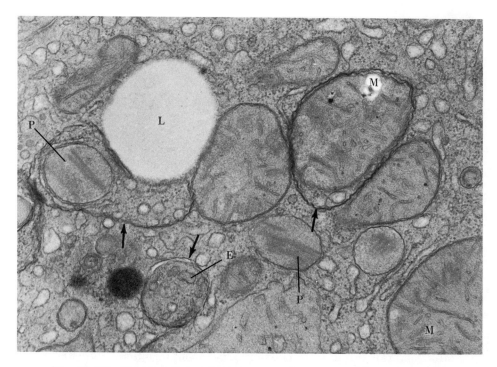

Fig. II-55 *Region of cytoplasm in a rat hepatocyte. Three autophagic vacuoles are present; their delimiting membranes are indicated by the arrows. Within one vacuole a mitochondrion is present (M), a second contains a peroxisome (P), and the third, fragments of ER (E). Other organelles, not included in autophagic vacuoles, lie nearby (M, P). L indicates a lipid droplet in the cytoplasm near one of the vacuoles. × 30,000. (Courtesy of L. Biempica.)*

To what extent does autophagy account for the destructive phases of normal steady-state intracellular turnover, as defined in the introduction to Part 2 (p. 41)? This question is still being debated. Lysosomes are the major intracellular sites of most of the common enzymes capable of extensively degrading macromolecules, so one might expect them to participate centrally in turnover. The *half-life* (time required for half the molecules of a population to be replaced) of hepatocytic mitochondria is about 5 or 6 days. For peroxisomes the figure is about 1 to 2 days, and hepatocyte ribosomal RNA has a half-life of about 5 days. The formation in each hepatocyte of one autophagic vacuole containing one mitochondrion every 10 minutes or so could account for the known rates of mitochondrial turnover. This seems not to be excessively rapid, although the actual rates of autophagy are not known. As might be expected for partially non-selective autophagic processes the turnover of intracellular macromolecules is at random with respect to age. A recently made molecule has much the same probability of being degraded as an older one of the same type. This is seen, for example, by giving cells brief exposure to radioactive amino acids and following the subsequent fate of proteins made during that period. Radioactivity disappears from various populations of proteins with an exponential (random) time course.

Since the degradation of an organelle within an autophagic vacuole seems to involve essentially simultaneous destruction of all visible components of the organelle, if turnover depended simply upon autophagy of whole organelles, all macromolecules of a given type of organelle should show essentially the same turnover rates. This does seem to be the case for several proteins of peroxisomes. In contrast, the turnover rate of mitochondrial outer membrane proteins is higher than that of inner membrane proteins. Membrane lipids of ER and other organelles turn over at different rates from the proteins of the same structures. Various soluble enzymes of the hyaloplasm turn over at very different rates. Differences in turnover rates among components of a given organelle are tentatively considered to reflect dynamic exchanges and replacement mechanisms by which a large structure can gain and lose individual macromolecules or arrays of molecules without extensive destruction. This ties in with the information about membrane formation and changes discussed in earlier sections (Sections 2.4.4 and 2.7.4).

There is no contradiction between molecule-by-molecule replacement and autophagy; both could occur simultaneously and contribute to turnover. And it should be noted that if molecule-by-molecule turnover of an organelle occurs, one still must account for the eventual degradation of the molecules. Is there some kind of "microautophagic" process by which individual molecules are taken into lysosomes and degraded? Selective microautophagy would also be useful in explaining the turnover behavior of soluble enzymes. Various investigators have speculated that microautophagy might occur by a pinocytosis-like incorporation of cytoplasm during events in which vesicles bud from the lysosome surface into the interior (see below and Fig. II-50). But present techniques have yet to provide a decisive test of the concept.

Some puzzling correlations have been noted between turnover rates and the sizes of proteins, or their susceptibility to attack by proteolytic enzymes. These are observed with procaryotes (see Section 3.2.5), which lack lysosomes, as well as with the lysosome-containing eucaryotes. While still unexplained, the observations have led to speculation that nonlysosomal enzymes or other molecules may play important roles in turnover, perhaps by altering organelles or macromolecules so as to induce their uptake by lysosomes. In studies of the turnover of blood glycoproteins it was observed that treatment of the proteins with the enzyme neuraminidase greatly enhanced the rate of endocytosis and destruction of the proteins by hepatocytes. The removal of sialic acid by the enzyme exposes the sugar galactose in the carbohydrate chains of the glycoproteins. Hepatocytes possess receptors that bind or respond to these altered chains. The search is now on for analogous effects in normal intracellular turnover, although, as yet, nothing is known about possible receptors on lysosome membranes.

Although autophagy is a normal physiological phenomenon proceeding, probably at low rates, in normal cells, many more autophagic vacuoles are found when cells are under metabolic stress (when they are undergoing extensive developmental remodeling or responding to hormonal stimulation or to injury). Autophagy may be a means by which cells digest bits of their own cytoplasm, surviving periods of starvation, stress, or metabolic urgency without self-destruction. It is not uncommon for cells that are soon to die to show an abundance of autophagic vacuoles. This is true in many cases of cell death associated with the normal modeling of organs and tissues during development. But this should not be construed to mean that autophagic vacuoles lead to the death of the cells. Indeed, it may be that the cells die, under these circumstances, *despite* a "defense" mechanism involving autophagy.

How do portions of the cytoplasm come to be enclosed within the membranes of autophagic vacuoles and how do acid hydrolases enter the vacuoles? Figures II-50 and II-56 outline one hypothesis which suggests that the endoplasmic reticulum, particularly in the Golgi zone, is involved in forming the delimiting membrane of the vacuole and perhaps in directly providing lysosomal hydrolases. It also has been proposed that the delimiting membranes of some autophagic vacuoles derive from the Golgi apparatus. Fusion of other lysosomes with autophagic vacuoles has been demonstrated in hepatocytes; presumably this contributes hydrolases to the vacuoles.

Several additional autophagic phenomena should be mentioned. In cells of some endocrine glands the membranes surrounding secretion granules can fuse with lysosomes leading to a degradation of the granule contents. This process, called *crinophagy*, increases markedly when exocytic release of the secretions is suddenly decreased and there is a consequent temporary imbalance between synthesis of the secretions and their export from the cell. This raises a fascinating problem, What features of lysosomes or secretion granules are altered so that crinophagic fusion is promoted and excess secretory materials do not accumulate?

Lysosomes often tend to diminish in size as digestion proceeds and digestion products diffuse out. This involves a decrease in lysosome surface area, which can occur by the invagination of the lysosome delimiting membrane and the budding off into the interior of small vesicles

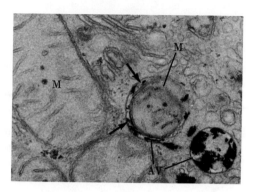

Fig. II-56   *Two autophagic vacuoles (AV) in a hepatocyte from a section of rat liver that was incubated to show sites of* acid phosphatase *activity. The reaction product appears black. The vacuole at the left contains a mitochondrion (M). Arrows indicate appearances suggesting that the membranes surrounding this mitochondrion may be forming by the flattening and transformation of an acid phosphatase-containing sac as proposed in Figure II-50. A second mitochondrion (M) at the left of the micrograph is not within an autophagic vacuole.* × *30,000.*

or tubules (Fig. II-50). Lysosomes known as *multivesicular* bodies result from such processes. Multivesicular bodies are prominent in the disposal of pinocytosed materials. It is likely that the membranes internalized within them include some that originally were part of the cell surface, entered the lysosome surface by fusions of endocytic vesicles, and then passed into the lysosome interior by invagination. This may be one of the routes important for the turnover of cell surface macromolecules. On the other hand, in protozoa large endocytic vacuoles sometimes seem to shrink by the budding of vesicles in the opposite direction, that is, into the cytoplasm. There is much current interest in the possibility that some vesicles formed in this way may rejoin the plasma membrane, restoring part of the cell surface withdrawn by endocytosis.

2.8.4   CELL INJURY   It was suggested soon after the discovery
AND CELL DEATH   of lysosomes that abnormal conditions such
as the uptake of injurious materials might
lead to the leakage of lysosomal enzymes into the cell cytoplasm and to
consequent cell injury or death. Although unequivocal evidence for this
suggestion is very difficult to obtain, it remains a potentially important
mechanism that may aid in understanding cell pathology. For example,
some investigators believe that in miners' disease, *silicosis*, silica particles, taken up by macrophages of the lung into phagocytic vacuoles that

fuse with lysosomes, act upon the vacuoles' membranes to make them "leaky." There is little reason to doubt that after a cell has died, its lysosomal hydrolases are released and participate in destroying cellular and extracellular materials.

When certain compounds, including drugs and hormones, are administered to cells, changes are observed in the membranes of lysosomes subsequently isolated. Some agents, such as vitamin A, appear to *labilize* the lysosomes (that is, make them less resistant to disruption) and others, such as cortisone, seem to stabilize the organelles. It is believed by many that in the cell, too, labilizers promote leakage of lysosomal hydrolases and that stabilizers have the opposite effect. Such substances might also affect the fusion of lysosomes with endocytic vacuoles. However, definitive evidence is still being sought.

In inflammations and diseases such as arthritis, hydrolytic enzymes are released from phagocytes and other cells into surrounding tissues where they may do damage. Sometimes this release results from cell death. In addition, hydrolases may sometimes be released from lysosomes of living cells. This occurs, for example, under some experimental circumstances in which lysosomes fuse with forming phagocytic vacuoles that have not completed their separation from the cell surface. The normal developmental remodeling of cartilage and bone involves extensive degradation of extracellular materials which is mediated, in part, by hydrolases secreted by the cells involved.

2.8.5   ***STORAGE OF*** Many indigestible substances, such as iron,
INDIGESTIBLE finely dispersed gold, and some large car-
RESIDUES IN bohydrate polymers, experimentally admin-
LYSOSOMES istered to cells, accumulate within the
lysosomes. Among the most interesting natural accumulations are those that occur with aging in several tissues of man, and those that apparently result from genetically caused absence of lysosomal hydrolases. The frequency of granules called *lipofuscin* pigment granules increases with age in human nerve, heart, and liver cells. The granules are secondary lysosomes (residual bodies) in which lipids and other materials accumulate. Many congenital "storage diseases," fortunately rare but often fatal early in childhood, show abnormally large lysosomes. Most of those thus far studied appear to result from the inherited defect of a gene required for the synthesis of a specific hydrolase normally present in lysosomes.

The first studied of these "lysosomal diseases" is a generalized *glycogen* storage disease, called Pompe's disease, in which the missing enzyme, normally found in lysosomes of various tissues, is involved in the degradation of glycogen. In the absence of this enzyme the liver lysosomes become engorged with glycogen. Some 30 storage diseases are now known. Each reflects a deficiency in a specific enzyme, and each shows a characteristic pattern of accumulation of *mucopolysaccharides, lipids,* or related compounds in tissues such as nerve cells, muscle, spleen, or liver. Interestingly, when cells (fibroblasts) from a patient with a storage disease are grown in tissue culture along with normal cells (or cells from a patient with a different storage disease) the abnormal deposits may disappear. Evidently the normal cells release hydrolases into the medium and the lysosomes of the abnormal cells obtain the missing enzyme through endocytosis. Chapter 5.2 will discuss some diagnostic and therapeutic aspects of storage diseases.

Not all abnormally stuffed lysosomes, however, need arise from enzyme deficiency. Diseases are known in which the lysosomes apparently contain a normal complement of enzymes but cannot cope with materials that enter them. This may result, for example, from the presence of abnormally large amounts of hard-to-digest molecules in the cell. A suggestion along these lines has been advanced for one of the factors that may contribute to the lipid deposits found in blood-vessel walls in *atherosclerosis*.

The disorders being discussed raise an important unresolved question: What is the usual fate of a lysosome? In protozoa, residues of digestion are released from the cell by fusion of the membrane of a residual body with the plasma membrane, that is, by exocytosis. The extent to which similar processes occur in mammalian cell types is not certain. Maturing red blood cells seem to lose many of their cytoplasmic organelles by autophagy followed by exocytosis. But the existence of storage diseases implies that many cell types have, at best, limited capacities to rid themselves of lysosome contents. Some, such as macrophages, are known to retain molecules in their lysosomes for very prolonged periods. The fact that new digestive structures can fuse with older ones, and that lysosomes can grow smaller as digestion proceeds, raises the possibility of a kind of recycling system in which the size and number of lysosomes in a cell remains more or less constant despite the slowness or absence of exocytosis.

c h a p t e r **2.9**
PEROXISOMES (MICROBODIES, GLYOXYSOMES)

Peroxisomes were seen in rodent kidney and liver in the early 1950s, when electron microscopy of sectioned biological materials was in its infancy; they were then called microbodies. One of the peroxisome enzymes, urate oxidase, was among the first enzymes to be studied in isolated fractions of rat liver cells. The distribution of this enzyme in cell fractions was markedly similar to, but not identical with, that of the lysosomal enzyme acid phosphatase. It proved to be situated in peroxisomes rather than lysosomes.

A number of procedures have been developed to separate peroxisomes from other organelles, and cytochemical methods are now available for light microscope and electron microscope visualization of peroxisomes (see Figs. I-19 and II-57). It is now established that peroxisomes are distinctive organelles of widespread occurrence both in plants and animals.

2.9.1 OCCURRENCE AND MORPHOLOGY

The peroxisomes of rodent liver and kidney, and those of various plant tissues, have diameters ranging from 0.5 to more than 1 μm. They are delimited by a single membrane, about 65–80 Å thick, and they contain a finely granular matrix. In a number of tissues, including mammalian liver (man is an exception) and diverse plant cell types (Fig. II-57), a core or nucleoid is present. The structure of the nucleoid is characteristic of the cell type and species; for example, in rat liver the nucleoids consist of straight tubules arranged in parallel in a crystallike array while in mouse and hamster liver they have more the appearance of twisted strands.

Microscopists initially referred to these organelles as "microbodies." Work with cell fractions from rat liver demonstrated the presence, in microbody-rich fractions, of enzymes catalyzing reactions that involve hydrogen peroxide; hence the name *peroxisomes* came into use. It was originally thought that peroxisomes were present only in liver and kidney of mammals, in a few other animal cell types, such as cells of the fat body of the butterfly *Calpodes*, and in certain plant cells (Fig II-57 A and B).

The advent of a cytochemical method for the demonstration of catalase (Figs. I-19 and II-57E), one of the peroxisomal enzymes, changed this view. Peroxisomes were found in virtually all cell types of mammals; the red blood cells are the chief exception known at present. For example, in the absorptive cells lining the small intestine (Fig. III-19) such organelles are present in large numbers (Fig. II-57D and E). They had been overlooked by microscopists largely because they lack nucleoids or other immediately striking morphological features. Peroxisomes similar to those of the intestine are present with greatly different frequencies in many different mammalian cell types. These are often

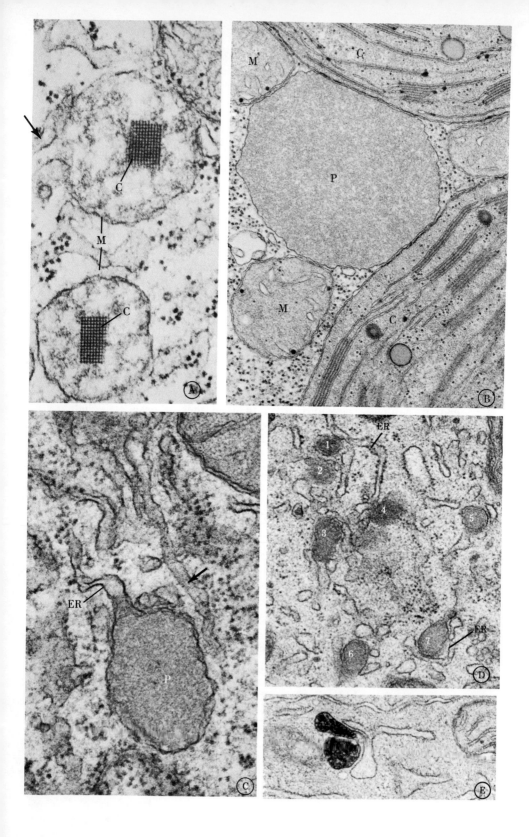

referred to as *microperoxisomes*. Usually they are quite small, measuring 150–250 nm in length.

An important feature of peroxisomes, illustrated in Figure II-57, is their close association with the endoplasmic reticulum. In fact, continuities between ER and the delimiting membranes of peroxisome membranes are commonly encountered. Such continuities are particularly numerous with microperoxisomes.

2.9.2   ***ENZYME*** A familiar pattern of cell organization, the
ACTIVITIES AND location of several metabolically related
FUNCTIONAL enzymes within one organelle, is well
SIGNIFICANCE illustrated by peroxisomes. The enzymes
                                    initially identified in rat liver fractions included several that produce hydrogen peroxide (urate oxidase, *D*-amino acid oxidase, α-hydroxy acid oxidase) and catalase, which decomposes H_2O_2. Peroxisomes in different plant and animal cells can vary considerably in their enzymatic makeup, but they do tend to contain at least some peroxide-producing enzymes and all (with one or two possible exceptions that still require verification) contain catalase. Urate oxidase is found only in the nucleoid-containing peroxisomes. It is possible that in some cases, such as rat liver, the nucleoids are actually crystals of this enzyme.

Our knowledge of the metabolic participation of peroxisomes is far more complete for plant cells than for animals. In many green leaves

◄ *Fig. II-57   A. Two peroxisomes from a cell of the grass plant* Avena *show crystal-like nucleoids. The delimiting membranes* (M) *are continuous with those of the ER (arrow).* × *75,000.* **B.** *Portion of a cell of tobacco leaf showing a peroxisome* (P) *closely associated with chloroplasts* (C) *and mitochondria* (M). × *40,000.* **C.** *Portion of a hepatocyte in rat liver. The peroxisome* (P) *was sectioned in a plane that did not include its nucleoid. The continuity of the peroxisome membrane with the ER is evident.* × *60,000.* **D.** *Portion of an absorptive cell in guinea pig small intestine. A cluster of microperoxisomes (1–7) is present in this small area. ER, mostly smooth, is present nearby. The ER is quite tortuous, and the delimiting membranes of the microperoxisomes are irregular, so the continuities between the ER and peroxisomes are difficult to see. They can be demonstrated through use of a device that permits one to rotate and tilt a section in an electron microscope, permitting the selection of the most favorable angle from which to view a structure.* × *37,000.* **E.** *A portion of a cell like that in* **D,** *but incubated by a cytochemical method utilizing diaminobenzidine (DAB) at alkaline pH (see Fig. I-19). The density of the peroxisomes is due to the reaction product (oxidized DAB) resulting from the action of catalase.* × *44,000. A and B courtesy of Frederick, Newcomb, Vigil and Wergin; C, courtesy of Biempica.*

there is a light-dependent pathway resulting in oxygen consumption, known as *photorespiration*. This involves the interplay of chloroplasts, mitochondria, and peroxisomes. The rate of photorespiration is increased by high light intensity, low CO_2 concentrations, and high O_2 concentrations. Under these conditions excess molecules are produced of a 2-carbon compound, *glycolate*, a major photosynthetic by-product. Apparently glycolate passes out of chloroplasts and gains access to the adjacent peroxisomes. Isolated leaf-cell peroxisomes have been shown to possess the enzymes of the "glycolate" pathway, among others. Through reactions in which oxygen is consumed, these convert glycolate to compounds such as the amino acid glycine which, in turn, enter the mitochondria and are further metabolized there. The close associations often observed among chloroplasts, mitochondria, and peroxisomes (Fig. II-57B) are thus apparently correlated with metabolic interplay among the organelles.

Seeds of castor beans and other plants store fats in tissues known as endosperm. During early development of the embryos, this fat is transformed into sugars (gluconeogenesis) which are used for ATP-generating metabolism. The transformation depends upon a sequence of enzyme reactions known as the *glyoxylate* cycle (glyoxylate should not be confused with glycolate, just discussed—the two are, however, metabolically interconvertible). The responsible enzymes are localized within peroxisomes that also possess catalase as well as enzymes that oxidize fatty acids with the production of H_2O_2; some plant scientists prefer to call these organelles *glyoxysomes*. The involvement of endosperm peroxisomes in lipid metabolism probably accounts for two morphological features, (1) the peroxisomes are found quite close to the surfaces of lipid storage droplets in the cytoplasm, and (2) the number of peroxisomes increases dramatically as lipid breakdown begins.

Peroxisomal roles in animal cells are still unclear. The catalase in the organelles presumably protects the cell against the peroxide generated by other peroxisomal enzymes; H_2O_2 is quite toxic. Perhaps the catalase also decomposes peroxides generated elsewhere in the cell. The roles of the oxidases are mysterious. Urate oxidase may be important in degrading purines (the class of nitrogen-containing compounds of which the adenine and guanine bases of nucleic acids are examples). In frogs and chickens two other enzymes of purine catabolism accompany this oxidase in hepatic peroxisomes. In various mammals peroxisomes (particularly microperoxisomes) are very numerous in cell types where lipid absorption, transport or storage is prominent, and peroxisome number or size sometimes changes in animals administered drugs that affect lipid metabolism. This has led to the proposal of a role of peroxisomes in lipid metabolism, perhaps in gluconeogenesis, as in plant seeds. But there is no firm evidence to substantiate this.

2.9.3   ***FORMATION***   From their morphological relationships, peroxisomes seem to arise as outgrowths or swellings from the ER (Fig. II-57). Perhaps they bud off in a manner similar to the formation of secretion granules. However, especially with microperoxisomes, the continuities with the ER are so common as to raise the possibility that some peroxisomes may maintain long-term, or even permanent, functionally important connections with the reticulum. Yet it is premature, and perhaps even incorrect, to conclude that peroxisome formation simply involves the movement of enzymes and other molecules from polysomes into ER cisternae and thence into the forming organelles. In the case of catalase it appears that the molecules entering the peroxisomes are not completed enzymes, but rather subunits. Apparently the formation of the functional enzyme molecules occurs within the peroxisomes themselves through combination of the four subunits of catalase plus an iron-containing coenzyme (heme). Some data obtained by exposing hepatocytes to radioactive amino acids and isolating cell fractions at intervals thereafter raise the possibility that catalase subunits can reach the peroxisomes by presently unknown routes of transport perhaps through the hyaloplasm or through some labile structure that does not survive tissue fractionation.

c h a p t e r **2.10**

CENTRIOLES, CILIA, AND FLAGELLA

Structures with the characteristic morphology of centrioles occur in most animal cells and in some plant cells. They participate in cell division as will be discussed in Chapter 4.2. Of their other roles, the best established is their relationship to *cilia* and *flagella*, the long motile structures that extend from the surface of many unicellular organisms, from the sperm of a wide variety of plants and animals, and from other cell types of multicellular organisms. Cilia and flagella invariably form in association with *basal bodies*, organelles widely regarded as alternate forms of centrioles.

2.10.1   ***STRUCTURE***   Centrioles usually occur in pairs. In typical
OF CENTRIOLES   nondividing cells there is one pair per cell (Fig. II-58). Often this pair is located close to the Golgi apparatus. Characteristically, the centrioles are cylindrical

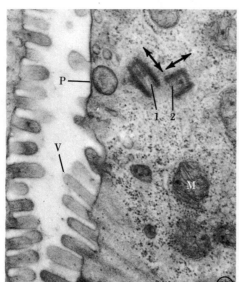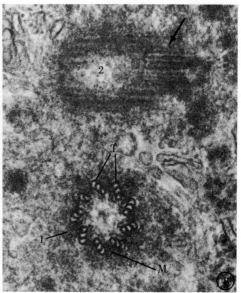

Fig. II-58   *A. Portion of the absorptive surface of a cell of the intestine (chick embryo) showing* centrioles. *Both centrioles of the pair (1, 2) have been cut longitudinally. The long axes of the centrioles (arrows) are almost perpendicular to one another. M indicates a mitochondrion; P the plasma membrane; and V, microvilli. × 35,000. (Courtesy of S. Sorokin and D. W. Fawcett.) B. Portion of a cultured cell from the Chinese hamster showing a centriole pair sectioned at a different angle from A. Centriole 1 is cut transversely; it shows the characteristic arrangement of tubules (T) as 9 triplets embedded in a cylinder of dense matrix (M) surrounding a less dense central region. Centriole 2 is cut obliquely thus obscuring the pattern of tubules. Part of one tubule cut in longitudinal section is seen at the arrow. Approx. × 70,000. (Courtesy of B. R. Brinkley and E. Stubblefield.)*

structures approximately 0.15 μm in diameter and 0.3–0.5 μm long. This is just at the limit of resolution of the light microscope, so that little was learned of their detailed structure until electron microscopy developed.

Centrioles have a characteristic appearance in the electron microscope. The two members of the pair often lie with their long axes perpendicular to each other (Fig. II-58A). Each is made of nine sets of tubulelike structures, arranged as the walls of a cylinder; each set is a *triplet* composed of three closely associated tubular elements (Fig. II-50D). These triplets are embedded in an amorphous matrix. At one end of the centriole a "cartwheel" structure is often found (Fig. II-62). The presence of this structure at only one end defines a polarity of the centriole which will be outlined later (Section 3.3.3).

2.10.2   STRUCTURE OF CILIA AND FLAGELLA   In typical ciliated cells each cell contains large numbers of cilia about 2–10 μm long. In flagellated cells there are usually only one or two flagella per cell; they are often as long as 100–200 μm. The diameters of both cilia and flagella usually are less than 0.5 μm.

Cilia and flagella are of similar structure. They resemble centrioles in having nine sets of tubules arranged in a cylinder 0.15–0.2 μm in diameter (Fig. II-59) and referred to collectively as the *axoneme*. Unlike centrioles, an additional pair of tubules is found in the center of the cylinder. Also, the peripheral sets are "doublets" of two tubular elements each, rather than the triplets of centrioles; "arms" extend from the doublets as shown in Figure II-59. The pattern of cilia and flagella is thus 9 + 2 instead of 9 + 0 like that of the centrioles. In addition, whereas centrioles lie within the cytoplasm and have no surrounding membrane, mature cilia and flagella are bounded by a membrane that is an extension of the plasma membrane.

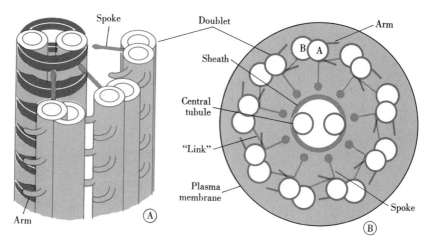

Fig. II-59   *Structure of a cilium or flagellum.* **A.** *An interpretation of the arrangement of tubules and associated material as seen from outside.* **B.** *The structures seen in a cross-section. Each doublet consists of an A and a B tubule; two arms are attached to the A tubule. Note that the A tubule is a complete circle in cross section, whereas part of the wall of the B tubule is shared with the A. Spokes seem to occur at intervals along the length of each doublet; perhaps they connect the doublets to the sheath surrounding the central tubules. Some investigators believe that fine links, made of the protein "nexin," attach adjacent doublets to one another; there may also be bridges between the two central tubules. (After I. Gibbons, P. Satir, and many others.)*

2.10.3   ***CILIA AND***   Cilia beat by producing an *effective stroke*
                   FLAGELLA;   *(power stroke)* followed by recovery, that is,
               FUNCTIONAL   a return to the original position (Fig. II-60).
                   ASPECTS   The rate of motion is rapid enough that the
                              tip may attain velocities of several thousand
micrometers per minute, and a given cilium can beat a dozen to several
dozen times a second. Flagella move with an undulating snakelike motion (Fig. II-60). For both cilia and flagella the overall movement pattern, as seen in three dimensions, sometimes has helical or circular
aspects (Figs. II-60, III-11). For protozoa and sperm, cilia or flagella
endow the cells with motility. Movement of cilia of the cells lining the
gills of molluscs induces circulation of water past the gills. In ciliated
epithelia such as those lining the trachea of mammals, collective movement of several hundred cilia on each cell provides a strong upward
current that helps carry mucus and trapped dust out of the lung.
(Cigarette smoking inhibits normal ciliary action in the respiratory
tract and thus affects the normal mechanisms for removal of foreign
matter.)

  Cilia typically beat in coordinated waves, so that at any given
moment some cilia of a cell are in the effective positions of the cycle and
others are in the recovery positions (Fig. III-11). This assures a steady
flow of fluid past a surface. The limited available information on ciliary
coordination will be discussed in Section 3.3.3.

  Much progress has been made in recent years toward relating the

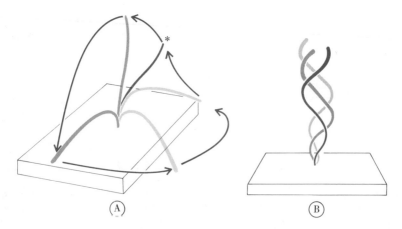

Fig. II-60   *Successive stages in the motion of a cilium (A) and a flagellum
(B). The effective stroke of the cilium begins at *. For many cilia, especially
in protozoa, recovery of the cilium occurs in a different plane from the
effective stroke. As Figure III-11 shows, ciliary motion is often somewhat
more complex than indicated here. (After B. Parducz and P. Satir.)*

structure of cilia and flagella to their motion, and it is to be anticipated that a detailed molecular picture will be available before long. Cilia or flagella that have been detached from the cell can beat if ATP is added to the suspension medium. This indicates that cilia and flagella contain the machinery for motion, rather than being passively moved by activities elsewhere in the cell. If the tubules of the axoneme contracted so as to shorten by about 10 percent of their length, much of the motion could be accounted for. However, there is little evidence for tubule contraction. Instead, careful observations of cilia during bending indicate that there is a cycle of tubule sliding (Fig. II-61) in which the doublets move longitudinally with respect to one another. Thus the situation resembles, to some extent, the contraction of striated muscle which, as the next chapter will outline, depends on the sliding of fine filaments along one another. It was, however, necessary to demonstrate that the sliding of ciliary tubules is unlikely to be simply the *result* of bending of the cilium: Isolated flagella were fragmented, their membranes removed, and they were briefly digested with proteolytic enzymes. They were then exposed to ATP and studied by microcinematography (p. 20). The structures elongated greatly, indicating that sliding had taken place without bending. This strongly suggests that sliding is a primary event. Presumably sliding is harnessed to generate bending by interaction of the tubules with some other ciliary components that are disrupted in the experiment just described, perhaps the "spokes" shown in Figure II-59 or links that some suspect to exist between adjacent doublets.

The roles of the central tubules require elucidation. Because certain nonmotile cilia lack the central tubule pair (see, for example, Sections 3.6.3 and 3.8.1), it has sometimes been suggested that they are essential for motility. Especially since early studies indicated that the effective stroke of many cilia is oriented at right angles to the plane defined by the central pair, it has been proposed that they have some important mechanical function related to the pattern of the ciliary beat. This still may be borne out although careful reevaluation has cast some doubt as to whether the effective stroke is always oriented at right angles to the central plane. Further, sperm and other cells of a few species, and mutants of some unicellular organisms, have cilia or flagella in which the central tubule arrangement is very different from that of Figure II-59. A single modified tubule may be present, there may be more than two, or the tubules may be absent. The deviant forms are often nonmotile but some are not, indicating that the pattern of paired central tubules is not an absolute requirement for motion.

The molecular morphology of the axoneme is currently under intensive investigation. The tubules themselves are made of molecules of the *tubulin* class, similar to those of other cellular microtubules to be

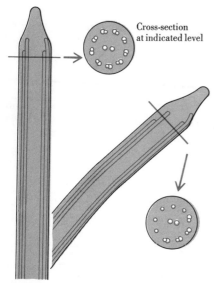

Cross-section
at indicated level

Fig. II-61 *The sliding of doub-lets during motion of a cilium in the gills of a mussel. One of the two tubules of each doublet protrudes further into the tip of the cilium than does the other. Thus, cross sections at an appropriate constant distance from the tip will show a change in tubule pattern during bending, from all doublets to doub-lets and single tubules. The entire doublets slide with respect to one another; but the two members of each doublet retain their positions relative to each other. (After P. Satir.)*

discussed in the next chapter. In an important experiment, axonemes isolated from the protozoan *Tetrahymena* were treated with salt solutions that selectively extract the central tubules and the arms on the peripheral doublets. "Ghosts" consisting of armless peripheral doublets are left behind. From the salt extract the protein *dynein* can be isolated. When purified dynein is added to the extracted ghosts, the arms reappear. Dynein is an ATPase and is likely to be involved in making energy available for sliding. One hypothesis suggests that under the influence of ATP, the arms (dynein) are able to form transient bridges between adjacent doublets, analogous to the bridges formed in filament sliding in muscle (Section 2.11.2).

2.10.4   BASAL BODIES;
THE FORMATION
OF CILIA AND
FLAGELLA

A basal body is present at the base of each cilium or flagellum (Fig. II-62). This is true from the onset of formation of a given cilium or flagellum. Often fibers (*rootlets*) extend from the basal body deep into the cytoplasm, perhaps as part of a system that anchors the basal bodies (Figs. II-62 and III-24A).

Each basal body has the same 9 + 0 structure as a centriole. The nine peripheral doublets of the associated cilium or flagellum appear almost as extensions of two of the three tubules of the basal body's triplets. That basal bodies and centrioles are closely related, probably as alternate forms of the same organelle, is indicated by several lines of

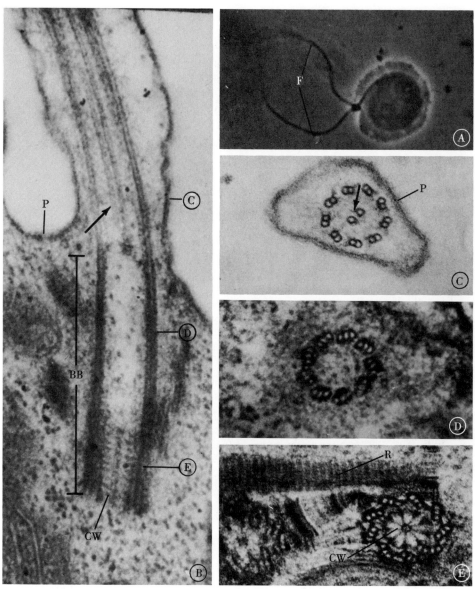

Fig. II-62   *Flagella and basal bodies of the unicellular organism Naeg-*
laria during its flagellate phase. A is a phase contrast micrograph; the two
flagella are seen at (F). B to E are electron micrographs. B is a longitudi-
nal section and C, D, and E are transverse sections at the levels indicated
by the corresponding letters in B. P indicates the plasma membrane, CW the
"cartwheel" structure at one end of the basal body (BB), and the arrows
show the central pair of tubules of the flagellum. In E a cross-striated
rootlet (R) is seen. Such structures are often found extending from basal
bodies into the cytoplasm; perhaps they help to anchor the basal body. A ×
2000; B-E approx. × 90,000. (Courtesy of A. D. Dingle and C. Fulton.)

evidence. (The distinction between the two is functional and derives from the preelectron microscope era when the identity in structure could not be known.) For example, in some developing sperm cells a flagellum grows in association with a centriole during division, while the centriole is still in its customary position for dividing cells at the poles of the mitotic spindle (Section 4.2.7). Thus the same body can function simultaneously as both a centriole (in cell division) and a basal body (in flagellum formation).

It has been proposed that most of the growth of cilia and flagella depends upon the assembly of structural subunits which are "preformed," in the sense that subunit synthesis is a separable process from subsequent assembly into structures. If the two flagella of the unicellular organism *Chlamydomonas* (Fig. II-46) are broken off by mechanical agitation or other means, new ones regenerate. Preventing protein synthesis by the addition of inhibitors (such as *cycloheximide*) after amputation does not prevent regeneration, although the regenerates do not reach full length. If only one flagellum is removed, the other shortens for a while as regeneration occurs, then both return to normal length. When the numerous cilia at the surface of sea urchin embryos are removed, regeneration occurs even in the presence of inhibitors of protein synthesis.

Presumably free subunits are usually present in the cells studied in these experiments and are available for assembly. Thus although the subunits are presumed to be chiefly protein, newly made proteins are not needed for regeneration of considerable lengths of cilia and flagella, and the inhibition of protein synthesis does not prevent regeneration. The experiment in which one flagellum is removed is interpreted as indicating that subunits can be freed from one flagellum to support the growth of another.

The basal bodies in some way control assembly of ciliary and flagellar subunits. Of interest is the fact that the 9 + 0 structure of the basal body can control the formation of the somewhat different 9 + 2 structure. In one investigation cells were fed radioactive amino acids at different times during regeneration of flagella and were then studied by autoradiography. Results of the study indicated that as a flagellum grows it adds much of its new protein (identifiable by radioactivity) at the tip rather than at the base. Apparently the basal body initiates ciliary tubule formation, and the tubules then grow by continued addition of subunits at the end opposite the basal body. Such assembly takes place in the test tube when purified tubule proteins are added to isolated basal bodies or to portions of cilia (Fig. II-63).

Analysis of the molecular details of basal body roles must await more adequate information about their composition (see Section 2.10.5). Recent studies on isolated basal bodies suggest that they are rich in proteins resembling those that constitute the ciliary tubules.

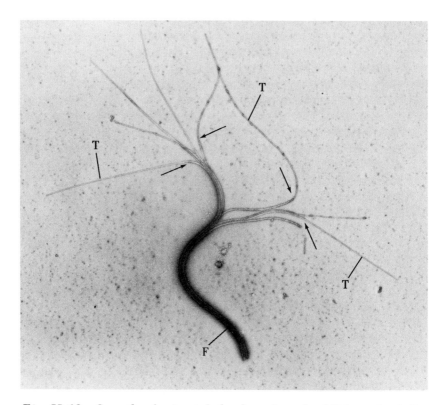

Fig. II-63 *Growth of microtubules by oriented addition of tubulin molecules. A flagellum (F) from* Chlamydomonas *(see Fig. II-46) was detached from the cell, and its membrane removed. It was then placed in a solution containing tubulin (purified from pig brain) under conditions favoring microtubule formation. The original ends of the doublets of the flagellum are indicated by arrows. Continuous with the A tubules (Fig. II-59) of the doublets are the microtubules (T) formed from the pig brain tubulin.* × *7000. (From C. Allen and G. Borisy,* J. Molecular Biology, *90:381. Courtesy of the authors.)*

2.10.5 **CENTRIOLE** The study of centriole duplication faces
 DUPLICATION two main difficulties. First, centriole chemistry is poorly understood; the organelles
are too small for adequate cytochemical study, and they have only recently been isolated in a form pure enough to permit biochemical study.
Second, the mode of formation may vary from cell to cell. When ciliated
protozoa, such as *Paramecium*, divide by binary fission, the precise
pattern of basal bodies near the surface (Fig. III-10) is transmitted to
both daughter cells. This involves the duplication of each basal body in
coordination with cell division. In other cases, however, basal bodies
appear at one point in the cell life cycle despite the previous absence of

microscopically visible centrioles or basal bodies. This is true of the unicellular organism *Naeglaria* (Fig. II-62). This organism may exist in a nonflagellated form in which centrioles, basal bodies, or flagella have not been found, or it may become a flagellated cell with prominent flagella and basal bodies. This sudden appearance of basal bodies in cells that previously lacked visible centrioles or basal bodies also occurs in the sperm of some plants. It remains to be determined whether some self-duplicating structure is present that gives rise to basal bodies; it is possible that such structures have not been recognized in the microscope because of their small size or lack of distinctive appearance.

The possible presence of nucleic acids in centrioles and basal bodies is an unsettled matter. Staining and autoradiography have demonstrated DNA in preparations of isolated portions of the cortical region of ciliated protozoa (the region just below the cell surface, where numerous basal bodies are aligned). But in some cases this DNA has been attributed to artifactual contamination of the preparations by material from other cell regions disrupted during isolation of the cortex. RNase treatment seems to remove some electron-dense material from basal bodies and centrioles, and some staining studies have also been interpreted as suggesting the presence of RNA. At present the evidence for centriolar RNAs is less widely doubted than that for DNAs but clearly more information is needed.

In many cells centrioles appear to duplicate by the growth of a new *procentriole* near the "old" centriole. The procentriole often contains the 9 + 0 pattern but is much shorter than the old centriole. It is characteristically oriented at right angles to the old centriole (Fig. II-64) and eventually grows into a complete centriole. This occurs in cells where centriole duplication is part of cell division, as well as in some cells where many basal bodies form. In some of the cells forming basal bodies, several procentrioles appear more or less simultaneously, radially arranged around the old centriole. Procentriole formation appears to involve neither fission nor direct budding from preexisting centrioles; how it does occur remains to be explained.

In cells of the respiratory and reproductive tracts of mammals, only two centrioles are present early in development; eventually large numbers of basal bodies are formed and produce the cilia that line the tracts. In the respiratory system, some of the basal bodies have been seen to form near the centrioles, but many arise from small aggregations of amorphous or finely fibrillar material with no obvious resemblances to centrioles. Although these aggregates are considered by some to be "generative" structures, originally formed by centrioles, it is possible that they arise from some unrecognizable precursor material that originates elsewhere. In the cells that develop into the sperm of certain ferns and cycads, large compound bodies are formed from an unknown source and then fragment to produce numerous procentrioles which

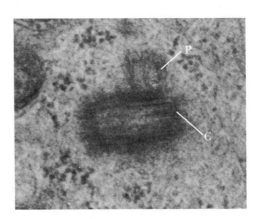

Fig. II-64 *Portion of a human cultured cell (*HeLa, *see Chapter 3.11) showing a centriole (*C*) and an associated procentriole (*P*). It is thought that the procentriole grows at right angles to the centriole eventually producing a configuration resembling Figure II-58.* × *64,000. (Courtesy of E. Robbins.)*

ultimately give rise to the basal bodies of the many flagella found on each cell.

"Self-duplication," in the sense that an old structure participates in the formation of a new one, more or less directly, may well be involved in the instances where procentrioles form near old centrioles. But how formation of centrioles or basal bodies occurs in cells initially without centrioles is unclear. This is related to the more general question as to how the distribution and assembly of microtubules is controlled by the cell (p. 170-171).

c h a p t e r **2.11**

MICROTUBULES AND MICROFILAMENTS

2.11.1   MICROTUBULES   As preparative methods for electron microscopy improved, microtubules became evident as a regular component of the cytoplasm of most cells. Microtubules are best preserved by fixatives like glutaraldehyde or formaldehyde, and only in the past decade have electron microscopists used these agents routinely to treat tissue before it is exposed to osmium tetroxide (Chapter 1.2) which, used alone, often disrupts the tubules. Microtubules are thin cylinders, approximately 200–300 Å in diameter (Figs. II-65 and IV-19); it is not known whether there are any limits to possible lengths since beyond several micrometers the measurements are very difficult to make in the thin section used for electron microscopy. They are not membrane delimited. From images such as Figure II-65A it is concluded that the microtubule wall is made of some form of repeated basic structural building block. In cross section a microtubule

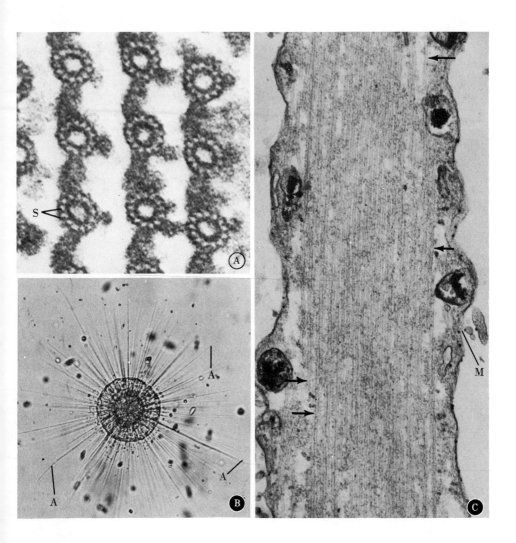

Fig. II-65 **A.** *Several microtubules from a protozoan, viewed in transverse section after staining with tannic acid. Tannic acid brings out details of the tubule substructure; each tubule appears as a ring of subunits (S). ×*
450,000. **B.** *The protozoan* Actinospherium *shows many modified pseudopods (axopods), (A) that extend radially from the body of the cell. ×*
100. **C.** *A longitudinal section of an axopod like the ones in* **B**. *Its membrane (M) is part of the plasma membrane. Numerous microtubules (arrows indicate four) are found longitudinally arranged inside the axopod. Transverse sections would show many circular outlines (see Figs. I-10 and III-32). × 45,000. (From L. G. Tilney and K. R. Porter; A is from J. Cell Biology,* **59**:*267.)*

has the appearance of a circular array of about a dozen subunits (13 are very commonly reported). In longitudinal view, particularly with negative staining, the walls sometimes appear to be made of parallel filaments, again numbering about a dozen.

Microtubules are constructed chiefly of proteins called *tubulins*. Some other proteins that may be present have yet to be studied intensively. Tubulins have molecular weights of approximately 50,000–60,000. Purified tubulin preparations from several sources consist of mixtures of roughly equal amounts of two proteins, differing slightly in size and composition. Precisely how these are arranged in the microtubule is uncertain. However, the present consensus is that they are associated as "dimers" of two tubulin molecules, one of each type. The model diagrammed in Figure II-66 illustrates the structure proposed from microscopy. The microtubule wall is thought to be made of globular subunits 50 Å in diameter (this is the size of a tubulin molecule). The subunits are helically arranged around the wall and also fall into rows running parallel to the tubule's long axis. The longitudinal rows of subunits are probably the structures that appear as filaments in the wall. There are hints that variations in composition or structural details sometimes occur among microtubules of different cells or at different sites in a cell, but the general organizational principles usually seem quite similar in widely disparate species and cell types.

Two major interrelated roles have been assigned to microtubules. Together with other structures, they participate in maintenance of cell shape (a "cytoskeletal" role), and they are important in intracellular motion. They often are distributed as might be expected for cytoskeletal structures. For example, microtubules are longitudinally disposed along the elongate processes of cells such as nerve cells (Fig. III-26) or the rigid cytoplasmic extensions of some protozoa (Fig. II-65). The red blood cells of some fish have a flattened disc-like shape and microtubules are arranged as a circular band, just under the edge of the disc. Microtubule participation in motion is suggested by their involvement in cilia and flagella already discussed (Section 2.10.3). Moreover, microtubules often are arranged along pathways of oriented movement of other structures in cells. Thus in cell division, chromosome movement is mediated by the mitotic spindle in which microtubules are quite prominent (Section 4.2.8 will discuss this in detail). Rapid changes in the color of fish result from responses of pigment cells such as *melanophores* to environmentally triggered physiological changes. Within these cells, pigment granules migrate along paths delineated by microtubules, spreading throughout the cell or concentrating in a small area. The corresponding changes in light absorption by the cells alter the appearance of the fish's surface.

Microtubules can be reversibly disrupted by exposing cells to high pressure, low temperature, or drugs like colchicine and vinblastine, which bind to tubulin molecules. (Cilia and flagella are exceptional in this regard; their tubules resist such perturbations.) When the pressure is elevated or temperature lowered, the microtubules of the pseudopods shown in Figure II-65 disappear, and the pseudopods retract into the cell. With the return of the cell to normal conditions, both microtubules and pseudopods reappear. Tubule-disruption experiments of this type confirm the importance of the microtubules for aspects of cell shape and motion. Colchicine seems to bind specifically to tubulin.

Paralleling their behavior in cells, tubules can disassemble and reassemble readily in the test tube. Assembly can be induced in solutions of purified microtubule proteins containing tubulin dimers plus some other material such as certain non-tubulin proteins. Microtubules will form in such solutions if the temperature and ionic environment are appropriately adjusted (for example, Ca^{++} must be low) and ATP or GTP (guanosine triphosphate) is added. Tubulin molecules themselves have GTP molecules or derivatives attached, but how these, or the ATPs or GTPs added to the solution, contribute to assembly is still not known. Small discs or rings, roughly 300 Å in diameter, found in the assembly solutions may be important for initiating the assembly process or as intermediates.

Assembly and disassembly of microtubules presumably provides a flexible control of cell shape and oriented motion. What factors

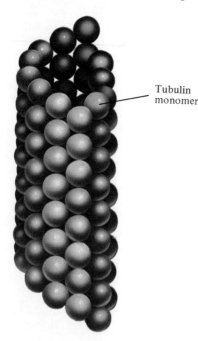

Tubulin
monomer

Fig. II-66   Interpretation of the structure of a microtubule as an assembly of globular subunits. In line with recent evidence, the structure has been drawn as a left-handed helix, but the helix directionality is still being debated. Also to be definitively determined are the relative positions of the two members of each tubulin dimer (p. 169) within the tubule wall. Perhaps they alternate in the longitudinal direction (that is, within the filament-like longitudinal rows of subunits). (After L. Amos, H. P. Erickson, A. Klug, J. Randall.)

influence the assembly process and microtubule distribution in the cell? Circumstantial evidence implicates ions such as Ca^{++} and agents such as cyclic-AMP. Altered intracellular levels of such components are accompanied by changes in microtubule assembly or distribution. However, Ca^{++} and cAMP have so many effects upon cells that it is difficult to disentangle specific direct actions upon microtubules. In the preceding chapter we outlined evidence that basal bodies are sites of initiation of microtubule assembly in cilia or flagella. Other microtubules frequently radiate from "satellite" bodies lying close to centrioles. These bodies have an undistinguished amorphous structure. Many investigators tentatively conclude that such bodies represent an important type of "microtubule organizing center" or "nucleating center." These "centers" might serve to initiate microtubule assembly in many cell types. In part this conclusion stems from the fact that many plant cells that lack centrioles possess microtubules, often terminating in aggregates of amorphous material. Evidently, centrioles *per se* are not needed for tubule assembly.

A major unresolved question is whether microtubules themselves are responsible for generating the movement of other structures, or whether some other system produces the motion and the microtubules orient it. Many investigators believe that molecules at the surfaces of microtubules interact with adjacent material. Sometimes there appear to be special associations between microtubules and membrane-delimited vesicles or other structures moving in the directions defined by the tubules' orientation. Apparent attachments are seen that usually are ill-defined, but sometimes resemble short bridges; perhaps these are involved in producing movement. Structures resembling fine bridges are also occasionally seen to extend between adjacent members of arrays of microtubules. The interlinking of tubules could be important for stabilizing three-dimensional arrays or perhaps for generating sliding of tubules with respect to one another.

The possibility has been entertained that the interior of a microtubule is hollow and that small molecules move through the cavity. There is no strong evidence for this.

2.11.2   **FILAMENTS AND MOTION: STRIATED MUSCLE**   The most thoroughly studied microfilaments are those of striated muscle (Figs. II-67 and II-68). The striated muscle fiber contains within it numerous elongate myofibrils, which are made of regularly arranged fine filaments and which form the basis of muscle contraction. Each myofibril may be regarded as being composed of repeating contractile units or *sarcomeres* (Fig. II-67). Within the sarcomere there is a regular

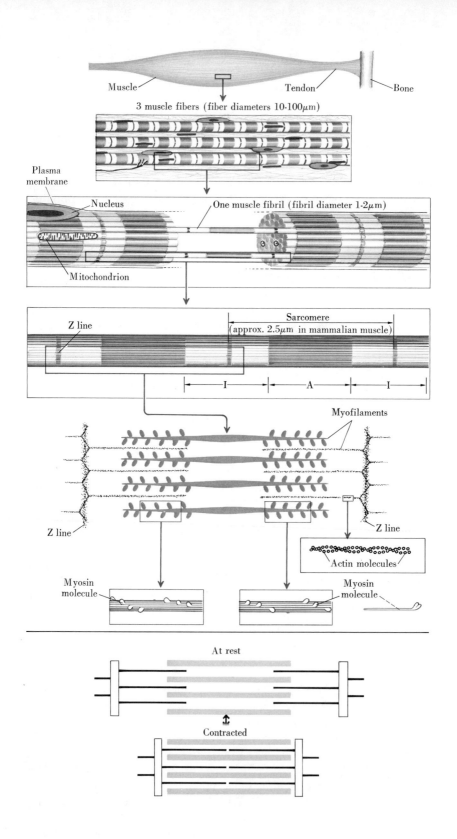

Muscle   Tendon   Bone

3 muscle fibers (fiber diameters 10-100μm)

Plasma
membrane

Nucleus

One muscle fibril (fibril diameter 1-2μm)

Mitochondrion

Z line

Sarcomere
(approx. 2.5μm in mammalian muscle)

I   A   I

Myofilaments

Z line

Z line

Actin molecules

Myosin
molecule

Myosin
molecule

At rest

Contracted

pattern of thick (100 Å) and thin (60 Å) filaments as shown in Figure II-68. This alignment of filaments gives these muscle fibers their cross-banded (striated) appearance. The zones of thin filaments (I bands) can readily be distinguished from those of thick filaments (A bands) by light microscopy. (The individual filaments, however, are below the limit of resolution of the light microscope.)

The polarizing microscope (Section 1.2.2) shows that a regular organization of longitudinal elements is present in living muscle as well as in fixed preparations. The interference and phase microscopes can be used for measuring the amount of material present in different regions of the muscle cell. The comparison of muscle fibers in their normal state with fibers that have been treated with solutions that extract the muscle protein *myosin* shows that only the A bands lose mass. This study and immunohistochemical ones (see Fig. II-71 and p. 22) demonstrate that myosin is found in the thick filaments and that another muscle protein, *actin*, is present in the thin filaments.

Figure II-67 diagrams the generally accepted theory about the arrangement and functioning of protein molecules in the filaments. According to this theory, muscle contraction takes place when the two sets of thin filaments of each sarcomere slide toward the center, past the central set of thick filaments. The sliding of filaments shortens the sarcomere. If the muscle is stretched and held at fixed length, the extent of the overlap of thick and thin filaments is reduced, and the force the muscle can generate is correspondingly reduced. This indicates that the interaction of the two sets of filaments is responsible for the force of contraction.

Although many details of muscle contraction have been explained, some still remain obscure. The sliding itself has been attributed to the action of bridges between thick and thin filaments (Fig. II-68) which

◄ *Fig. II-67* *The structure of vertebrate striated muscle (see also Fig. III-31). The unit comparable to a more conventional cell is a* fiber. *It is surrounded by a single plasma membrane but contains several to many nuclei. Within the fiber,* myofibrils *are present. These consist of filaments which are ordered in such fashion as to produce the appearance of alternating A and I bands along the fibrils. Each filament is of several hundred protein molecules. The diagram illustrates the probable arrangement of* myosin *molecules in the thick filaments and of* actin *molecules in the thin filaments. Many mitochondria are present throughout the fiber; they are found between adjacent myofibrils.*

The region between two successive Z lines along a myofibril is referred to as a sarcomere. *The sarcomere can be thought of as a repeating unit of function. The sliding of filaments within a sarcomere during contraction is illustrated at the bottom of the diagram. (Modified from Loewy and Siekevitz; after the work of H. Huxley and others.)*

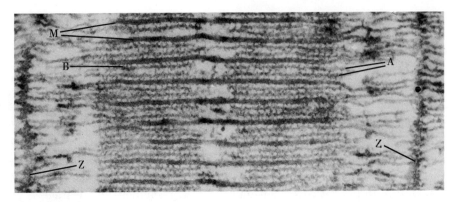

Fig. II-68 *A* sarcomere *from a rabbit muscle. The thin filaments (A) extend inward from the Z line; in the center they overlap with the thick filaments (M). Indications of fine bridges between thin and thick filaments are seen in the zone of overlap (B). × 100,000. (Courtesy of H. Huxley.)*

might serve to pull filaments past one another. As Figure II-67 indicates, the myosin molecules in one half of the thick filament appear to be oriented with opposite polarity to the molecules in the other half. In their interactions with myosin, the thin filaments also reveal a polarity (Fig. II-71). The result is that the two halves of the thick filament "pull" thin filaments "in" toward the center; the thin filaments associated with one end of a thick filament move in the opposite direction to those associated with the other end. It is known that myosin can split ATP and that actin has ATP or derivatives bound to it. However, the mechanism by which the chemical energy of ATP is converted into the mechanical energy of muscle contraction remains unclear.

Models have been proposed that translate filament interaction into molecular terms. It is proposed that the bridges are parts of thick filaments that protrude from the filaments in a helical arrangement (Fig. II-67). Studies of isolated myosin molecules indicate that each molecule has a portion referred to as the "heavy meromyosin portion," which contains sites that can bind to isolated actin molecules. The heavy meromyosin portion is able to split ATP, and presumably it participates in the bridges between actin and myosin filaments.

From immunohistochemical staining (p. 22) two proteins, in addition to actin, have been identified in the thin filaments, *troponin* and *tropomyosin*. These proteins participate in the control of contraction by Ca^{++} ions. Current thinking is that in the absence of Ca^{++} the proteins inhibit the myosin ATPase and the interaction of actin and myosin, and that this inhibition is reversed by Ca^{++} ions. As will be seen in Chapter 3.9, the triggering of muscle contraction by a nerve impulse is mediated by the specialized endoplasmic reticulum (sarcoplasmic reticulum) of the muscle fiber, which controls the concentrations of Ca^{++} in the fiber's cytoplasm. The elongate tropomyosin molecules are thought to run along the thin filaments, perhaps lying in or near a groove between the two chains of actin molecules (Fig. II-67). Attached to the tropomyosins, probably near their ends, are the more globular troponin molecules which thus occur at regularly spaced intervals (approximately 400 Å) along the thin filament.

It has recently been suggested that a protein known as *actinin* is concentrated at the Z-line.

2.11.3   **MICRO-** The existence of microfilaments in many
             FILAMENTS cell types has long been recognized. Among
                          the earliest to be visualized were the bundles within the microvilli of the absorptive cells of the intestine (Fig. III-20). These filaments merge with another zone of packed filaments just below the microvilli, the terminal web. In epithelial cells (Section 3.5.1) of skin and other tissues, bundles of parallel fine filaments in the cytoplasm sometimes terminate near special regions of the plasma membrane, where adjacent cells are associated by structures called *desmosomes* (Section 3.5.2). In diverse cell types filaments are found in groups of variable sizes and distributions; often they form parallel bundles in the cytoplasm (Fig. II-70) or networks near the plasma membrane. Many of the filaments are 50–70 Å wide; some propose that the name "microfilament" be reserved for filaments of this size. The filaments may play some cytoskeletal roles, but most interest is now focused on their probable participation in many types of cellular motion.

It is likely that most organelles in most cells are in constant motion, both random and nonrandom, with the effect of speeding molecular transport and facilitating exchanges and interactions among organelles. We referred earlier in this chapter to the movement of pigment granules along microtubule-delineated channels. Comparable highly oriented movement of cytoplasmic structures is probably commonplace: pinocytosis vesicles pass from the cell surface where they form toward the region of the Golgi apparatus; materials flow down the elongate processes of nerve cells (Section 3.7.2); the cytoplasm streams rapidly around the central vacuole of many plant cells (Fig. II-69); in the phenomenon known as "capping," antibodies or certain other proteins that initially are bound uniformly around the surface of lymphocytes (p. 245) and some other white blood cells become congregated at one pole of the cell prior to endocytosis. Cytoplasmic flow is also intrinsic to cellular mobility. Amebae, white blood cells, macrophages, tissue culture cells, and the tips of growing nerve cells can "crawl" along surfaces by extending pseudopods. In this way amebae may move at rates greater than 100 μm per minute. As pseudopods form, a marked bulk flow of cytoplasm occurs in the direction of the ameba's movement. This involves a continual formation and dissolution of a gel-like outer region (cortex) of the cytoplasm, as diagrammed in Figure II-69. From experiments with

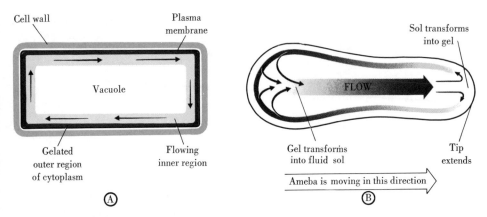

Fig. II-69 *Two important types of cytoplasmic motion.* **A.** Cyclosis *in a plant cell—the rapid streaming of cytoplasm around a central vacuole.* **B.** *One type of* ameboid motion. *The extension of a pseudopod by an ameba is accompanied by constant flowing of cytoplasm in the direction of extension and probably by continual transformation from gelated ectoplasm or* cortex *(outer region of cytoplasm) to fluid* endoplasm *at the posterior end with the reverse transformation at the anterior end.*

tissue culture cells, ameboid movement is thought also to involve the formation of transient firm attachments between portions of the cell surface and the surface over which the cell is moving.

In many cases the distribution of filaments suggests their involvement in movement: they abound at the interface between stationary and moving cytoplasm in plant cells; close associations are sometimes seen between filament bundles and membrane-delineated cytoplasmic structures; nerve processes contain many filaments, longitudinally arranged; filaments are numerous near the surface or deeper in the cytoplasm of various ameboid cells. Given their prominent role in movement of muscle cells, it is not surprising that most hypotheses implicate microfilaments as generators of other types of motion. Evidently they work in concert with microtubules, and possibly with membrane macromolecules (see also pp. 269, 329).

In theory the filaments could exert their influence in several ways. If suitably anchored, individual filaments, or groups, might move other materials along their lengths through interactions, perhaps resembling the "bridge" formation and "sliding" processes of muscle cells. Or, they might somehow pull structures to which their ends are attached. The localized contraction, or assembly and disassembly of a filament network, might change the shape or resistance to deformation of a cell region, affect the passage of other structures through the region, or alter the distribution of macromolecules in, or the geometry of cell surfaces to which the network might be attached. For motion of amebae it is proposed that the bulk flow of cytoplasm results from pressure generated by contraction of the cortex at the rear, or by contraction or some other process at the front end that pulls the cytoplasm forward.

Under proper conditions, the drug *cytochalasin B* interferes with several types of cytoplasmic movement. In some but not all cases this seems to be correlated with disruption of microfilaments. Beclouding the interpretation of these observations is the fact that the drug can affect other cell activities, notably transport of sugars and other molecules across the plasma membrane. Thus there is debate as to the nature of cytochalasin B effects on motion.

A promising recent finding is that the 50 Å microfilaments of many nonmuscle cells are made of proteins identical to or very closely resembling the "muscle protein" actin. In a few cases, such as the microvillar core, this has been established through direct biochemical analysis of isolated proteins; in others (including some plants) it is inferred from antibody staining procedures and allied methods such as the binding of myosin fragment (Figs. II-70 and II-71). Proteins resembling myosin are also widespread, and filaments on the order of 100 Å in diameter have been identified in some cell types. However, these "thick" filaments, when present, are relatively sparse, and the amounts of myosinlike proteins generally are small when compared with actin. Also, the interdigitating filament arrays seen in striated muscle are not seen in other cells. Thus if actin–myosin interactions are involved in nonmuscle motility, they must differ in important details from those in muscle. Interestingly, influences of Ca^{++} ions on motility, and the presence of proteins resembling actinin and tropomyosin seem widespread.

c h a p t e r **2.12**
THE HEPATOCYTE: A CENSUS

This is a convenient point to return to the hepatocyte for a summary of its content of major cytoplasmic organelles.

It is difficult to obtain quantitative information about the numbers and volumes of organelles in a cell as large and complex as the hepatocyte. Estimates of the relative numbers of different organelles can be made from light microscope preparations such as those shown in Figures I-11 through I-19, but they are limited by several factors, such as the resolving power of the light microscope (Section 1.2.1). Sections for the electron microscope are thin, and a given section often includes only

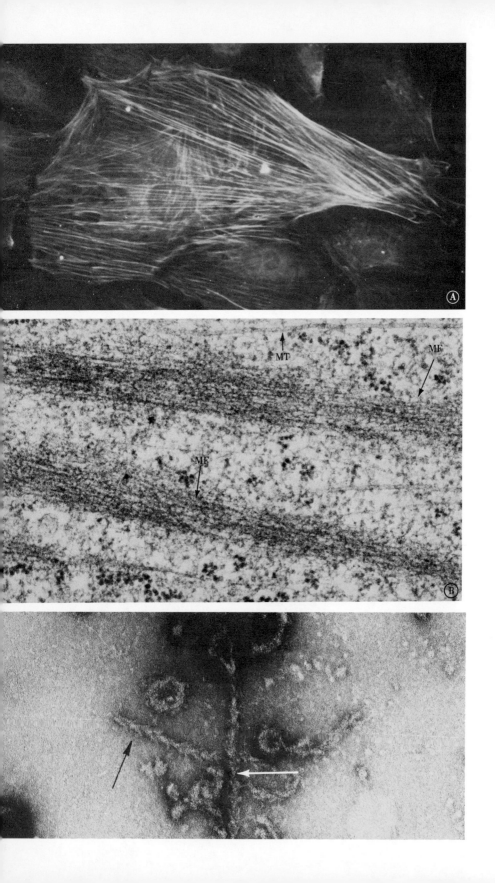

◄ *Fig. II-70 Microfilaments in cultured mouse fibroblasts ("3T3 cells") A. This cell was photographed in the fluorescence microscope following "staining" with antibodies to actin purified from muscle. Molecules of fluorescent dye had been linked to the antibodies. The dye produces bright fluorescence where sufficient dye was bound. This reveals the presence of actin (which binds the anti-actin antibody) in numerous elongate cytoplasmic "fibers." B. Electron micrograph of a 3T3 cell showing bundles of microfilaments (MF). These correspond to the "fibers" in A. MT indicates a microtubule. × 64,000. (A. Courtesy of E. Lazarides and K. Weber. B. From R. Goldman,* Exp. Cell Res., **90**:333, 1975.)

a small portion of a given structure. Organelles rarely have simple shapes; thus treatment by the usual geometric formulas is imprecise. For organelles such as the lysosomes, the variety of morphological types complicates estimates: some types may be difficult to identify in the microscope for census purposes. A section may be regarded as a sample of the cell, the contents of which depend on the size, number, shape, and distribution of organelles and on the section thickness and angle. Statistical formulas are often used to relate the information from sections to the cell as a whole; this approach is called *morphometry*.

Isolated cell fractions are of limited use, unless extreme care is taken to ensure that the fraction is reasonably pure, and unless it is possible to determine what proportion of the total cell content of the organelle is present in the fraction studied. Sometimes this proportion can be approximated by comparing the fraction with the unfractionated cell in terms of the amounts of characteristic components, such as the mitochondrial respiratory enzymes or nuclear DNA.

◄ *Fig. II-71 Negatively stained preparation (Fig. II-39) showing microfilaments (arrows point to two) isolated from cultured hamster cells ("BHK-21 cells") and exposed to a solution containing "heavy meromyosin" (HMM) prepared from muscle myosin. HMM is thought to be the portion of the myosin molecule that binds to actin; it can be prepared from myosin by gentle protease treatment; In this micrograph numerous HMM's have bound to each filament producing an "arrowhead" or "herringbone" pattern along the length of the filament; this is seen most clearly at the left-hand arrow. The formation of this pattern suggests that the microfilaments are made of actin. Note that the arrowheads all point in the same direction on a given microfilament indicating a polarity in arrangement of the actin molecules. × 112,000. (From R. Goldman,* Exp. Cell Res., **90**:333, 1975.)

Even the cells of a single type may vary considerably within an organ. Within a hepatic lobule (see Chapter 1.1), for example, the hepatocytes near the periphery are exposed to blood that is richer in oxygen and nutrients than the cells near the central vein; by the time the blood reaches the center of the lobule it has undergone considerable exchange of material with the peripheral cells. Probably related to this is the fact that mitochondria, lysosomes, ER, and glycogen differ in both size and amount when peripheral and central cells are compared. For such reasons, the estimates that have been made are very rough and can convey only an overall impression of an "average hepatocyte."

A rat hepatocyte averaging 20 μm in diameter has a surface area on the order of 3000 square micrometers (μm^2) and a volume of roughly 5000 cubic micrometers (μm^3). The nucleus occupies 5–10 percent of the cell's volume, mitochondria occupy 15–20 percent, peroxisomes 1–2 percent, and lysosomes less than 1 percent. (The figure for lysosomes includes residual bodies, some pinocytic digestion vacuoles, and a small number of autophagic vacuoles; the vesicles that may be primary lysosomes are excluded, since their number and volume are exceedingly difficult to estimate.) Glycogen varies greatly in amount, but may occupy 5–10 percent of the cytoplasmic volume. The ER occupies roughly 15 percent of the cytoplasmic volume and has a surface area of 30–60,000 μm^2; rough ER and smooth ER make approximately equal contributions (estimates vary, depending in part on details of tissue preparation). Perhaps 300 ribosomes are present per square micrometer of rough ER surface. The Golgi apparatus is sometimes said to occupy 5–10 percent as much of the cell as the ER, but the diversity of components of the apparatus makes such estimates uncertain. On the order of half the volume of the cytoplasm is accounted for by hyaloplasm plus free ribosomes, microtubules, and microfilaments.

The number of mitochondria per cell is usually thought to be about 1000–2000. A few hundred peroxisomes (perhaps 400–500) are present per cell, and, excluding small vesicles, there are possibly one half as many lysosomes as there are peroxisomes. There are several million ribosomes, one pair of centrioles, and a large but unknown number of microtubules and microfilaments.

FURTHER READING

Brady, R.O. "Hereditory Fat Metabolism Diseases." *Scientific American*, **229**: no. 2, p. 88, Aug. 1973.

Branton, D., and R.B. Park, Eds. *Papers on Biological Membrane Structure.* Boston: Little, Brown & Co., 1968, 311 pp. A collection of articles describing research on cellular membranes including some of the key early work.

Busch, H., Ed. *The Nucleus*, 3 vols., New York: Academic Press, 1974. A collection of review articles covering modern research on many aspects of nuclear function.

Capaldi, R.A. "A Dynamic Model of Cell Membranes." *Scientific American*, **230:** no. 3, p. 327, Mar. 1974.

Cold Spring Harbor Symposium on Quantitative Biology 37: Mechanism of Muscle Contraction, Cold Spring Harbor, N.Y.: Cold Spring Harbor Press, 1973. A collection of research articles.

Cohen, C. "The protein switch of muscle contractions," *Scientific American*, **233:** no. 5, p. 36, Nov. 1975

Dingle, J.T., and others, Eds. *Lysosomes in Biology and Pathology*, several vols. Amsterdam: North Holland Publ. Co., 1969 and thereafter. A collection of articles reviewing research on many aspects of lysosomes.

Fawcett, F. *The Cell, An Atlas of Fine Structure*. Philadelphia: Saunders, 1966, 448 pp. An extensive collection of electron micrographs of organelles with explanatory material.

Fox, C.F., "The Structure of Cell Membranes," *Scientific American*, **262:** no. 2, p. 30, Feb. 1972.

Gibor, A. "Acetabularia: A Useful Giant Cell," *Scientific American*, **215:** no. 5, p. 118, Nov. 1966.

Govindjee, and R. Govindjee. "The Absorption of Light in Photosynthesis," *Scientific American*, **231:** no. 6, p. 68, Dec. 1974.

Holtzman, E. *Lysosomes, a Survey*. Vienna: Springer-Verlag, 1975. A concise discussion of present views and research.

Kroon, A.M., and C. Saccone. *The Biogenesis of Mitochondria*. New York: Academic Press, 1974, 551 pp. A collection of research articles.

Levine, R.P. "The Mechanism of Photosynthesis." *Scientific American*, **221:** no. 6, p. 58, Dec. 1969.

Munn, E.A. *The Structure of Mitochondria*. New York: Academic Press, 1975.

Murray, J.M., and A. Weber. "The Cooperative Action of Muscle Proteins." *Scientific American*, **230:** no. 2, p. 58, Feb. 1974.

Miller, O.L. "The Visualization of Genes in Action." *Scientific American*, **228:** no. 3, p. 34, Mar. 1973.

Neutra, M., and C.P. Leblond. "The Golgi Apparatus." *Scientific American*, **220:** no. 2, p. 100, Feb. 1969.

Nomura, M. "Ribosomes." *Scientific American*, **221:** no. 4, p. 28, Oct. 1969.

Nomura, M., A. Tissieres, and P. Lengyel, Eds.. *Ribosomes*. Cold Spring Harbor Laboratory, N.Y., 1974. A collection of research articles.

Novikoff, A.B., and J.M. Allen, (Chairmen). "Symposium on peroxisomes." *J. Histochemistry and Cytochemistry*, **21:** 941–1020, 1973. A collection of short articles describing recent research.

Racker, E. "The Membrane of the Mitochondrion." *Scientific American*, **218:** no. 2, p. 32, Feb. 1968.

Reinert, J., and H. Ursprung, Eds. *Origin and Continuity of Cell Organelles*. New York: Springer-Verlag, 1971, 342 pp. A collection of articles reviewing research on many organelles.

Satir, B. "The Final Steps in Secretion," *Scientific American*, **233:** no. 4, p. 28, Oct. 1975.

Satir, P., "How Cilia Move." *Scientific American*, **231:** no. 4, p. 44, Oct. 1974.

Soifer, D. Ed. *The Biology of Cytoplasmic Microtubules*, Annals of N.Y. Acad. Sci., **253:**1, 1975. A collection of research articles.

Watson, J.D. *The Molecular Biology of the Gene*, 2nd ed. New York: Benjamin, 1970, 662 pp. A comprehensive introduction to modern molecular biology that also covers much basic biochemistry.

Zelitch, I., *Photosynthesis, Photorespiration and Plant Productivity*. New York: Academic Press, 1971, 347 pp. A discussion of the biochemistry of photosynthesis and related processes.

Consult the introductory texts, collections of articles, and monograph series listed at the end of Part 1 for additional references.

part 3

CELL TYPES: CONSTANCY AND DIVERSITY

The diameters of known cells range from 0.1 μm for the simplest procaryotes to 1–100 μm for most protozoa, algae, and cells of multicellular organisms; occasional specialized cells, such as some egg cells, have diameters in the millimeter or centimeter range. This diversity in size is matched by a diversity in morphology and metabolism. The simplest cells are not much more complicated in structure than a mitochondrion or a plastid. The most complex show intricate patterns of thousands of organelles.

Each cell type is characterized by a particular organization of its organelles and macromolecules. Structural differences among cell types are related to differences in function. Sometimes this relation is obvious; for example, chloroplasts are present in plant cells which are capable of photosynthesis, and absent in animal cells which are incapable of photosynthesis. Often, however, the relations between structure and function are more subtle. Thus in some insect flight muscle fibers the thick filaments extend throughout the length of the entire sarcomere, almost from Z line to Z line. The sliding of thick and thin filaments during contraction involves only short, probably oscillatory, movements producing length changes of less than 5 percent. (These muscles control the extraordinarily rapid wing motion.) By contrast, in the skeletal muscle of vertebrates where movements are slower, much greater length changes occur, often exceeding 20 percent. The thick filaments occupy only part of the sarcomere length, near the center. Presumably this is associated with the extensive filament sliding that takes place (see Section 2.11.2).

While different organization characterizes cells doing different things, cells doing similar things generally show similar organizations. For example, most animal cells engaging in the extensive synthesis of *proteins* that are secreted at the cell surface have large nucleoli and extensive rough ER. The Golgi apparatus in these cells is large and usually located between the nucleus and the region of the cell surface where the secretion is released. In contrast, a variety of cells producing steroids have abundant smooth ER.

These similarities permit one to speak of cell types (protozoa, algae, absorptive cells, gland cells, and so forth). Findings made in the study of one cell can often be extrapolated to others of similar structure, but this must be done with caution. Although similarities often result from common evolutionary origin, evolutionary changes do not stop once a particular organization develops; consequently, cells with general features in common often differ in detail. Also, comparable organization can be achieved from very different evolutionary starting points (*convergent evolution*).

The great diversity of cell types, with different organization and functions, has provided favorable experimental materials. Experiments on nuclear function described earlier (Chapter 2.2) were possible because the alga *Acetabularia* is a large unicellular organism, relatively simple in structure, easy to maintain in the laboratory, and capable of surviving drastic microsurgery. The chloroplast biogenesis experiments on *Chlamydomonas* (Section 2.7.4) were facilitated by the availability of mutants in which chlorophyll synthesis is dependent on light. Studies of the extraordinarily large neurons of lobsters and squids have supplied much information about nerve impulse conduction. And bacteria have been used extensively in studies of heredity and biochemistry because of their relative simplicity, their rapid growth in media of simple and readily controlled composition, and the availability of many mutants.

Partial catalogs of cell types exist. Thus histology textbooks (*histology*: the study of tissues) may describe virtually all cell types of man or of other organisms; a few such books are listed at the end of this part. This text will not attempt any such catalog. Rather, a number of cell

types have been selected to illustrate major differences in the organization and features of procaryotes and eucaryotes, animal cells and plant
cells, and specialized cells of multicellular animals. In the last group,
mammalian cells have been chosen more often than others because they
have been most extensively studied. Some discussion of specialized
cells of plants and invertebrates has been included, as well as a brief
consideration of the viruses. The latter are not cells and their evolutionary status in the biological world is still uncertain. There has been much
controversy about whether viruses are truly "living," for they depend on
cells for their reproduction, synthesis of macromolecules, and most
other activities. However, because of their simplicity, viruses have been
of incalculable value in the study of many important biological problems. Like the bacteria, they have been used to provide much of the
present insight into the molecular mechanisms of heredity.

c h a p t e r **3.1**
VIRUSES

Viruses have none of the organelles discussed in Part 2 and,
strictly speaking, they have no metabolism. From one point of view they
represent the ultimate in specialization. For their duplication they depend entirely on their ability to enter cells of living organisms and to
redirect the metabolism of the cells (by substituting their nucleic acid
for the cell's DNA). Different viruses enter different cells, from bacteria
to man. In extreme cases the host cell very rapidly becomes diverted
from its normal functions to the production of new viruses. In these
cases the host cell ultimately dies, releasing the viral progeny to infect
other cells. Their dependence on cells for reproduction argues against
the consideration of viruses as *primitive* in the evolutionary sense, that
is, as the ancestors of cellular life. Perhaps the simplest notion is that
they arose as a sort of parasite at the same time as, or subsequent to, the
origin of cellular life; possibly some are degenerate forms which have
evolved from structures that were once capable of independent life.

3.1.1   **STRUCTURE;**   A great variety of viruses are known (Fig.
ENTRY INTO   III-1). Many are extremely small, about the
CELLS   size of a ribosome; the largest have maxi-
mum dimensions of 0.1–0.3 μm.

Basically, viruses consist of a nucleic acid core (DNA or RNA)
wrapped in a coat made of fewer than a hundred to several thousand

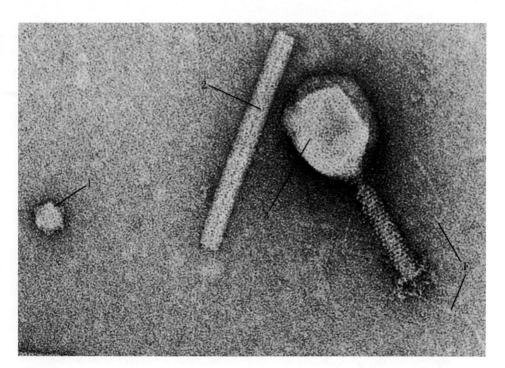

Fig. III-1   *Negatively stained preparation (Fig. II-39) showing three different viruses. 1 is φX-174, a single-stranded DNA virus infecting bacteria. 2 is a* tobacco mosaic virus *that infects tobacco plants. 3 is the bacteriophage* T$_4$; *the head and tail are clearly visible but the tail fibers are difficult to preserve and only some of them are seen (F; see Fig. IV-4). × 250,000. (Courtesy of F. Eiserling, W. Wood, and R. S. Edgar.)*

protein molecules (see Figs. III-2 and IV-4). In the more complex viruses some lipid and polysaccharide components are also present. Thus, certain viruses are bounded by lipoprotein membranes. The RNA viruses (for example, polio virus and type A influenza virus) are the only biological systems known in which RNA and not DNA is the hereditary material. Often the RNA is present in its usual single-stranded form, but some viruses have a core of double-stranded RNA similar in properties to DNA. The viral protein coat protects the nucleic acid during the extracellular phase of the virus life cycle. Also, some of the proteins in the coat probably bind the virus to the cell surface prior to the entry of viruses into cells. Some viral surface proteins include enzymes that aid in penetrating surface layers of cells. The viral core may also include a few protein molecules; for example, some RNA viruses carry enzymes for RNA transcription and certain ones possess enzymes that generate DNA copies of RNAs (Section 3.12.3).

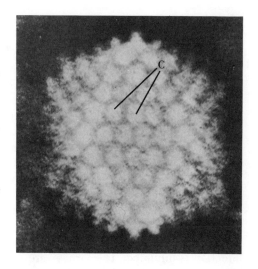

Fig. III-2 Negatively stained adenovirus, a DNA virus that infects human and other animal cells. The micrograph shows the capsomeres (C), protein subunits composing the "coat"; these have a polyhedral arrangement. (Careful study indicates that the coat consists of 252 capsomeres arranged as an icosahedron). × 650,000. (Courtesy of R. A. Valentine and H. G. Pereira.)

Figure III-1 illustrates the diversity of viral form. In the simpler viruses the coat consists of a number of protein units, of one or a few types, geometrically arranged around the nucleic acid. In many small viruses the protein subunits surround a DNA (or RNA) core to form a polyhedron (Fig. III-2). In the case of tobacco mosaic virus (TMV), the core of RNA is arranged as an elongate spiral and is surrounded by protein units in a helical pattern (see Fig. IV-4); the overall structure is an elongate rod. Some viruses have a much more complex structure involving several different kinds of protein. For example, T_4, one of the *bacteriophages* (viruses that attack bacteria), has a head composed of DNA surrounded by protein subunits; in addition, it has a "tail" made up of several sets of other types of subunits including a set of fibers at the end (see Fig. IV-4). The tail attaches the virus to the cell surface; enzymes in the tail help digest the bacterial surface. The DNA passes through the tail and then through the hole in the bacterial wall produced by the tail enzymes; a contraction of the tail apparently aids the entry of the DNA into the cell, leaving the rest of the virus outside.

For viruses infecting animal cells, the initial step of entry into the cells is attachment of the virus to the cell surface. Different viral strains have evolved coats with different specificities in terms of the cells to which they can attach and the particular surface components to which they bind. The steps following binding probably vary considerably for different viral strains; they are incompletely understood. Eventually viral nucleic acids are uncoated and appear in the cytoplasm. Some investigators believe that for many viruses the coats can interact with the cell surface so as to permit direct penetration of the viral core across the plasma membrane. Others feel that membrane penetration occurs after the virus has been taken up in an endocytic vacuole. Commonly, viruses are seen to be endocytosed after attachment to the cell surface. But by itself this observation does not establish that endocytosis is a *necessary* step in viral penetration; a few viruses might enter by a different, less readily detected route, and it might be these that are more important for multiplication. Probably both escape from endocytic vacuoles and direct penetration of the plasma membrane can occur, with one or the other mechanism predominating for a given strain of virus. The events of membrane penetration are not understood. In a few cases, such as some of the membrane-enveloped viruses, fusion of the viral coat with the hosts' plasma membrane or endocytic vacuole membrane is thought to be involved.

If it is to release viable viral cores, escape from endocytic vacuoles presumably must occur prior to fusion with lysosomes. In fact, when viruses are coated with antibodies before adding them to cell cultures, infection is inhibited. The viruses are endocytosed, but the antibodies seem to render them incapable of leaving the vacuoles and they are degraded within lysosomes. One virus, *Reovirus*, apparently shows exceptional behavior in these regards. It is taken up into lysosomes via endocytosis, but it is not degraded. In this instance the virus has evolved in such fashion that the lysosomal hydrolases accomplish its uncoating, but the viral core (double-stranded RNA plus some proteins) is resistant to hydrolysis. Present thinking is that while still within the lysosomes the core initiates the sequence of macromolecule replication and synthesis which leads, by incompletely understood steps, to the formation of viral progeny outside the lysosomes.

Virus *release* from cells may occur with the death and rupture of host cells (*lysis*). Sometimes host cell destruction is less rapid, and the virus is released by inclusion in a bud pinched off at the cell surface. The bud remains around the virus. It is generally believed that in this way membrane-enveloped viruses obtain components of their membrane directly from the host cell surface membrane and associated structures. However, in a few cases some of the proteins, and perhaps other constituents in the membrane surrounding the virus, have been shown to be specific molecules that are made only in infected cells, under the influence of the virus genome. Apparently these are incorporated in the host's membranes prior to viral budding.

3.1.2   **EFFECTS ON**   Within the cell different viruses concentrate
             THE HOST   in different regions. Some types are found
                               in the nucleus, some in the endoplasmic
reticulum (Fig. III-3), and others elsewhere in the cytoplasm. The viruses somehow evade or overwhelm mechanisms that tend to protect

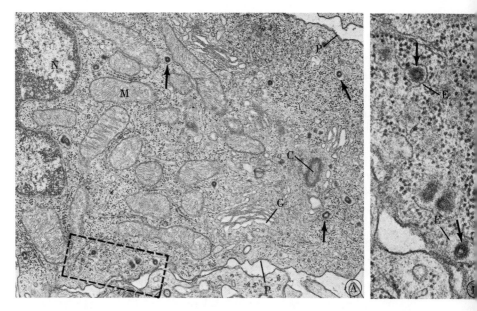

Fig. III-3   *Portion of a cell in the spleen of a leukemic mouse. In **A** (× 18,000) N indicates the nucleus; P, the plasma membrane; C, a centriole, and G, a part of the Golgi apparatus. The electron-dense bodies within the endoplasmic reticulum (arrows) are enlarged in **B** (× 56,000). The delimiting membrane of the ER is seen at E. The bodies within the reticulum are viruses. (Courtesy of E. de Harven.)*

the cell. The resistance of Reovirus RNA to lysosomal hydrolysis is mentioned in Section 3.1.1. Some of the bacteriophages have DNA modified in such a manner that they resist degradation by nucleases in the host bacteria; unusual DNA bases are sometimes present including some with glucose molecules attached. Effects on cell metabolism vary greatly. Certain viruses, such as some bacteriophages and the viruses causing polio and influenza, drastically alter cell metabolism and may lead to cell death. Host cell functions may be entirely diverted to the production of new viral components, for which the nucleic acid of the infecting virus acts as a template. In the case of some viruses with single-stranded RNA as their genetic material, a special "replicating form" of RNA is synthesized. The virus first brings about formation of a complementary copy of its RNA, and then uses this in turn to produce new RNAs which arc complementary to this copy. This results in the production of RNAs that are identical to the original RNA. Viral-infected cells also often synthesize a variety of enzymes not normally present in cells but involved in the production of specific viral constituents. For example, normal cells contain no enzymes capable of using RNA templates rather than DNA templates for synthesis of RNA, but cells infected with RNA viruses do have such enzymes. The viral nucleic acid specifies production of these enzymes as well as production of the viral coat proteins. Host cell ribosomes and other components are used in the synthesis. With poliovirus the viral RNA is translated into one very large polypeptide chain which subsequently is cleaved by proteases into several smaller protein molecules. Among the cleavage products are protein molecules destined for incorporation in the viral coat and, probably, enzymes for RNA replication.

As viral components accumulate, assembly of viruses begins in the cell. Reconstitution experiments with some of the simpler viruses suggest that the nucleic acids and coat proteins can simply associate spontaneously to form the viruses. However, as Section 4.1.5 will discuss, more complex processes may be involved.

In some cases viruses enter into a coexistence with the host cell and remain reproducing in long-term residency more or less in synchrony with the cell. The extreme form of this is found in some bacteria where certain (*lysogenic*) viruses become closely associated with the bacterial DNA. The viruses behave very much like part of the host chromosome; viral and bacterial DNA replicate at the same time. Such long-term relationships may break down under some circumstances. The virus becomes virulent, duplicates much more rapidly than the host cell, and soon kills the cell. It is speculated that some diseases of higher organisms are due to such activation of latent viruses. Interestingly, in the bacteria that cause diphtheria, the toxin responsible for the disease is coded for by genes of bacteriophages present in the bacteria.

As discussed later (Section 3.12.3) viruses are known that cause tumors when injected into animals or that bring about rapid and abnormal growth when added to cell cultures.

Nonviral infective agents may also influence cell behavior. Thus a strain of the fly, *Dropsophila* is known in which the viable offspring are almost exclusively female; males die as embryos. This trait is due to the presence of an infectious microorganism, a spirochete, transmitted through the ova. Infective particles are found in the cytoplasm of some strains of *Paramecium*. Bacterial infection of higher organisms influences cell activities in diverse ways—witness the varieties of bacterial diseases.

3.1.3   ***VIRUSES AS***   Since genetic information is generally
EXPERIMENTAL   stored as DNA base sequences, the amount
TOOLS   of DNA in a cell provides an initial basis
for estimating the genetic information available to that cell. Typical proteins consist of a few dozen to a few hundred amino acids; 300–400 is often used as the figure for an "average" protein, although this is somewhat arbitrary. It requires three

RNA bases to code for a single amino acid, and since RNAs are transcribed from only one of the two complementary DNA strands, this means that it requires three DNA base *pairs* to code for an amino acid. Thus a *very* rough measure of the number of different proteins for which a cell might carry information is obtained by dividing by three the number of base pairs in its DNA. The reasons the estimate is highly imprecise are (1) not all DNAs carry information for proteins (or for special RNAs such as the tRNAs or rRNAs) and (2) some DNA occurs in multiple repetitive copies (Sections 2.3.4 and 4.2.6). Cells of vertebrates and higher plants contain hundreds of millions to billions of base pairs (10^8–10^9), bacteria on the order of 10^5–10^7, and viruses between 10^3 and 10^5. In TMV the RNA is a single strand with slightly more than 6000 bases and poliovirus RNA has 7500 bases. T_4 bacteriophage has roughly 200,000 DNA base pairs specifying about 25–50 different proteins. Several viruses are known that contain only a few thousand base pairs, enough to specify only three or four proteins, including the coat proteins and those required for function in a host cell. The DNA of the *polyoma* virus (Section 3.12.3), whose coat is of 72 similar protein subunits, is a circular molecule 1.6 μm long. There are approximately 3000 base pairs per micrometer of DNA double helix, so this viral DNA contains about 5000 base pairs. By contrast, the circular DNA of the bacterium *Escherichia coli* is about 1000 μm long, and a single chromosome of a eucaryote cell may contain enough DNA to make a double helix that is millimeters or centimeters in length. A single mitochondrion may contain circular molecules of DNA only 5 μm in length (Fig. II-43). (To convert these numbers into molecular weights use is made of the fact that the molecular weight of DNA is 660 per base pair.)

These comparisons dramatize the relative simplicity of viruses. One can now hope to identify the function of every gene in the viral "chromosome" (nucleic acid molecule) and to identify every protein for which the nucleic acid codes—a perspective that is presently unrealistic for chromosomes of cells. As simple "self-reproducing" structures, viruses have been of great value as experimental tools. The demonstration that only the DNA of some bacteriophages enters bacteria was an important piece of evidence in establishing the primacy of nucleic acids in heredity. Reconstitution of certain viruses from their isolated components (Section 4.1.5) is relatively easy. This is being exploited to unravel the mode of formation of more complex biological structures. The involvement of some viruses in the transformation of normal cells into cancer cells is providing insight into the role of nucleic acids in cancer (Section 3.12.3). As it is better understood how in viruses nucleic acid molecules many thousands of Å long are packed into spaces a few thousand Å or less in length, valuable clues will be gained to the rules governing nucleic acid arrangement in plant and animal cells.

Another interesting point is that viruses show evidence of temporal control mechanisms—certain of the viral proteins may be synthesized early in infection and others only later. The molecular bases for this are being sought, in part since they may shed some light on the more complex temporal changes in gene expression in eucaryote development (Chapter 4.4). Some of the regulation depends on the initial synthesis of viral products that influence the course of subsequent events. These may include enzymes that introduce new synthetic capacities in the cell and other proteins that interact with the viral nucleic acids or host cell enzymes to alter their properties.

c h a p t e r **3.2**
PROCARYOTES

The small size of procaryotes and, until recently, difficulties in preserving their structure for electron microscopy have hampered extensive analysis by microscopy. Today the structure of some bacteria and blue-green algae is known in fair detail, and a beginning has been made in the study of procaryotes such as the *mycoplasmas*. The situation is quite different with respect to biochemistry and genetics; some bacteria, *Escherichia coli* and related forms, are probably the most intensively studied and best understood organisms. Because of the wealth of such biochemical information, many questions about the relations of structure and function may be studied to advantage in bacteria.

The overall cellular dimensions of procaryotes are usually on the order of a fraction of a micrometer to a micrometer or two. Evidence is appearing for a higher degree of intracellular organization than was once thought to be present in procaryotes, but this organization is simpler than that of eucaryotic cells. In some cases the cells contain moderately elaborate arrangements of membranes, but these usually are not organized as discrete organelles with a special surrounding membrane separating them from the rest of the cell.

c h a p t e r **3.2A**
MYCOPLASMAS

Mycoplasmas are the simplest known cellular organisms. They were earlier known as pleuropneumonialike organisms (PPLO) because some produce respiratory diseases. The procaryote class to which they belong is now known as *Mollicutes*. The diameters of the smallest mycoplasmas are 0.3–0.4 μm; the largest nonfilamentous forms measure almost a micrometer. Filamentous forms, consisting of chains of attached cells, reach more than 40 μm in length. Some mycoplasmas appear to be motile, but the basis of their movement is not understood.

Thus in size some mycoplasmas overlap with the largest viruses and with the smallest bacteria. Mycoplasmas produce diseases in animals and plants including respiratory and other diseases in humans. The possible pathogenic (disease-causing) roles of recently described viruses inside of mycoplasmas needs to be clarified. Vaccines have been produced against *Mycoplasma pneumoniae* and large-scale testing of their effectiveness in reducing human respiratory illness caused by the organism is underway.

3.2.1 STRUCTURE AND FUNCTION Figure III-4 is based largely on the best studied species. In all mycoplasmas studied thus far, a circular DNA molecule is contained in a nuclear region that is not separated from the remainder of the cell. Ribosomes are usually distributed randomly in the hyaloplasm, but sometimes they show helical arrays. A plasma membrane of the usual three-layered structure and thickness (Chapter 2.1) delimits the cell. As in other cells the membrane consists of lipids and proteins and some proteins, including one glycoprotein thus far identified, have part of their polypeptide chain exposed at the external surface of the plasma membrane. Unlike bacteria, no elaborate cell wall surrounds the membrane. Whether the DNA molecule is attached to a part of the plasma membrane as in bacteria (Fig. III-5) is not fully settled, but some investigators think that DNA replication involves sites on the membrane.

The enzymes present in the cell include, among others, those required for DNA replication, for the transcription and translation involved in protein synthesis, and for the generation of ATP by anaerobic breakdown of sugar along pathways similar to those described earlier (Chapter 1.3). Mycoplasmas can live and grow by themselves in artificial growth media; unlike viruses, they do not require host cells for duplication. Of course they can, and do, multiply in host cells, often with pathogenic effect. Special pains need to be taken to prevent contamination by mycoplasmas of cells grown in culture (Chapter 3.11).

3.2.2 MACRO-MOLECULES AND DUPLICATION The smaller of the mycoplasmas contain DNA with about 750,000 base pairs, and a few hundred ribosomes. (By comparison, bacteria contain roughly 10 times as many DNA base pairs and about 50-100 times as many ribosomes.) Such cells probably can make no more than some 500-1000 different kinds of proteins. The largest mycoplasmas can make about twice as many.

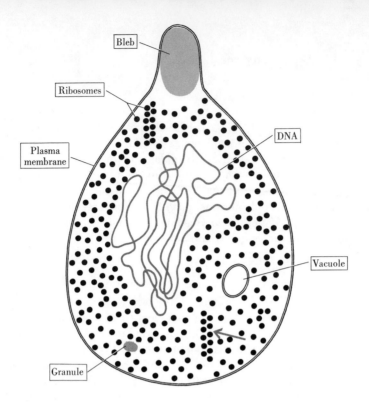

Fig. III-4 *Diagram of a mycoplasma. (After the work of D. R. Anderson and J. Maniloff and H. J. Morowitz.) Of the great variety of mycoplasmas only a few have been studied by electron microscopy. These differ in cell shape, arrangement of DNA fibrils, patterns of ribosome distribution (in one study, many of the ribosomes were found associated in long helical arrays; arrow in diagram), and other features. Some of these differences may reflect intrinsic differences among different mycoplasma types. Others probably result from differences in growth conditions and in methods of preparation of cells for microscopy.*

    The mycoplasmas that have been studied all show the presence of a delimiting plasma membrane (with a triple-layered structure as diagrammed) and of ribosomes and DNA fibrils (a single circular double helix of DNA per cell is thought to be present). They have no cell wall and no nuclear envelope or other elaborate intracellular membrane systems. The granule and vacuole represent structures seen in some types and are of uncertain significance. For example, some investigators claim that the vacuole is actually a deep infolding of the plasma membrane that appears free from the plasma membrane in some preparations only because connections were not included in the thin section being studied. Blebs also have been noted only in some types of mycoplasmas.

Because of their small size and other difficulties of observing mycoplasmas, little is firmly established even about matters such as cell division. In most cases binary fission appears to be the mechanism of cell division, although budding reportedly occurs in some species and growth into filaments which fragment may characterize other species. In *Mycoplasma gallisepticum* a bleb is present at one end of the cell

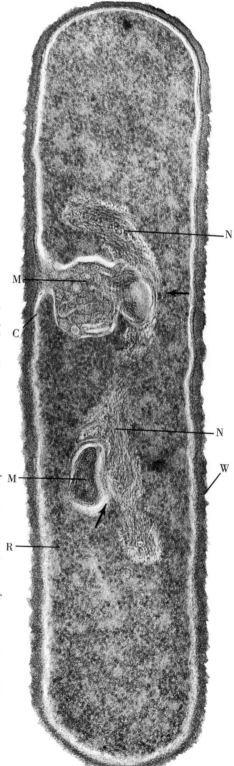

Fig. III-5 *A Gram-positive bacterium,* Bacillus subtilis, *showing the prominent wall (W) surrounding the cell, mesosomes (M), and nuclear area (N). The mesosomes are infoldings of the plasma membrane (C) and also are closely associated (arrows) with the nuclear region. The small dense granules (R) in the "cytoplasm" are ribosomes.*

As is true of procaryotes in general, the arrangement of DNA in the nuclear region cannot readily be analyzed by microscopy of sectioned cells. Many short fibrils are seen in thin sections and one cannot determine visually whether they are part of one folded molecule or represent some other, unknown arrangement. Study of the structure of procaryote chromosomes (or "genophores" as they sometimes are called to distinguish them from the chromosomes of eucaryotes) has thus depended heavily on investigation of material isolated from the cell. Approx. × 75,000. (Courtesy of A. Ryter.)

(Fig. III-4). Preparatory to cell division a second bleb forms at the end of the cell opposite to the original bleb. The cell then pinches into daughter cells, each with nuclear region, ribosomes, and bleb.

Mycoplasmas are probably far more complex than the primitive self-duplicating systems from which life arose. Yet some may be close to the minimum size and complexity required for independent cellular life and reproduction under present-day conditions. Knowledge of their metabolism and reproduction may shed light on control mechanisms and other aspects of macromolecular synthesis and structural duplication in more complex cells.

c h a p t e r 3.2B
BACTERIA

There are a great many different types and species of bacteria, but only a few have been thoroughly studied with modern techniques. One of these is *Escherichia coli*, (*E. coli*) found in the mammalian digestive tract. *E. coli* is a favorite experimental organism for geneticists, biochemists, and molecular biologists. An *E. coli* cell measuring 1 by $2\mu m$ may contain roughly 5000 distinguishable components, ranging from water to DNA and other macromolecules. Its genetic information is carried by a single DNA molecule; enough information is present (roughly 3×10^6 base pairs) to code for several thousand different proteins; on the order of 20,000 ribosomes are present. A single cell, by doubling at rates of more than once every 15–30 minutes, can give rise to millions of essentially identical cells in a short time. Growth and reproduction can occur in a chemically defined growth medium, containing only glucose and inorganic salts. This indicates that the cell contains all the necessary enzymes to synthesize both metabolic precursors and macromolecules from very simple molecules. The combination of relative genetic simplicity, ease of experimental manipulation (for example, production and isolation of mutants), metabolic versatility, and the ability to grow on simple media and to produce large uniform populations makes *E. coli* an outstanding experimental tool. This is exemplified by work with *E. coli* on the synthesis of the enzyme *β-galactosidase*. As outlined below, study of the production of this enzyme has progressed to the point where specific proteins controlling the transcription of the pertinent mRNAs have been identified. Recently the specific sequences of nucleotides in the DNA regions with which these proteins interact have been mapped.

Bacterial metabolism is extraordinarily diverse. Many bacteria have pathways that differ only in some details from those of higher organisms. Aerobic and anerobic pathways for sugar breakdown and mechanisms of electron transport and ATP formation show many features in common with comparable processes in eucaryotic plant and animal cells. But some metabolic pathways are essentially unique to certain bacteria; an example is the utilization in photosynthesis of H_2S and other sulfur compounds rather than H_2O. The varied biochemistry of different bacteria is correlated with survival of different bacteria in virtually every type of environment. This variety has provided opportunity for insight into metabolic controls of many kinds.

3.2.3   **STRUCTURE**   Electron microscopy shows that bacterial cells possess a plasma membrane under a cell wall, ribosomes and polysomes, and one or several nuclear regions (Fig. III-5). Under most growth conditions *E. coli* contains two or more nuclear regions (nucleoids), each containing apparently identical copies of the one chromosome—a circular structure made of a thin fiber about 1000 μm long. As far as is known, the fiber is a single continuous double-stranded DNA molecule. In contrast to eucaryotic cells, no histones are associated with the DNA, although *polyamines* (small organic molecules that contain amino groups) may be bound to some of the phosphate groups of the bacterial DNA. As has been stressed before (p. 3), the procaryote "nucleus" is not separated from the "cytoplasm" by a special membrane system. But this does not imply that the DNA is in a random or disorganized configuration. In fact, one can isolate the nucleoid in the form of a compact, coiled, and folded array. Work has now begun on the functional geometry of the nucleoid and the manner in which enzymes, ribosomes, and other components interact with it. Section 2.3.5 pointed out that polysome formation and translation of a bacterial mRNA may commence as the RNA molecule is being transcribed. Correspondingly one finds RNA molecules with attached ribosomes associated with the DNA. The RNAs may somewhat resemble those in Figure IV-18.

Though the differences have no known major functional significance, bacterial ribosomes are slightly smaller than the ribosomes of the cytoplasm of eucaryotes (they sediment at 70 S as opposed to 80 S) and contain slightly more RNA (60 percent) than protein (eucaryotic cell ribosomes contain 40–50 percent RNA). No endoplasmic reticulum or Golgi apparatus is present.

The extracellular wall that surrounds the cell is usually a moderately rigid structure responsible for the maintenance of cell shape. Its removal leads to alteration of the cell into a fragile "protoplast," which may burst from water influx unless osmotic conditions in the surrounding medium are carefully adjusted (see Section 2.1.1). Bacteriologists have long used *Gram stains*, which enabled them to distinguish two different major bacterial classes by light microscopy. Many bacteria, such as the pneumonia-producing organism *Pneumococcus*, are colored by the Gram stains and are referred to as Gram-positive. Other bacteria, such as *E. coli*, are Gram-negative. The cell walls of Gram-positive bacteria contain a *murein* ("peptidoglycan") network. This consists of polysaccharide chains with attached peptide chains (of amino acids) that link the polysaccharides to one another, producing a strong network surrounding the cell. The walls of Gram-negative bacteria contain a similar network but have in addition an outer layer of lipids, complexed with protein and polysaccharides and often appearing as a three-layered membrane in the electron microscope. Some of the outer layer components in Gram-negative cell walls are toxic to animals and may account for certain effects of bacterial infection.

The differences in cell wall chemistry between Gram-positive and Gram-negative bacteria are associated with differences in sensitivity to important antibacterial agents. For example, Gram-positive bacteria are usually more sensitive to penicillin. Pencillin interferes with the interlinking of cell wall macromolecules during cell growth and thus reduces the rigidity of the wall and the resistance of the cell to osmotic rupture. The polysaccharide-splitting enzyme *lysozyme* also affects Gram-positive species more readily than Gram-negative ones; in the latter, the lipid-containing layer probably acts as a barrier to penetration of the enzyme into the wall. When white blood cells phagocytose bacteria, cytoplasmic granules rich in lysozyme fuse with the phagocytic vacuoles, as do the lysosomes (Fig. II-51). Some bacteria survive uptake by white blood cells and other phagocytes, probably because their cell walls remain intact and protect the cell itself from osmotic and enzymatic disruption (Section 2.8.2).

Certain bacteria can transform into dormant spores that are metabolically inactive and extraordinarily resistant to extremes of temperature, dehydration, and other drastic environmental conditions. The spores contain a single nuclear region and relatively little cytoplasm. Each is enclosed in a special thick extracellular coat. The controls of spore formation and germination are being actively studied; these processes are analogous to the differentiation of specialized cells of the multicellular eucaryotes. During spore formation new cell products are made. During germination, synthetic machinery that previously was inactive is turned on; a similar process occurs upon fertilization of egg cells (Chapter 4.4). Problems being investigated include the identification of the control mechanisms that operate to produce orderly synthesis of the right amounts of the right components at the right time, ensuring bacterial transformation into spores under some conditions and conversion back into active cells under other conditions.

3.2.4   ***SEGREGATION*** Figure III-6 shows a multienzyme complex
                OF FUNCTIONS; that has been isolated from *E. coli*. Several
                CELL DIVISION dozen protein subunits are associated in a
                structure that carries out a number of
sequential steps in pyruvate metabolism (Chapter 1.3). The complex
can be dissociated into three distinct enzymes and can then be reconstituted in the test tube. The ability of the complex to form spontaneously
from mixtures of its components is of great interest for understanding
the mode of formation of biological structures (Chapter 4.1). Similar
complexes are present in mitochondria, and other multienzyme complexes are found in various eucaryotic organelles.

The existence of multienzyme complexes in bacteria indicates the
high degree of organization that underlies the apparent morphological
simplicity. Respiratory enzymes and others, some probably in the form
of multienzyme complexes, are attached to the plasma membrane or to
special internal membranes thought to derive from the plasma membrane. For example, there are a variety of photosynthetic bacteria. *Bac-*

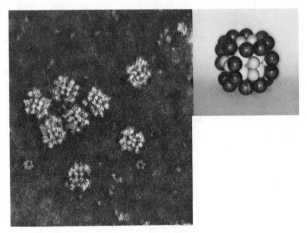

Fig. III-6 *Pyruvate dehydrogenase multienzyme complexes isolated
from* E. coli *and negatively stained. The diameter of the complex is 300 Å,
roughly twice that of a ribosome (see Fig. III-5). Each complex contains
several molecules of three types of enzymes which in turn are made of
subunits (probably a total of 72) arranged as indicated in the model at the
upper right.* × *300,000. (Courtesy of L. J. Reed, R. M. Oliver, and D. J.
Cox).*

teriochlorophyll and other photosynthetic pigments differ somewhat from the comparable pigments of eucaryotes. (Some bacteria utilize hydrogen sulfide in photosynthesis and do not evolve oxygen; the H_2S is oxidized producing sulfur as higher plants oxidize H_2O producing oxygen.) The photosynthetic pigments and enzymes are associated with internal membranes that are arranged as lamellae, tubules, or vesicles in different species; these are not separated from the rest of the cell as a discrete chloroplast by a surrounding membrane. (See Fig. III-7 for a cell with comparable organization.)

Interest has centered on another membranous structure, the *mesosome*, formed by infolding of the plasma membrane (see Fig. III-5). These are found chiefly in Gram-positive forms, although some Gram-negative forms show similar but simpler infoldings of the plasma membrane. Mesosomes, or similar special regions of plasma membrane, may play roles in bacterial duplication and division. Such division involves DNA replication, growth, and separation into two cells by formation of a *septum* across the cell. The septum grows in from the surface; it consists of a plasma membrane and cell wall. The central questions is: How is DNA behavior controlled so that daughter cells receive the proper share of nuclear regions? By analogy with eucaryotic cells (Chapter 4.2) it might be expected that the chromosome is anchored to another structure that controls the chromosome's position during division. In electron micrographs the DNA of each nuclear region appears to be attached at one point to a mesosome, and the mesosomes are often seen near the forming septum and attached to daughter nucleoids, as if controlling separation. Circumstantial evidence suggests that attachment of the nucleoid to the plasma membrane may be important in DNA replication. For example, when *E. coli* nucleoids are isolated by gentle methods it is observed that cells actively replicating DNA yield nucleoids with some membrane fragments attached, whereas cells not replicating their DNA yield membrane-free nucleoids. Presumably this reflects corresponding alterations in nucleoid-membrane associations in the cell.

Certain strains of the bacteria called *Halobacteria* show specialized patches in their plasma membranes. These regions are known as "purple membranes" since their composition is dominated by a protein-pigment similar to the rhodopsin of vertebrate retinal photoreceptors (Section 3.8.1). The protein molecules are arranged in regular arrays and seem to span the thickness of the membrane, penetrating through the lipid bilayer. An important set of proposals under investigation suggests that the purple membrane regions control the passage of ions such as H^+ across the plasma membrane, that this is affected by light and that the result of the ion movements is the generation of concentration gradients that are involved in formation of ATP (Sections 2.6.2 and 2.7.3) or other metabolic processes. Thus, it is hoped that work on the purple membranes will help elucidate the relations of membrane structure to ion transport and permeability and also shed light on the functioning of chloroplasts, mitochondria, and photoreceptor cells.

Phagocytosis, pinocytosis, and lysosomes have not been observed in bacteria. However, some bacteria produce *exoenzymes* that act outside the plasma membrane and break down macromolecules into smaller units that can pass into the cell. Such enzymes can facilitate passage of bacteria into or within tissues of organisms they infect, as well as providing nutrients for the bacteria. The release of extracellular lytic enzymes has been likened to the emptying of lysosomal enzymes into a phagocytic vacuole of an animal cell; a phagocytic vacuole is, after all, a bit of the enternal milieu that has been enclosed within a membrane and taken into the cell. However, the release of enzymes in bacteria does not, as far as is known, involve fusion of a membrane-delimited enzyme-containing vacuole with the plasma membrane. Enzymes can enter the extracellular medium when some of the individuals in the population die and liberate their contents. But this seems an unlikely general mechanism for release. One proposal is that bacterial exoenzymes pass directly through the plasma membrane. Possibly the polysomes that synthesize the exoenzymes are attached to the plasma membrane, and newly synthesized enzymes leave the cell by mechanisms like those responsible for passage of proteins from bound polysomes into the ER cavities of eucaryotic cells (see Section 2.4.1). It should be noted that controlled breakdown and resynthesis of regions of the extracellular wall is a necessary step in bacterial growth and division. Part of the wall must be dismantled to permit an increase in cell volume and new wall must be made to separate daughter cells. Presumably, enzymes released locally from the cell participate in such remodeling. However, the extent to which localized breakdown of wall structure occurs and the controls and mechanisms of remodeling are still incompletely understood, though several alternative proposals are being investigated.

There is no evident attachment of ribosomes and polysomes to other structures in the bacterium, perhaps with the exception of the plasma membrane. However, the possibility must be left open that some organization of ribosomes exists, which is either broken down when cells are fixed for microscopy or is not readily recognizable in the microscope.

Many bacteria, including *E. coli*, have flagella, but these are quite unlike the flagella of eucaryotic cells. There may be many flagella per cell. Each consists of a single fiber, 100–200 Å thick and several micrometers long, protruding through the cell wall from a basal structure anchored in the plasma membrane. The fiber is made largely of the protein *flagellin*. Evidently, rotatory movement of its flagella move the bacterium. Much shorter fine filaments called *pili* also protrude from the surface of some bacteria. Many investigators believe that special *pili* function in the association of bacteria during *conjugation*, a form of sexual reproduction (Section 4.3.3), but in general the significance of the filaments is not understood.

3.2.5   METABOLIC CON-   Study of *enzyme induction* and *repression*
TROLS; PLASMIDS   in bacteria has elucidated mechanisms by which genes can be turned on or off. Many *E. coli* enzymes are *constitutive*, that is, they are synthesized irrespective of the growth medium. *Inducible* and *repressible* enzymes are produced in some growth media but not in others. The most intensive work on induction has been done on an *E. coli* metabolic pathway responsible for steps in the metabolism of certain sugars and referred to as the *β-galactosidase system*. In the absence of the sugar *lactose* (or other "inducing" compounds of similar molecular structure), the bacteria make very few molecules of the enzymes that metabolize lactose. When inducing compounds are added to the growth medium, extensive and coordinated production of three related enzymes starts: *β-galacotosidase* which splits lactose into simpler sugars—galactose and glucose; a *permease* involved in entry of lactose into the cell; and an *acetylase* involved in certain reactions of sugars. The proteins which show these enzymatic activities are coded for by genes (DNA nucleotide sequences) that are adjacent to one another along the chromosome; a single mRNA molecule transcribed from this DNA contains the information for all three proteins. Several thousand *β*-galactosidase molecules rapidly accumulate in the cell following exposure to lactose.

The explanation advanced for these findings proposes that DNA sequences responsible for the control of transcription are located adjacent to the DNA (*structural gene*) that actually codes for the enzymes, and that the control involves the binding of specific proteins to these sequences. The RNA polymerase that transcribes a given DNA sequence into an mRNA first attaches to the DNA at a *promoter* sequence near the sequence to be transcribed. For constitutive enzymes it is suspected that the relative rates of transcription of different genes may reflect, in part, differences in the associated promoters leading to different affinities for RNA polymerases. For inducible enzymes such as *β*-galactosidase there are additional controls. Adjacent to the promoter sequence is an *operator* DNA sequence. A separate regulatory gene codes for a *repressor* protein that can bind specifically to this operator sequence. Such binding inhibits transcription, possibly by blocking access of the polymerase to the promoter. But the affinity of repressor protein for inducers is much greater than for DNA, so that when inducers are present they bind to the protein, freeing the operator and promoter. This "depresses" transcription.

Another regulatory DNA sequence linked to the β-galactosidase promoter is "recognized" by a protein, sensitive to cyclic AMP. When cAMP levels are high, the protein interacts with this site in such a way as to permit transcription. When cAMP levels drop, transcription is prevented. When the cells are supplied with glucose, rendering metabolism of other sugars metabolically superfluous, intracellular cAMP falls and β-galactosidase is not synthesized, irrespective of the presence of inducers. The three control DNA sequences governing the β-galactosidase system (promoter, operator, and cAMP site) are included in a stretch of 122 nucleotide pairs located between the sequence coding for repressor protein and that coding for the β-galactosidase enzyme.

Several other metabolic pathways have also been shown to be under comparable controls. For the β-galactosidase system, which is involved in *breaking down* components of the growth medium, enzyme synthesis depends on the presence of inducers. For other pathways, such as one responsible for *synthesizing* the amino acid, histidine, production of the enzymes is repressed if the end product molecule resulting from operation of the pathway is provided in the growth medium. The evolution of such mechanisms has provided flexibility in metabolism; the synthesis of certain enzymes is adjusted to the particular environmental conditions. Since β-galactosidase may contribute as much as 3 percent of the total protein in a bacterium, it is clearly advantageous that its synthesis is halted when appropriate substrates are not available.

A different sort of control mechanism is known as *feedback inhibition*. This controls enzyme *activity*, rather than enzyme *amount*. If enzyme E catalyzes the production of b from a, and b is further metabolized by other enzymes to produce d, then in many cases d can be shown to inhibit the activity of E. Thus as excess d builds up, further synthesis of b and d is slowed down. The simplest explanation for this is that d binds to E and alters its enzymatic properties. It is likely that enzymes subject to feedback control have special binding sites that may be separate from the "active sites" at which they bind and act upon their substrates. Binding of an inhibiting compound to the protein at the special site might result in change of the three-dimensional shape of the protein molecule, altering the three-dimensional arrangement of amino acids at the "active site," and thus inhibiting enzymatic activity. Such effects of binding at one site upon the properties of another site on the same molecule are known as *allosteric* effects.

Feedback inhibition is a well-established phenomenon. The involvement of shape changes in proteins (allosteric effects) has been proposed more recently. Suggestive findings of shape changes associated with inhibition of enzyme activity have been made for only a few enzymes, but the proposal is a plausible one from what is known of protein structure and function.

There have been some interesting observations on regulation of the degradative phase of macromolecule turnover in bacteria. For example, in rapidly growing bacteria there is only a very limited breakdown of proteins. However, when the cells are starved for carbon or nitrogen by placement in deficient media, they stop growing and they increase intracellular protein degradation. This makes small molecules available for energy metabolism, the synthesis of new enzymes, or other activities. It is also found that bacteria can selectively degrade severely abnormal proteins, synthesized as a result of mutations or the presence of abnormal metabolites (such as certain "amino acid analogs", compounds that substitute for normal amino acids). In these studies a general correlation has been noted between the rates at which a given protein turns over in the cell, and the ease with which the protein is degraded by proteolytic enzymes in the test tube. Comparable correlations are sometimes reported for eucaryotes. Thus elucidation of the mechanisms by which bacteria control the breakdown of their proteins may have broad implications.

A final example of important phenomena encountered in bacteria is the existence of *episomes*. These are DNA structures with the unusual ability to exist either free in the cytoplasm or in close structural and functional association with the bacterial chromosome. One example was discussed in Section 3.1.2; lysogenic bacteriophages can replicate, essentially as part of the bacterial chromosome, or they may become independent of the chromosome, undergo rapid replication, and kill the host cell. Another episome is the bacterial sex factor known as *F*. This is a circular DNA molecule which can replicate independently in the bacterial cytoplasm or as part of the bacterial chromosome. In the latter case the addition of *F* to the bacterial chromosome results in the ability of the bacterium to transfer parts or all of its chromosome to another cell as part of a mechanism of sexual reproduction. This takes place during conjugation when cell-to-cell continuity is established by membrane fusion, as will be briefly outlined in Seciton 4.3.3.

Episomes are a category of *plasmids*, DNA molecules, generally circular, that can replicate and maintain an independent existence in the bacterial cytoplasm, separate from the chromosome. Methods are now available for isolating plasmids from bacteria, incorporating other DNA molecules into them, and then reintroducing the DNA into bacteria. The added DNA will replicate along with the rest of the plasmid and the bacterial chromosomes. Thus, in principle, one can obtain almost limitless duplicates of any isolable DNA molecules by "growing" them in bacterial "hosts." This could turn out to be an exceptionally useful technical trick for study of hard-to-purify genes. But it also involves potential hazards. Through inadvertence or accident a bacterial strain carrying genetic information harmful to humans or other organisms could conceivably be created and distributed into the environment. There is a continuing debate in the biological community about needed safeguards and requirements for self-restraint among researchers in the creation of new microbial strains. Even the more difficult ethical question has been raised, whether research with exciting prospects but such harmful potential should be restricted by concerted agreement of scientists.

c h a p t e r **3.2C**
BLUE-GREEN ALGAE

Blue-green algae occupy a wide variety of habitats, and some are of considerable ecological or economic significance. In the context of this book, their most interesting characteristic is that the blue-green algae are *procaryotes*, resembling bacteria in many regards. A few grow as single separate cells, and many form filamentous multicellular colonies.

The metabolism of blue-green algae is based on photosynthesis. They are the most primitive plants to possess chlorophyll and in which photosynthesis produces oxygen. In addition to chlorophyll, these algae contain unique pigments, collectively called *phycobilin*; one of these pigments (*phycocyanin*) is blue and another (*phycoerythrin*) is red. The variability in color of different species of blue-green algae usually results from differing amounts of green (chlorophyll), blue, and red pigments.

The algae show an adaptive response to light. When grown in red light they form more pigment that absorbs red light and consequently they look blue. When grown in blue light they produce more blue-absorbing pigment and thus look red. These pigments are combined with polypeptides in the cell to form biliproteins. The energy they absorb can be transferred to chlorophyll for use in photosynthesis. Thus, the cell can efficiently utilize a broader spectrum of wavelengths than would be true with chlorophyll alone (Section 2.7.3). The phycobilin pigments along with other special metabolic features, and the ability to form spores that are quite resistant to the environment, faciliate growth of the algae under conditions of temperature, water salinity, dryness, and light intensity that would preclude survival of most higher plants.

As in all procaryotes the circular DNA (as usual, a double-stranded helix, probably circular) is not segregated into a nucleus separated from the cytoplasm by a nuclear envelope. No endoplasmic reticulum, Golgi apparatus, or mitochondria have been observed. The ribosomes and polysomes are "free" and are similar to those of bacteria. The photosynthetic apparatus (Figure III-7) is not segregated into a membrane-delimited chloroplast. The cell walls resemble the walls of bacteria in containing lipoproteins, lipopolysaccharides, and extensively interlinked polymers involving sugars and amino acids. These polymers are like those of some Gram-negative bacteria (Chapter 3.2B), and like the walls of many bacteria, under some circumstances the algal walls can be dissolved by the enzyme lysozyme to produce "protoplasts." (It is widely thought that bacteria and blue-green algae have evolved from a common ancestral form.) An additional gelatinous sheath often surrounds the cell wall of the algae.

The photosynthetic apparatus of blue-green algae is somewhat more highly organized than that of photosynthetic bacteria. The pigments are present in flattened sacs, called *lamellae* or *thylakoids*, which are arranged in parallel array. In usual preparations their mem-

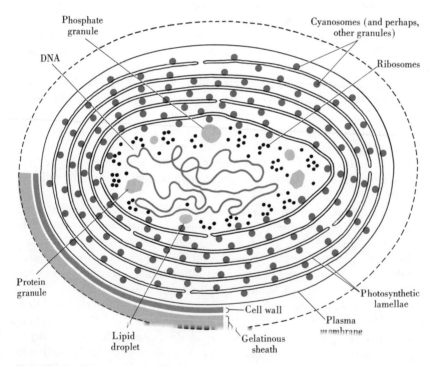

Fig. III-7   Diagram of a blue-green alga cell. (Based on the work of M. Lefort-Tran and others.)

branes show the typical three-layered structure. The lamella closest to the periphery of the cell may be continous with the plasma membrane.

Granules, about 300–400 Å in diameter, are observed between the thylakoids, with at least some apparently attached to the outer membrane surface. These granules may be seen by electron microscopy of thin sections and of freeze-etched perparations. The granules are often referred to as *cyanosomes* or *phycobilisomes*; they are the sites of phycobilin pigments. The cells can be disrupted (using a device known as a "French press") and fractions prepared by density gradient centrifugation. When this is done, the cyanosomes are separated into a fraction found to contain phycobilins, while the lamellar membranes are found in the fraction containing chlorophyll and other pigments (such as yellow *carotenoids*). If the cells are fixed in glutaraldehyde prior to disruption and centrifugation, then fragments are obtained which consist of membranes with cyanosomes still attached. Such cyanosome-membrane fragments retain photosynthetic functions that neither membranes nor cyanosomes possess alone. Negative staining shows the phycobiliproteins to consist of subunits. To what extent the variation in subunit number and size reported by different investigators is due to different procedures and different species used is uncertain. Still unclear also are the functions of the subunits.

The cells divide by inward growth of plasma membrane and wall producing two cells of equal size as in bacteria. Mesosomes or similar structures have yet to be observed and the mechanisms for separation of daughter "chromosomes" are not known.

Although they have no obvious specialized motion-related organelles such as cilia or flagella, blue-green algae are capable of movement. Gliding and rotatory motions have been observed whose mechanisms are unknown.

c h a p t e r **3.3**
PROTOZOA

Protozoa are single-celled eucaryotic organisms with the same basic functions and with the same organelles as cells of higher organisms. Many different types occur (amebae, ciliates, flagellates, and so forth), each with particular characteristics associated with its own mode of life. Unlike the cells of multicellular organisms, each protozoon must carry out all the functions required for both survival and duplication of the whole organism. For this reason, although they usually are thought of as low on the evolutionary scale, the protozoa include the most complicated and diversified types of known cells.

Pinocytosis and phagocytosis are used by most protozoa in feeding. Amebae often have elaborate surface coats (see Fig. II-7) that adsorb material to be taken in by phagocytosis and pinocytosis, and many other protozoons have special "mouth" regions where food is ingested. The vacuoles and vesicles that form acquire lysosomal enzymes which digest the contents. Finally, egestion, or defecation, of indigestible residues occurs by fusion of the vacuole membranes with the cell membrane.

Interesting structural specializations are found in various protozoan species. For example, some secrete a hard, extracellular shell containing calcium or silicon. Others possess *trichocysts*, cytoplasmic bodies that fuse with the plasma membrane and extrude filamentous structures used in defense and trapping of food. (Similar, though more complex bodies found in *Hydra*, called nematocysts, will be discussed in Section 3.6.2.) Most fresh water forms and some that live in salt water contain *contractile vacuoles* (water expulsion vacuoles). Especially in fresh water, the organisms are not in osmotic equilibrium with their environment, and they constantly tend to take in water. In bacteria and higher plants living in water, this tendency is counteracted by the presence of a rigid cell wall which prevents swelling. In protozoa the water that comes in ultimately is removed by the contractile vacuoles which fuse with the plasma membrane and expel the water at the cell surface. In some organisms, for example, *Paramecium*, the vacuoles are large, round, membrane-delimited bodies surrounded by an elaborate arrangement of small vesicles and tubules that carry fluids to the vacuoles. Contractile vacuole functions apparently are energy dependent since they can be inhibited by cyanide or other inhibitors of respiration. The vacuole filling–expulsion cycle may be completed in less than one minute; in one hour an organism can expel a volume of water several times its own volume.

Some flagellate protozoa contain *kinetoplasts*, unique structures having sufficient DNA for detection by light microscope cytochemistry, but having as well the typical mitochondrial membrane configurations. As the cells multiply, these structures divide. Study of kinetoplast DNA may produce insights into the role of nucleic acids in more typical mitochondria.

In addition to the kinetoplasts, other specialized features of protozoa are particularly interesting for the understanding of cell organization and function. For example, a few species have giant "centrioles," rodlike structures tens of micrometers long which, like ordinary centrioles, appear to be involved in formation of the cell division spindle. The limited information available suggests that the structure of the giant centrioles is quite different from ordinary centrioles, although their function is apparently similar. Their study may therefore shed light on centriole function.

Some protozoa can be conveniently grown as large populations in a medium of controlled and well-defined composition. Many are large enough for microsurgery.

3.3.1 MACRONUCLEUS AND MICRONUCLEUS

In ciliates and some other protozoa, two types of nuclei are regularly present in the same cell. One of these often is quite large (Figs. III-8 and III-9), contains much DNA, and shows the presence of nucleoli, These *macronuclei* are considered to participate actively in producing RNA used in cell metabolism. The smaller *micronuclei* contain much less DNA. Their most evident role is in sexual reproduction. They duplicate by mitotic division similar to that of ordinary eucaryotic cells (Section 4.2.1) and they can participate in meiosis (Section 4.3.1). During sexual reproduction (*conjugation* in protozoa) two individuals exchange micronuclei through temporary cytoplasmic bridges. Meanwhile the macronucleus of each degenerates to be replaced later by alteration and growth of a micronucleus. This micronuclear transformation involves the synthesis of much DNA. In

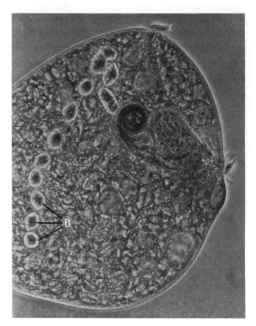

Fig. III-8 Portion of the protozoan Stentor. *The macronucleus appears as a chain of beads (B). The micronuclei are too small to be well demonstrated in this photograph.* × *100. (Courtesy of L. Margulis.)*

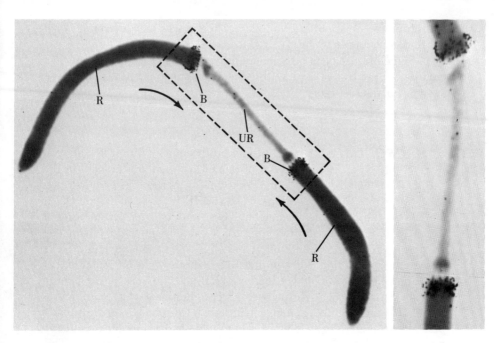

Fig. III-9 *Autoradiogram of the macronucleus of the ciliated protozoan Euplotes. The cell had been exposed to tritiated thymidine for 20 minutes before fixation; the DNA synthesized during this interval is therefore radioactive. Grains are found only over two short lengths of the nucleus; these are the replication bands (B). The bands reflect a wave of replication passing toward the center of the nucleus (arrows); the wave has not yet reached the zone at UR. The chromatin in the region between the bands and the ends of the nucleus has already been replicated (R). × 600. Insert at right is an enlargement of the area in dotted lines. × 1000. (Courtesy of D. Prescott.)*

some species of protozoa amicronucleate strains are known that are asexual but survive and function with only a macronucleus. In other strains with both types of nuclei, removal of the macronucleus causes cell death even if a micronucleus is present.

The simplest explanation for these observations is that the macronucleus contains many copies of the genetic information needed for the day-to-day functions of the cell. It seems to represent an "amplification" mechanism whereby numerous replicates of DNA, derived initially from the micronucleus, are made available for functions such as RNA production. This might explain the observations made on macronuclear regeneration in the ciliate *Stentor*. As Figure III-8 shows, the macronucleus of this organism has the form of a string of beads. All but one of the beads can be removed without killing the cell. Fragments of cells provided with a single "bead" can regenerate complete cells. If only one or two copies of given genes were contained in the macronucleus, it would be difficult to see how a complete nucleus, capable of sustaining cell life, could be regenerated from a small fragment that could not contain more than a small fraction of the original macronuclear DNA.

Direct analysis of the DNA of macronuclei and micronuclei has begun, using techniques such as DNA–RNA hydridization (Fig. II-10) and DNA reannealing (Section 4.2.6). One interesting finding is that in some ciliates, such as *Stylonychia*, many of the DNA sequences present in micronuclei are not found in macronuclei. As the macronucleus forms, extensive fragmentation and selective breakdown of DNA occurs, as well as repeated duplication of some portions of the DNA. While such differences in DNA sequences between macronucleus and micronucleus are absent or much less dramatic in other species, these observations are of great potential interest given the differences in roles of the two nuclear types. Conceivably they imply that the macronucleus can be selectively constructed so as to contain many copies of the DNA-encoded information needed for its function and few or none of superfluous information.

Interesting studies of macronuclear replication have been performed with the ciliate *Euplotes* (Fig. III-9). The macronucleus of this species is a long, sausage-shaped structure. Replication of the contents occurs in a progressive fashion, starting at both ends and moving toward the center. The actual points of replication can be distinguished in the microscope as a pair of bands whose appearance differs from the rest of the chromatin. This provides a most convenient system for studying duplication of nuclear material. The chromatin in the bands appears to transform from clumps and patches into dispersed fibrils (individual chromosomes have not been distinguished). Autoradiography shows that DNA synthesis occurs at the bands (Fig. III-9). On the other hand, RNA synthesis takes place everywhere but at the bands. Thus it is held that no synthesis of new RNA takes place on chromatin as it duplicates. From other evidence this seems true of other cell types as well. Newly synthesized proteins accumulate at the bands, suggesting that new chromosomal proteins become associated with DNA as it replicates. Newly labeled protein is also found in the nucleus, away from the bands. As the amount of DNA in the bands doubles, the histone content also doubles. The total duplication time for the *Euplotes* macronucleus is several hours. After duplication of its contents, the nucleus is pinched in two across its long axis as the cell divides.

Macronuclei appear to be one device whereby the metabolic capacities of the nucleus are expanded to serve a large volume of metabolically active cytoplasm (see Section 3.3.2). Polyploidy (Section 4.5.1), polyteny (Section 4.4.4), and selective amplification of specific genes (Section 3.10.1 and Fig. IV-18), and DNA repetitiveness (Section 4.2.6) probably have similar effects.

3.3.2   *NUCLEUS AND*   Amebae survive microsurgical experiments
**   *CYTOPLASM***   well which has facilitated several interesting
experiments. We referred in Section 2.5.3
to the influence on the Golgi apparatus, of removing the nucleus and subsequently replacing it. Another line of investigation is based on labeling the nuclear macromolecules of one ameba by growing it in appropriate radioactive precursors, then transplanting the nucleus into another (host) ameba which has not previously been exposed to radioactive molecules. It can be shown by autoradiography that proteins and RNAs of the transplanted nucleus pass into the host's cytoplasm; some of these molecules subsequently turn up in the host's nucleus. Although the experimental system is a quite artificial one, the results suggest that there is a class of proteins and perhaps also some RNA molecules that can pass back and forth between nucleus and cytoplasm. Could such molecules serve some regulatory or other functions in the interplay of nucleus and cytoplasm?

Other interesting experiments on the relations of nucleus and cytoplasm have explored the hypothesis that the initiation of cell division depends on the cell's reaching a certain size. One form of this hypothesis suggests that division mechanisms have evolved so that a rapidly growing cell tends to divide before it exceeds a critical volume. For a sphere, the surface is given by $4\pi r^2$, while the volume is $(4/3)\pi r^3$. For a given increase in r, the volume increases proportionally more than the surface, since cubes of numbers increase faster than squares. While cells are not usually spheres, this sort of relationship of surface increasing at a slower rate than volume is a general one; as cells get larger, there is a *relatively* smaller surface available for exchange of material with the environment. Similarly, a fixed amount of DNA services an increasing mass of cytoplasm.

If one repeatedly cuts off portions of cytoplasm from growing amebae and thus prevents them from attaining a certain size, the cells do not divide. This suggests that one set of factors controlling cell division is related to cell size. However, the relationship is not a simple one. The growth of amebae may be curtailed by placing the cells in a nutrient-poor medium. Division is slowed, but does occur despite the fact that the cells are smaller than normal. Thus while growth and division may be under some common controls, these are flexible. Perhaps some key compound must reach a critical level before division occurs, and the normal rate of production of this component is such that it usually accumulates to the appropriate level as the cell reaches a certain size. Or perhaps a much more subtle or complex control is responsible. One experimental approach to this question is the creation of new nuclear-cytoplasmic combinations by transplantation or other procedures; we will discuss such experiments in Section 4.2.3.

3.3.3   BASAL BODIES, CENTRIOLES, CILIA, AND FLAGELLA

In ciliated protozoa like *Paramecium*, the cell surface and the cortex (the cytoplasm just below the membrane) are highly organized (Fig. III-10). The basal bodies of the cilia are arranged in a precise geometrical fashion. In a series of investigations, portions of the cortex were experimentally reoriented so that the basal body pattern was reversed and the cilia beat in the opposite direction from their neighbors. Such changed orientations persist through repeated cell duplications; some have been followed for 700 generations. This is one of a number of experiments that lend weight to the idea that much of the surface pattern depends for its reproduction on (unknown) local determinants rather than on the nucleus. Electron microscopy of duplicating protozoa shows that new basal bodies form in close association with old ones, suggesting some mode of self-duplication (see Section 2.10.5).

The basal bodies of ciliated protozoans show clearly the presence of a "cartwheel" structure within the 9 + 0 tubule pattern (see Fig. II-62). This is restricted to the end of the basal body opposite to the end attached to the cilium or flagellum. Observations on protozoa and other cells have raised the possibility that this structural polarity may reflect a functional polarity; perhaps one end of the organelle is involved in producing cilia or flagella, and the other in basal body duplication or

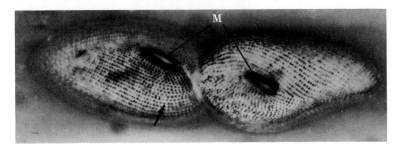

Fig. III-10   A dividing Paramecium *stained to show the basal bodies and related structures in the cell cortex (the cytoplasm just below the plasma membrane). The basal bodies are arranged in rows which appear as lines of granules (arrow); the two daughter cells have identical patterns of basal bodies. The "mouths" of the daughters are indicated by M. (Courtesy of R. V. Dippel and T. Sonnenborn.)*

growth. In basal body duplication in ciliated protozoa, a new procentriole (Section 2.10.5) forms near the cartwheel end of the old body. Cartwheels appear early in developing procentrioles.

Many of the ciliates and other organisms show elaborate systems of microtubules and fibers associated with the basal bodies; often these

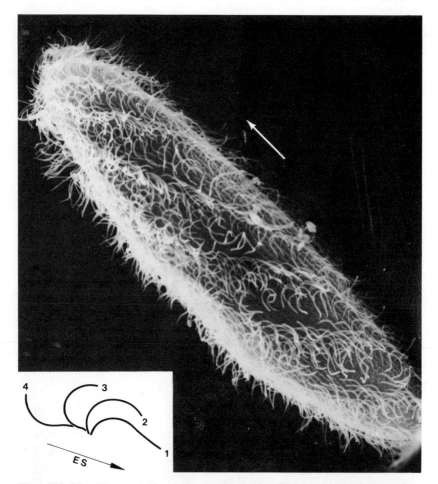

Fig. III-11   *Micrograph of the surface of a* Paramecium *taken with a scanning electron microscope. The animal was fixed while swimming in the direction indicated by the large arrow. As the micrograph shows, the movement of the numerous cilia is coordinated so they beat in waves (called metachronal waves); at a given time some cilia are in phases of the effective stroke and others are recovering (see Fig. II-60). The insert diagram shows the movement pattern of an individual cilium, seen from above; the arrow indicates the direction of the effective stroke, occuring in a plane perpendicular to the page, and the numerals show successive stages in recovery. × 1000. (From S. L. Tamm, T. M. Sonneborn and R. V. Dippell, J. Cell Biology, **64**:107.)*

appear as a network that seems to link the basal bodies together. It is possible that such systems coordinate the cilia of a cell and maintain the wave of ciliary motion that passes along the surface (Fig. III-11). A few microsurgical studies indicate that interruption of the connections between basal bodies can affect coordination. However, the results of other studies are equivocal, and the extent to which various factors participate in coordination remains to be established. One proposal suggests that the plasma membrane may conduct some coordinating impulses analogous to nerve impulses. Another maintains that much of ciliary coordination depends on simple physical or mechanical interactions resulting from the close proximity of cilia to one another and that no special coordinating devices are needed. As evidence for this last suggestion, several students of protozoa cite observations on spirochetes that are found attached in large numbers to some protozoa. Spirochetes are elongate procaryotic organisms which have fibers helically wound around their cell body and which are capable of motion somewhat similar to that of a flagellum. Despite the absence of a common membrane or other obvious communication device, the spirochetes attached to one protozoon move in waves. It even appears that motion of attached spirochetes coordinated with the action of its own flagella can contribute to the motion of the protozoon itself.

Ciliates such as *Paramecium* can alter their direction of motion when they encounter an obstacle or when they are placed in an orienting environment (for example, an electrical field). Change in direction results from changes in the frequency and orientation of the ciliary beat. This can extend to a reversal of the direction of the effective ciliary stroke so that the organism swims "backward." Experimentally, the rate and direction of the ciliary beat are found to alter with changed ionic environments, especially changes in Ca^{++} ion concentration. A current hypothesis suggests that events analogous to the alterations in electrical potential and plasma membrane ionic permeability occuring in conduction of nervous impulses (Section 3.7.2), also occur in the protozoa when they contact an obstacle. Ca^{++} ions might serve as the direct regulatory agents. For further analysis of these phenomena use is being made of mutant strains of *Paramecium* that show abnormalities in their patterns of motion ("behavioral" mutants). (See also p. 261).

c h a p t e r *3.4*
EUCARYOTIC PLANT CELLS

Algae other than the blue-greens (Chapter 3.2C), and all higher plants are eucaryotes, and, with very few exceptions, their cells contain the organelles described in Part 2. Also described in Part 2 is one of the characteristic plant organelles, chloroplasts. Two others; the cell walls, and vacuoles are considered in this chapter. Additional topics of plant science discussed in this text can most easily be found by consulting the Index listing, "Plant cells." The reader is also referred to the references listed at the end of this part.

c h a p t e r **3.4A**
ALGAE

These plants, some unicellular others multicellular (some such as the giant kelps, are quite large), are classified partly on the basis of their color, determined largely by the nature of the pigments present in addition to the chlorophyll. For example, in red algae a major additional pigment is similar to the red pigment of blue-green algae (Chapter 3.2C).

As Figure III-12 indicates, the chloroplast membranes of algae are usually less elaborately developed than in higher plants (see also Fig. II-46). The plastids are membrane bounded, and the internal sacs or thylakoids are arranged as long parallel sheets. Most algae lack the stacked-coin-like grana characteristic of higher plants (Figs. II-46 and 47) but sometimes the stacks of closely-associated elongated thylakoids found in algae (Fig. II-46) are also referred to as grana. The plastids contain ribosomes and 25-Å fibrils, probably of DNA. A *pyrenoid* where starch is stored and probably synthesized, and a *pyrenoid sac*, which also stores polysaccharides, are part of the plastid.

The mitochondria may be like those in Figure III-12 or tubular as in many protozoa (Fig. II-38). The ER is relatively scanty; as usual, it is continuous with the nuclear envelope. The Golgi apparatus often lies near the nucleus in algae and images suggest that membrane from the nuclear envelope feeds into the outer surface of the Golgi stack, as portrayed in the diagram. Lysosomes and peroxisomes have not been isolated from algae, but both organelles have been reported by electron microscopists. Autophagic vacuoles increase in number in starved *Euglena*. In *Euglena*, too, the number of peroxisomes has been shown to increase greatly when the algae are shifted from a glucose medium to one with acetate or alcohol, suggesting to the investigators that the peroxisomes may function in lipid and carbohydrate metabolism as they do in seeds (Section 2.9.2). In contrast to most higher plant cells, paired centrioles are present in many algae.

Many algae have prominent cell walls as in higher plants, but some do not. In certain algae the cell walls show specialized plaque-like scales. In diatoms the walls are made largely of silica. Small vacuoles,

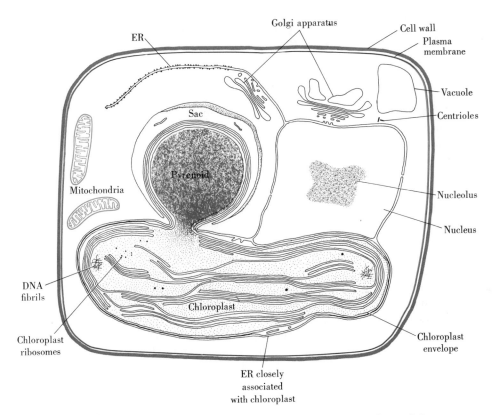

Fig. III-12 *Relationships among organelles in a hypothetical brown alga. Not included in this schematic diagram are free ribosomes and several types of cytoplasmic granules. Note that the flat sacs within the plastid are arranged in extended parallel arrays (compare with Figs. II-46 and II-47). (After G. B. Bouck.)*

derived from the Golgi apparatus, transport the scales or other materials to the forming cell walls. The cell walls may approach the complexity of those of higher plants.

In a large algae cell like *Nitella* (millimeters to centimeters in length) it is possible to suck out and thus to analyze directly the contents of the central vacuole that occupies much of the cell (as in Fig. II-69; Section 3.4.2). These cells also lend themselves well to the study of cyclosis (Fig. II-69); their size facilitates observation of the cytoplasmic streaming around the central vacuole.

Some unicellular algae have a single chloroplast which divides in synchrony with the cell, consistent with the "self-duplication" of chloroplasts (Section 2.7.4).

c h a p t e r **3.4B**
HIGHER PLANTS

Higher plants show a host of intriguing specializations of cell structure, chemistry, and physiology, some obvious and some subtle. For example, the varying properties of woods and plant-derived fibers such as cotton depend on the details of molecular arrangements in the cell walls of the tissues of origin. Another example is the group of plants known as *C-4 plants* because the initial products of their photosynthesis are 4-carbon molecules rather than the usual 3-carbon products (Section 2.7.2). The C-4 plants have a special leaf anatomy. The bundles of fluid-conducting vascular tissue are surrounded by two concentric layers of cells. The cells of the outer layer carry out the initial steps of C-4 metabolism. The products of these reactions pass into the inner layer where they are used, eventually, to produce starch or other carbohydrates. Thus, unlike C-3 plants, the overall photosynthetic process depends on metabolic cooperation between two separate sets of cells. The chloroplasts of C-4 plants show an unusual membranous reticulum (Fig. II-47) of unknown significance. Moreover in some C-4 species, the chloroplasts of the inner layer cells lack well-defined stacked-coin-like grana, thus somewhat resembling the plastids of some algae. C-4 metabolism is highly efficient in utilizing CO_2 which probably accounts for the fact that it is found in several agriculturally important plants such as corn and sugar cane. In addition, the efficient use of CO_2 permits the plants to function with smaller openings of their *stomata*, the structures through which leaves exchange gases with the atmosphere (Fig. III-13). This retards water loss from the leaves and is one reason that some C-4 plants can survive well in arid environments.

The present chapter will deal with several major features of cells of cells of higher plants. Fig. III-14 illustrates the dominant structural features found in many such cells—the presence of a prominent cell wall, and of a large central vacuole. The "further reading" lists should be consulted for discussions of the diversity of plant cells and of such important topics as the control of plant cell growth and differentiation by sensitive regulatory systems (for example, those involving phytochrome) and by plant hormones (p. 275).

3.4.1   **CELL WALLS**   Plant cell walls almost all contain cellulose, a polysaccharide made of glucose subunits. The cellulose molecules in the walls are in the form of multimolecular bundles or fibrils up to 250 Å thick and several micrometers long (Fig.

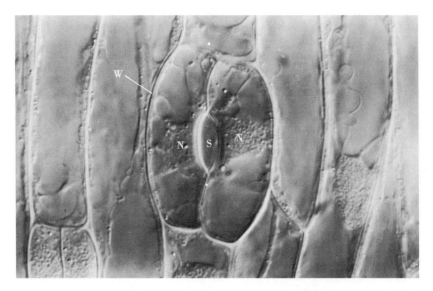

Fig. III-13   *Stomatal complex of an onion seedling photographed through a Nomarski interference microscope (Fig. I-6). The two guard cells (their nuclei are barely evident at N) control the extent of opening of the stoma, the space through which gas exchanges occur between the underlying tissues and the outside environment. W indicates the thin cell wall of one of the guard cells.* × *1500. (From B. Palevitz and P. Hepler,* Chromosoma, **46**:311, 1974.)

III-15). The polysaccharide chains run parallel to each other within the fibril. The chains are held together by many hydrogen bonds (p. 64), producing a highly cohesive structure. A matrix of other material surrounds the cellulose and contains additional polysaccharides, some more complex than cellulose, which crosslink the cellulose chains. Matrix compounds such as *lignin*, a complex polymer that imparts strength and rigidity to the cell wall, are prominent in woody plants. Polysaccharides known as *pectins* help bind adjacent cells together; they are abundant in the *middle lamella*, a layer formed between two adjacent cell walls. Removal of the pectins by enzymatic or other means permits cells with their walls to fall apart from one another, although the walls maintain their shape. Waxy substances in the cell wall help to protect many plants from drying out. Proteins may account for up to 10 percent of the cell wall. One of the proteins is unusually rich in hydroxyproline, the amino acid found in animal collagen; there are, however, chemical differences between the two proteins.

Wall morphology and chemistry vary greatly among the different plants and plant tissues and are responsible for many major features of the tissues. Openings present in the walls of many cells of multicellular

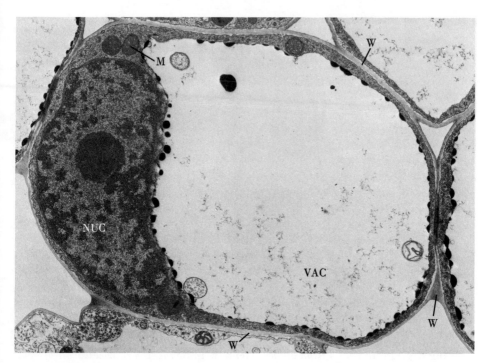

Fig. III-14 *A cell of the marsh plant* Limonium. *It shows the major features of many mature plant cells. The vacuole occupies much of the volume while the nucleus and organelles such as mitochondria (M) are found in a thin layer of cytoplasm surrounding the vacuole. The cell is separated from its neighbors by a cell wall (W).* × *9000. (Courtesy of M. Ledbetter.)*

plants permit passage of fluids. The fluid-conducting vascular system of higher plants is made partly of vessels composed of intercommunicating spaces bounded by complex cell walls left by dead cells. Cork owes its buoyancy to the closed and watertight compartments bounded by the cell wall, which contain air-filled spaces left when the cells die. (Cell walls were the structures seen by Robert Hooke when, in 1665, he used the term *cell* in reporting his investigations of cork and piths of several plants.)

In higher plants, cell division is restricted largely to specialized growing zones or meristems present in roots, shoots, and other organs. Cells produced in the meristems grow and differentiate into various tissues, each with a corresponding type of wall. The layers of the walls are produced sequentially by the cells (Fig. III-16). The outermost or *primary wall* is produced first—meristematic tissues possess only this layer. The initial wall fibrils laid down in the primary wall form a loose meshwork, and the wall is extensible, permitting considerable changes

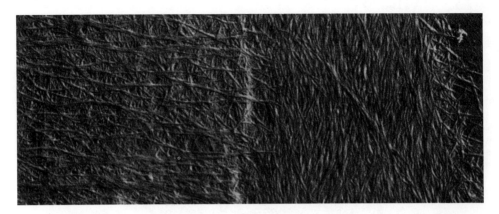

Fig. III-15 *A preparation of the wall surrounding a cell of a grasslike plant, the rush* Juncus. *A* shadowing *technique has deposited a thin layer of metal on the specimen from an angle forming a shadow analogous to that formed with light. Many* cellulose fibrils *in the layer of the wall seen at the left of the figure run roughly perpendicular to the direction of the fibrils in the adjacent wall layer, seen at the right.* × *38,000. (Courtesy of A. L. Houwink and P. A. Roelfson.)*

in shape as the cells grow. As the cell reaches its mature form it may produce an elaborate and rigid *secondary wall*. The *secondary wall* may contain distinct layers. In each layer the individual cellulose fibrils are oriented parallel to one another, but the fibrils of different layers are oriented at angles to one another (Fig. III-16). This layering contributes considerable strength to the wall; often adjacent layers have fibers at right angles to one another. A thin *tertiary* wall may also be present.

The Golgi apparatus plays a major role in the secretion of cell wall polysaccharides (Chapter 2.5). Figure III-17 illustrates the apparent movement of Golgi-derived vacuoles to the cell surface where their contents are emptied by exocytosis. This occurs when new cell walls form during cell division (Fig. IV-23) and at other times as well. Autoradiography of growing cells shows that new polysaccharides are added along the whole length of the wall—growth usually is not restricted to particular zones. There are, however, indications of complex control mechanisms. For example, the initial cellulose molecules made by the cells that produce cotton fibers are of varying length; they contain up to 5000 linked glucose molecules. When secondary wall formation begins, the number of glucose molecules per cellulose increases to almost 15,000 and the lengths of the newly made cellulose molecules become more uniform. With the labeled sugar, ^{3}H-fucose, autoradiography demonstrates that incorporation into cell walls varies among different cell types in corn roots. The root cap and epidermis show much incorpora-

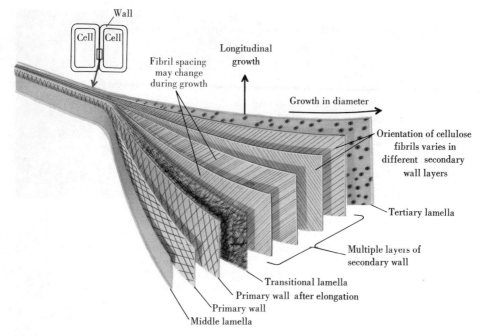

Fig. III-16   *Schematic representation of the various layers of the cell wall of a higher plant. The chief orientation of the cellulose fibrils in the various layers is indicated. This orientation or the spacing of fibrils may change somewhat during growth of some plants. (After K. Muhlethaler.)*

tion whereas other tissues show essentially none. With other sugars the labeling is uniform, suggesting that the fucose is incorporated in some specialized component.

In the case of cellulose, the fibrils do not form within the Golgi vacuoles but rather outside the cell, at or near the cell surface. The forming fibrils are often oriented parallel to the microtubules that lie under the plasma membrane but the mechanisms of cellulose deposition have not been established.

3.4.2   **VACUOLES**   The number and size of vacuoles varies in different cells and during development. In mature cells, a single vacuole is often present (Fig. III-14) and may occupy up to 80 percent or more of the cell volume. At earlier developmental stages, vacuoles occupy less of the cell and several small ones may be present. The vacuole is delimited by a three-layered membrane,

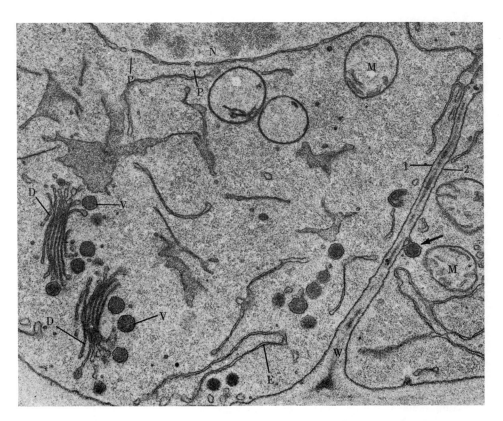

Fig. III-17 *Portions of two cells of corn root. The plasma membranes are seen at (1) and (2). N indicates a part of the nucleus and P nuclear pores. A mitochondrion is seen at M and endoplasmic reticulum at E. The Golgi apparatus (D) consists of numerous stacks of saccules. The large Golgi vesicles (vacuoles; V) contain an electron-dense material; apparently they migrate to the cell surface, fuse with the plasma membrane (arrow), and contribute their content to the cell wall (W). (Courtesy of W. G. Whaley, J. A. Kephart, and M. Dauwalder.)*

the *tonoplast*. As in algae, the cytoplasmic rim surrounding the vacuole of a cell such as that in Fig. III-14 may show extensive cyclosis (Fig. II-69).

The concentrations of salt, sugars, and other dissolved materials within the vacuole are very high. As a result there is an osmotic entry of water through the semipermeable tonoplast into the vacuole. This generates a pressure balanced by the mechanical resistance of the cell wall, so that the cytoplasm is pushed firmly against the wall. During cell elongation, which may be very rapid (cells can elongate at rates of 20-75 μm per hour), the cell walls are somewhat elastic. They stretch under the pressure generated by water uptake in the vacuole. Such elongation usually accompanies plant growth.

The vacuoles of many plants contain pigments such as the anthocyanins. The striking colors of petals, leaves, and fruit are due to such pigments; these colors are important in attracting insects and other organisms involved in pollination and seed dispersal.

Through cytochemistry and through isolation from tissue homogenates, vacuoles of some plant cells have been shown to contain acid hydrolases and thus to qualify as lysosomes. It has been suggested that vacuoles can engage in a type of autophagy in which a bud of cytoplasm protrudes into the vacuole and then pinches off and undergoes degradation. Recently some investigators have proposed that vacuole enlargement during cell differentiation and maturation may involve autophagy. They suggest that small areas of cytoplasm are enveloped by membranes to form autophagic vacuoles. The engulfed organelles are degraded and small vacuoles are formed; these merge to produce larger vacuoles, and finally the large central vacuole is produced. Some investigators believe that vacuoles may arise from the ER or Golgi apparatus.

When plant cells die, the tonoplast becomes interrupted and the hydrolases leak out into the adjacent cytoplasm. There, they participate in dissolution of the cell (autolysis). These events contribute to formation of the fluid-conduction system since the cell walls remain as hollow tubes. In addition autolysis is prominent during the withering and shedding (senescence) of flowers and leaves.

3.4.3 PLASMODESMATA

Holes, about 50–100 nm in diameter, are often found in the cell walls between adjacent cells in higher plants and many algae. Within these holes extensions of cytoplasm are present; these are called plasmodesmata. As shown in Figure III-18, the endoplasmic reticulum often is closely associated with the plasmodesmata.

The simplest explanation for the origin of many plasmodesmata is by the persistence of points of continuity or close contact between dividing cells which become more widely separated at the other points along their surface. In many dividing cells, regions containing remnants of the spindle responsible for the earlier separation of chromosomes may be seen to maintain continuity between daughter cells for some time after division of the rest of the cytoplasm has been completed (see Fig. IV-23). Division of the cytoplasm is retarded at these regions, and the plasma membrane of one daughter cell thus retains continuity with the membrane of the other daughter. Although this is their origin in many

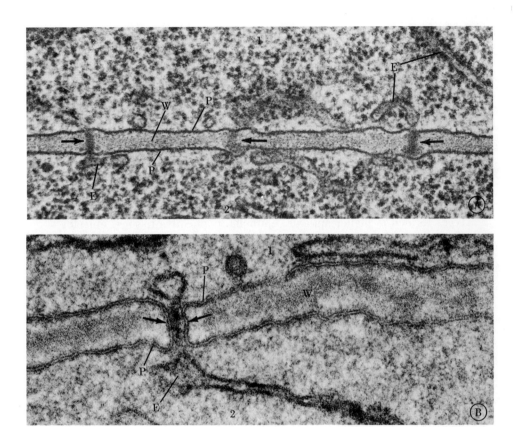

Fig. III-18 Plasmodesmata. *A. Portions of two adjacent cells (1 and 2) of* Arabidopsis, *a plant of the mustard family. The cell wall is seen at* W *and the plasma membrane at* P. *Arrows indicate plasmodesmata which, in this preparation, appear as dense regions that traverse the wall. Endoplasmic reticulum (E) is associated with the cell surfaces at the regions where the plasmodesmata are found. × 63,000. (Courtesy of M. Ledbetter.)* B. *A similar region showing two cells of corn. At the plasmodesma indicated by arrows it can be seen that the plasma membranes of the two cells are continuous, forming a channel connecting the cytoplasm of one cell with the other. The plasma membranes show a three-layered membrane structure. (Courtesy of H. Mollenhauer.)*

cells, it may be that plasmodesmata can also form in other ways; they probably are more dynamic structures than previously thought.

In some cases infecting virus particles apparently cross from one cell to the other via the plasmodesmata, which may be altered as a consequence. Whether large molecules generally can cross in this manner is not clear; organelles do not. However, the general opinion is that plasmadesmata provide a pathway for the movement of molecules from cell to cell. Interactions can thus occur even between cells separated by thick cell walls. There is great variation among cell types in frequency of plasmodesmata.

chapter 3.5
ABSORPTIVE CELLS

3.5.1   *TISSUES: THE SMALL INTESTINE*   Figure III-19 diagrams a cross section of the vertebrate small intestine; like many other organs, this contains the four major tissue types formed in higher animals: epithelia, connective tissues, muscle, and nerve. These tissues will be discussed in greater detail in subsequent chapters.

Many *epithelia* serve as covering tissues, at absorptive surfaces, in ducts, in skin, and elsewhere. They are continuous sheets of cells. Where the cells play a protective role, as in skin, the epithelium is often several cells thick (stratified). At absorptive surfaces the sheet usually is one cell thick. The absorptive surface of the intestine is thrown into a series of folds (Fig. III-19) that increase the absorptive area in comparison to a simple smooth-walled tube. (A much greater surface increase results at the cellular level, from the presence of microvilli (Section 3.5.3).) The intestine is lined by a single layer of epithelial cells that selectively absorbs material from the *lumen*, the space enclosed by the organ. The cells are referred to a *columnar* because they are taller than wide. The epithelium rests on a *basement membrane*, a mat of extracellular fibers and other material that may, in part, provide structural support. The cells at the tips of the villi are continually sloughed off into the lumen and replaced by others that migrate from further down. The population is maintained by the division of cells in the indentations (*crypts*) found at the base of the villi.

Gland cells also are classified as epithelial. The intestine contains three types of gland cells. *Goblet* cells scattered among the absorptive cells secrete mucus that lubricates and protects the lining. Other secretory cells are present at the base of the villi and in the crypts. Multicellular glands are present well below the absorptive surface in the first part of the intestine, near the stomach. Their secretion passes into the lumen via ducts. It is alkaline and probably, among other functions it neutralizes the acid in the stomach contents that enter the intestine.

Connective tissues serve to bind other tissues together, providing support and a framework within which blood vessels, lymphatic vessels, and nerves course. They are composed of cells surrounded by abundant extracellular materials, usually produced by the cells with which they

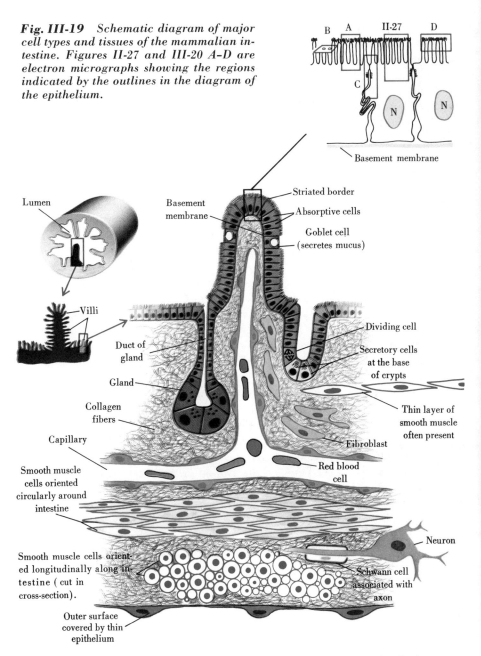

Fig. III-19 *Schematic diagram of major cell types and tissues of the mammalian intestine. Figures II-27 and III-20 A–D are electron micrographs showing the regions indicated by the outlines in the diagram of the epithelium.*

Basement membrane

Lumen

Basement membrane

Striated border

Absorptive cells

Goblet cell (secretes mucus)

Villi

Duct of gland

Dividing cell

Secretory cells at the base of crypts

Gland

Collagen fibers

Thin layer of smooth muscle often present

Capillary

Fibroblast

Smooth muscle cells oriented circularly around intestine

Red blood cell

Neuron

Smooth muscle cells oriented longitudinally along intestine (cut in cross-section).

Schwann cell associated with axon

Outer surface covered by thin epithelium

are associated. In fibrous connective tissues, widespread in the body, *fibroblasts* produce extracellular fibers, consisting mostly of the protein *collagen*. The fibers are surrounded by a matrix containing proteins and such polysaccharides as *hyaluronic acid*. Elastic connective tissues containing fibers of the protein *elastin* are found in the walls of arteries and elsewhere; such tissue provides resiliency to organs. Cartilage consists of cells surrounded by a stiff matrix of polysacharides and proteins; bone cells are surrounded by calcium salts in an organic matrix. Blood also is often classified as a connective tissue; large numbers of red blood cells and fewer white blood cells are carried in a protein-rich fluid, the *plasma*.

The connective tissue of the intestine is largely of the fibrous kind. In it, blood capillaries and lymphatics are present. These transport oxygen to the epithelium and other layers and remove nutrients absorbed from the lumen, as well as CO_2 and other waste products.

There are three major types of *muscle tissue*: the voluntary *skeletal* muscle (Section 2.11.2), the involuntary *cardiac* (heart) muscle, and *smooth* muscle. The first two types show cross striations. In the intestine a thin layer of smooth muscle cells underlies the surface epithelium and influences the overall shape of the epithelial layer. Separated from this by a thick region of connective tissues are two additional smooth muscle layers. In the inner one, the cells are oriented circularly around the intestine. In the outer, the cells have their long axes oriented longitudinally along the intestine. The presence of two outer layers of muscle cells with the cells in one layer oriented perpendicular to the cells in the other makes possible peristalsis and other complex movements of the intestine.

Nervous tissue consists of the impulse-conducting nerve cells (*neurons*) and certain associated cells called the *neuroglia* and *Schwann cells*. Some of the associated cells cover much of the surface of neurons and probably serve to modify neuronal activities.

One of the prominent groups of nerve cells in the intestine is disposed as a layer between the two outer muscle layers. These cells help to integrate the functions of the intestine with those of the rest of the body.

The outer layer of the intestine consists of a thin connective tissue layer within a covering of squamous (flat) epithelial cells. This outer epithelial layer is continuous with the *mesentery*, a thin sheet of tissue which attaches the intestine to the body wall.

3.5.2   **JUNCTIONAL**   Cells of epithelia are held together in sheets
          STRUCTURES   or clusters by special junctional structures.
                            There are several types of such structures; the particular ones present vary in different epithelia. In the absorptive epithelium lining the intestine, there are regions near the lumen where the outermost layers of plasma membranes of adjacent cells fuse, ob-

literating the intercellular space that usually separates adjacent cells (Figs. III-20 and 21). These regions are known as *tight junctions* and occur in a continuous band around each cell. Some distance below them are *desmosomes* (Fig. III-20). In contrast to the tight junctions, desmosomes occur in localized areas, or patches, rather than as a continuous band. In desmosomes the plasma membranes of the two cells are separated by an intercellular space of 100–200 Å or more. The extracellular material within the space may appear slightly denser than elsewhere, and sometimes it seems to be organized in closely associated layers. The adjacent cytoplasm shows a meshwork of filamentous material; additional filaments from other portions of the cytoplasm may also converge upon desmosomes.

In the intestine, and in some other tissues, the juntional structures near the lumen are known collectively as the *junctional complex* (Fig. III-20C); the tight junction is called a *zonula occludens* and the desmosome, a *macula adherens*. The region between desmosome and tight junction includes a *zonula adherens*; its structure generally is not as distinctive as the other regions but often somewhat resembles the structure of desmosomes in that the two plasma membranes are separated by about 200 Å and a band of filamentous material, just below the plasma membrane, encircles each cell.

Desmosomes are probably involved in anchoring cells together in many epithelia. Under experimental conditions (for example, shrinkage), in which cells tend to pull apart, they often remain attached to one another at desmosomes. Tight junctions are also widely found; they may help anchor cells together, but they also have somewhat different functions from desmosomes. It has been shown that tracer molecules—metals dispersed as fine particles—and proteins such as peroxidase (Section 2.1.4) do not penetrate into tight junctions, although they ordinarily pass readily through the spaces between cells. For this reason it is believed that tight junctions can seal off a surface by preventing material from penetrating between the cells; in the intestine, kidney, and other absorptive sites this makes movement through the cells themselves the only means of passage through the epithelium. Movement *through* cells is a more selective mechanism than movement *between* cells. In the latter, the size of a molecule as compared to the dimensions of the intercellular space is the chief limiting factor; in the former, many controlling devices determine movement (see Chapter 2.1).

Many junctions originally thought to be "tight," have been found to permit passage of certain small tracers. Often the tracer used is a suspension of lanthanum salts, containing particles smaller than 20 Å in diameter (Fig. III-21). These junctions are now recognized as a widespread type and are referred to as "gap" junctions (or *nexuses*). A gap of 20–40 Å is present between the adjacent cells. In face view of

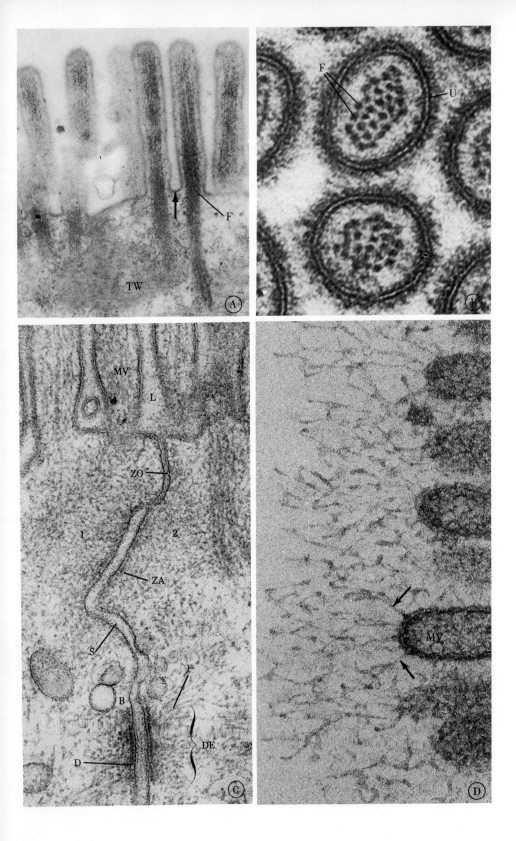

lanthanum-impregnated preparations the gap shows a regular arrangement (lattice) of particles or subunits (Fig. III-21); in freeze-fracture preparations (Fig. III-21) the membranes are seen to be rich in closely packed particles whose arrangement is similar to the lattice in the gap. One model for gap junctions interprets these observations as indicating a regular pattern of bridgelike contacts between the two plasma membranes, and suggests that the contacts might serve as channels across the 20–40-Å extracellular space. The extracellular space itself would thus have the form of a network of passages surrounding the "bridges." This view is prompted in part by the fact that cells associated by gap junctions are electrically "coupled" to one another. Thus the junctions are sometimes called *electrotonic*, or "low-resistance" junctions. An electrical current introduced in one cell passes rapidly into the other. This means that inorganic ions, the usual carriers of bioelectric currents, can cross from one cell to the other. In the absence of gap junctions this generally is not the case, since the plasma membrane is an insulator.

When gap junctions are present, certain dyes can be observed to pass from one cell to another. Evidently in some cases molecules of molecular weights of up to than 600 can traverse the junctions readily, and there may be some passage of even larger molecules. Many biologically important molecules fall within this range or not too far beyond it (for example, the molecular weight of nucleotides like cyclic AMP is less than 350). Therefore the junctions are potentially significant pathways for cell-to-cell communication. In South American electric fish the

◄ *Fig. III-20   Electron micrographs of the regions of intestinal epithelial cells indicated on the diagram in Figure III-19 (see also Fig. II-27). A. The microvilli are fingerlike extensions of the plasma membrane (arrow); each contains a core of filaments (F) that merges into a zone of filaments and amorphous material (the terminal web, TW) in the cytoplasm below the microvilli. × 50,000. (Courtesy of J. D. McNabb.) B. Cross section of two microvilli. The three-layered plasma membrane is seen at (U). The core filaments are seen transversely sectioned at F. × 275,000. (Courtesy of T. M. Mukherjee.) C. Junctional complex between two absorptive cells (1 and 2). MV is a microvillus on one cell and L is part of the lumen of the intestine. The desmosome (DE) and tight junction (zonula occludens, ZO) are seen. ZA indicates a zonula adherens (see text). At the desmosome, cytoplasmic filaments (F) radiate from a dense line (D) adjacent to the plasma membranes (the three-layered membrane appearance is barely evident at this magnification). An additional faint line bisects the intercellular space between the cells (B). × 96,000. (Courtesy of M.Farquhar and G. E. Palade.) D. Tips of microvilli from a section specially prepared to show the polysaccharide-rich surface coat. The coat consists of fine filaments some of which may be seen attached to the plasma membrane (arrow). × 160,000. (Courtesy of S. Ito.)*

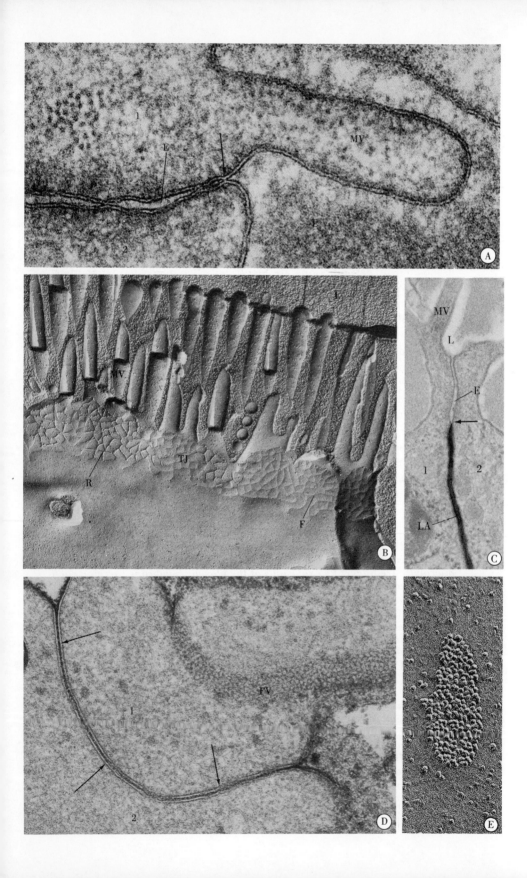

nerve cells in the spinal cord, controlling the discharge of the electric organ, are electrotonically coupled. Stimulation of one neuron causes all to fire and leads to simultaneous electrical discharge of the electric organ cells controlled by the neurons. Low-resistance junctions are also numerous in early embryos, suggesting they may have developmentally significant roles. They also are involved in coordinating contraction of some types of muscle cells (Section 3.9.2).

Certain junctional types other than gap junctions are sometimes suspected to be electrotonic, but there is still dispute about them since gap junctions may also be present on the same cells and account for the electrical properties. In insects, septate desmosomes are seen at which

◄ *Fig. III-21   Cell junctions. A. Region of tight junctions between two adjacent cells (1, 2) of the rat intestinal epithelium. MV indicates a microvillus; L, the lumen of the intestine; and E, the extracellular space between the cells (see Fig. III-19). The plasma membranes bordering the cells show the usual three-layered structure. At several points the adjacent plasma membranes of the two cells are closely apposed to one another and at one of these points (arrow) the membranes have apparently fused, completely obliterating the extracellular space (E). × 170,000. (Courtesy of G. E. Palade.)*
      B. Freeze fracture preparation of the rat intestinal epithelium. MV indicates microvilli, and L, the lumen. The fracture plane has passed through a region of tight junctions between adjacent cells (TJ) providing a face view of membrane structure at these junctions. Where thin sections show focal fusions (see A) freeze-fracture preparations show ridges (R) and furrows (F) in the membranes. (The ridges and furrows represent the complementary membrane faces mentioned in the legend of Figure II-4; the course of the fracture plane through the region is complex so that different membrane faces are seen at different points.) × 40,000. (From D. S. Friend and N. B. Gilula, J. Cell Biology, 53:758.)
      C. Tight junction (arrow) between two adjacent cells (1, 2) of the rat pancreas. MV indicates a microvillus, and L, the lumen into which secretions are released (see Fig. II-22). Electron-opaque lanthanum salts have been introduced into the extracellular spaces of the gland. Much lanthanum has penetrated into the space between the cells (LA) but its progress has been blocked by the tight junction and the extracellular space above the junction (E) is free of the tracer. × 40,000. (From D. S. Friend and N. B. Gilula, J. Cell Biology, 53:758.)
      D. Two rat hepatocytes (1, 2) associated by gap junctions. Each cell is bordered by a three-layered plasma membrane and the two adjacent membranes are separated by a small space or gap (arrows). Electron dense antimony salts (pyroantinomate) have been introduced into the extracellular space and have penetrated between the cells so that the gap appears dark. The area at FV is a gap junction so sectioned that a face view is seen. Here, small particles or subunits are outlined by the tracer. × 130,000. (From D. S. Friend and N. B. Gilula, J. Cell Biology, 53:758.)
      E. Freeze-fracture preparation of a gap junction from a rat ovary cell. The membrane shows the characteristic array of closely packed particles generally seen at gap junctions. × 100,000. (Courtesy of N. B. Gilula.)

the plasma membranes of adjacent cells are widely separated, but fine bridges (*septa*) seem to run from one to the other. (Somewhat similar looking junctions have also been seen in vertebrates.)

There have been a few reports that molecules up to the size of small proteins can traverse septate or gap junctions and thus pass from cell to cell, but some authors feel that the pertinent observations reflect artifacts induced by fixation.

Progress has been made in separating junctional regions from other portions of the cell surface and purifying them by centrifugation. The information that is beginning to accumulate on the specific proteins, lipids, and carbohydrates involved in the various types of junctions should eventually help provide molecular explanations for cell adhesion and communication.

The junctions between the *endothelial* cells lining capillaries are of importance in controlling passage to and from the blood stream of molecules such as sugars or proteins, which cannot diffuse as readily through the cells as water or gases probably can. In some tissues, such as brain tissue, adjacent endothelial cells are associated by tight junctions; in others, various less restrictive arrangements are found, and there are corresponding variations in the movement of material. For example, in the liver most molecules can pass readily through relatively large gaps in the endothelial lining of the modified capillaries, called *sinusoids* (see Figs. I-1 and I-2). This presumably facilitates the extensive exchanges that occur between the hepatocytes and blood stream. (An example of such an exchange is the passage of albumin, a major protein component of blood plasma, which is synthesized and secreted into the blood by hepatocytes.) In many endocrine glands, small interruptions ("fenestrations") are seen in capillary walls. Most muscle capillaries lack such "openings." As mentioned in Section 2.1.4, pinocytic vesicles are widely thought to contribute to the passage of molecules across some capillary walls.

3.5.3   **ABSORPTIVE**   All cells absorb small molecules through the
                CELLS;   plasma membrane by diffusion or by active
       MICROVILLI   processes of the types discussed in Chapter
                            2.1. Many, and perhaps most, eucaryotic cells can also take up macromolecules by pinocytosis. A few cell types show specializations at the luminal surfaces of tubular organs that are clearly related to absorption. Two of these cell types will be considered here, one in the kidney tubule and another in the intestinal tract.

In both cell types the absorptive area at the lumen surface is greatly increased by enormous numbers of microvilli, arranged in pre-

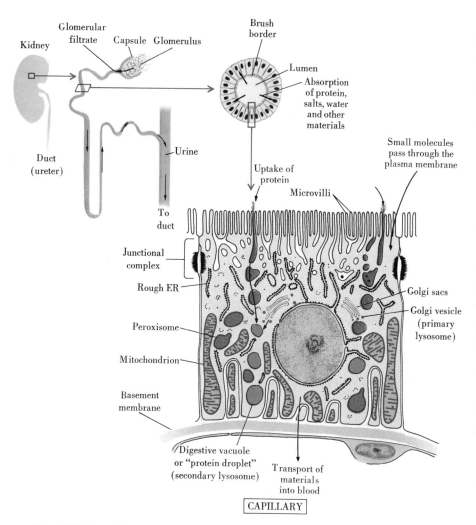

Fig. III-22 *Schematic representation of an absorptive cell of the rat kidney showing the path taken by protein absorbed from the lumen. Through selective absorption of some water, salts, proteins, and other components the filtrate of blood that enters the tubule from the glomerulus (a special capillary system) is modified into urine. Many of the salts and small molecules absorbed by the tubule cells are returned to the blood in the capillaries at the base of the cells. The diagram stresses the uptake and fate of protein which appears to be digested within the tubule cells.*

cise geometric array and collectively constituting the *brush border* or *striated border* (Fig. III-20 and III-22). For the intestine, it is considered that enzymes at the surface of the brush border hydrolyze various carbohydrates and other molecules. The surface is coated with filaments

25–50 Å thick that are expecially prominent in man, cat, and bat. They are attached to the outer surface of the plasma membrane and appear to be rich in acid mucopolysaccharide. The chief role suggested for the filaments is as a filter that keeps large particles from approaching the plasma membrane. However, some enzymes of the brush border may also be associated with the filaments.

The plasma membrane of microvilli in both the intestine and kidney tubules contains cytochemically demonstrable phosphatases (Section 1.2.3) for which roles in transport have been postulated.

A terminological convention should be noted. In absorptive epithelia the *base* of the cell is the region facing the capillaries, and the *free surface* or *apical region* is the opposite pole where absorption occurs. A comparable convention is used for exocrine secretory cells (Section 3.6.1); the apex of such cells is the zone where secretions are stored and released.

The occurrence of microvilli illustrates one evolutionary "solution" to the surface–volume "problem" raised in Section 3.3.2. The effective surface area is greatly increased by specialized cell shapes. Spherical shapes have minimal surface; alterations from the spherical increase the surface-to-volume ratio.

3.5.4   ***PROTEIN***   As Figure III-22 outlines, kidney tubules ***ABSORPTION***   receive a filtrate of blood. The modified capillaries that compose the glomerulus and the thin epithelial layer of the capsule permit water, ions, and small molecules, up to the size of small proteins, to enter the tubules while blood cells and most proteins are retained in the capillaries. Filtrate is modified into urine by the tubule cells which reabsorb water, salts, sugars, and other components and pass many of them back into the blood in the capillaries present near the bases of the cells. A variety of tracer proteins have been followed into kidney epithelium cells, mainly in the *proximal convolutions*, the initial portions of the tubules where much protein (especially in rats) normally is reabsorbed from the urinary fluid as it courses down the tubules. The tracers include radioactively labeled proteins detectable by autoradiography and the proteins ferritin and peroxidase that were discussed previously (Section 2.1.4).

Proteins enter the cell in pinocytic vacuoles that form at the ends of canalicular structures projecting into the cell between the bases of the microvilli (Fig. III-22). These vacuoles merge to form larger *apical vacuoles* that move toward the base of the cell. Perhaps from Golgi vesicles, the apical vacuoles acquire lysosomal hydrolases, thus becoming secondary lysosomes. Proteins and other macromolecules are broken down within the vacuoles, or *protein droplets* as they often are called. The soluble digestive products are probably utilized by the cell.

3.5.5   *TRIGLYCERIDE* Triglycerides are absorbed by the cells lin-
ABSORPTION ing the small intestine. The few pinocytic vacuoles that are seen at the base of the microvilli (Fig. III-20) play little, if any, role in the absorption of triglycerides. Tracer experiments show that the vesicles take up proteins and other macromolecules and deliver them to lysosomes, but since the intestinal lumen contains an abundance of digestive enzymes, such uptake probably is generally of minor importance for digestion of foods.) Triglycerides are digested in the intestinal lumen outside the cell and enter the cell, at the microvillous surfaces, in the form of small molecules: monoglycerides, fatty acids, and glycerol. These move into the ER where they are used in resynthesizing triglycerides and other lipids (see Fig. II-27). The lipids combine with specific proteins, producing lipoproteins within the ER. As outlined in Chapter 2.4, both triglycerides and phospholipids are thought to be made by ER, and protein is made by ribosomes attached to rough ER. In the apical region of the cell there are many continuities between rough and smooth ER (SER), so that materials can move readily between the two. An extensive network of SER is present a short distance below the microvilli. Lipoproteins are transported by SER toward the Golgi apparatus and possibly toward the lateral cell borders. Most of the lipoprotein is converted to droplets called *chylomicra*. The chylomicra apparently empty from the intestinal cells into extracellular spaces by fusion of the membranes that enclose them (Golgi vacuoles or SER) with the plasma membrane. From the spaces the droplets enter the lymphatic system through which they are carried to the blood stream.

It is interesting to note that after a fatty meal has been eaten, when a great deal of fat enters the cell, smooth ER becomes more abundant, possibly by loss of ribosomes from rough ER and by conversion of the cisternae and tubules into an interconnected meshwork and vesicles. This lends support to other evidence (Section 2.4.4) that the two systems, rough and smooth ER, may be interconvertible. After a fat-rich meal, the Golgi apparatus also contains lipoprotein resembling that seen in the ER; the significance of this remains to be determined. Conceivably it reflects addition of carbohydrate components to the chylomicra by the Golgi apparatus, and the packaging processes described in Section 2.5.2.

3.5.6   ***DISTRIBUTION***   In the kidney tubule cell (Fig. III-22)
OF   many elongate mitochondria are vertically
MITOCHONDRIA   aligned, in close relation to the plasma
membrane at the base of the cell, near the
capillaries. The plasma membrane shows deep infoldings. In addition,
the lateral surfaces of adjacent cells fit together by protrusions from one
cell fitting into indentations of the next; this *interdigitation* extends to
the bases of the cells so that a thin section of one cell shows mitochondria apparently in membrane-bounded compartments that are actually
extensions of neighboring cells not included in the section. The
mitochondria have a great many cristae and high levels of oxidative
enzyme activities, and they may be presumed to be producing much
ATP. The abundance of mitochondria and the complex folding of the
membrane are apparently devices that make energy available to a large
area of cell surface which is actively transporting ions and other substances (reabsorbed from the tubules) between the cell and the blood.

In the intestinal cell there are no basal interdigitations, and the
mitochondria show no special orientation at the cell base. The apical
mitochondria are more striking. They are very elongated and are
oriented lengthwise in the cell, parallel to the ER strands that are also
concentrated in the apical cytoplasm. Here, presumably, the mitochondria provide energy for the active processes involved in absorption of
various molecules and, perhaps, in lipid synthesis.

c h a p t e r **3.6**
SECRETORY CELLS

3.6.1   ***VARIETIES OF***   We have encountered several secretory
SECRETIONS   cells in the chapters on ER and Golgi
apparatus (Chapters 2.4 and 2.5). In pan-
creas (Figs. II-22 and II-23), pituitary (Fig. II-33), and intestinal glands
(Fig. II-31), as well as in others, proteins are manufactured on the
polysomes and are transported via the ER to the Golgi sacs (or sac-
derived vacuoles); here they are "condensed" into granules or viscous
fluid and "packaged" into membrane-delimited vacuoles. These vacuoles subsequently open to the surface by fusion of the vacuole membrane and the plasma membrane to discharge the secretion (exocytosis).
We further mentioned the likelihood that coats lining cell surfaces are
also secreted by exocytosis.

Cells that secrete protein have a well-developed ER and Golgi
apparatus. The same is true of many cells that secrete polysaccharide-

rich material, such as the mucus-secreting intestinal goblet cells and *chondroblasts*. The latter synthesize and secrete the mucopolysaccharide chondroitin sulfate, a major component of cartilage. In Chapter 2.5 we outlined evidence that steps in polysaccharide synthesis and "packaging" takes place in the Golgi apparatus, and we pointed out that in many (perhaps most) cells the secretion bodies formed by the Golgi apparatus are mixtures in which both carbohydrates and proteins are present in varying proportions and associations.

There are some obvious differences among secretory cells. *Exocrine* glands release their secretions into special duct systems; the secretions of *endocrine* glands (hormones) directly enter the blood stream. Not all secretions are released by simple processes of membrane fusion; for example, in the *sebaceous glands* (which secrete the oils that coat the skin and hair) the cells fill with secretion then die and disintegrate, releasing their content. Section 2.8.2 referred to the complex path taken by the secretions of the thyroid gland. The gland first releases *thyroglobulin* to an extracellular storage lumen, then takes it back into the thyroid cells, digests it, and releases the digestion products to the blood stream. Thyroglobulin is a glycoprotein to which iodines are attached through the action of a peroxidase enzyme (Fig. II-26A). An interesting possibility is that this enzyme is released to the storage lumen directly from the ER, through transport in small vesicles that fuse with the plasma membrane; in the lumen it can act to iodinate the thyroglobulin molecules.

The connective tissue components formed and secreted by fibroblasts include polysaccharides, such as hyaluronic acid, and the protein *procollagen*. The procollagen is synthesized on bound ribosomes and is transported in the ER to the Golgi apparatus; its presence in these sites can be demonstrated by immunohistochemistry (p. 22). Immunohistochemical techniques also permit the demonstration of the enzyme prolyl hydroxylase in the ER. This enzyme modifies the amino acid proline in the procollagen, converting many of them to hydroxyprolines, the form characteristic of the secreted protein. Procollagen gives rise to the extracellular collagen fibers of connective tissues. During or soon after its release from the cell it is cleaved by a proteolytic enzyme (peptidase) to form tropocollagen, which assembles into the fibers (Section 4.1.3). In our earlier discussion of insulin (Section 2.4.1) we pointed out that enzymatic cleavages occur as part of the intracellular processing of a number of proteins.

In some secretory cells the membrane-delimited granules formed by the ER and Golgi apparatus contain enzymes that subsequently participate in the synthesis of the cell's characteristic secretion. This is true of the adrenal medulla cells that produce epinephrine (adrenaline), a low molecular weight hormone that is synthesized from the amino acid tyrosine. (See Section 2.1.6 for the effects of epinephrine on hepatocytes.) Key steps in epinephrine synthesis occur in the secretion granules, and the hormone becomes bound there, complexed with ions and proteins; the complexes are released from the cell by exocytosis.

Some secretory cells, such as those secreting ions, have relatively little ER, and the Golgi apparatus is small. An important example is the *parietal cell* of the vertebrate stomach which secretes HCl into the stomach where the acid activates the digestive enzyme *pepsin*. The cell surface exposed to the lumen is enormously enlarged by a great many microvilli and by deep infoldings into the cell (*canaliculi*) (Fig. III-23). *Active transport*, in which the cell expends energy (Section 2.1.3), is involved in moving hydrogen ions out of the cell into the lumen. It is hardly surprising, therefore, that parietal cells contain many stout mitochondria with numerous cristae and high levels of oxidative enzymes. Within the cytoplasm, numerous tubules delimited by smooth membranes are present. Some believe that ions are secreted into the tubules and are stored there, bound to glycoproteins and other macromolecules bearing charges opposite to the charges of the secreted ions. Early literature, especially concerning the frog, suggested that the tubules could establish continuities with the plasma membrane, but recent studies have cast some doubt on this and the situation is pres-

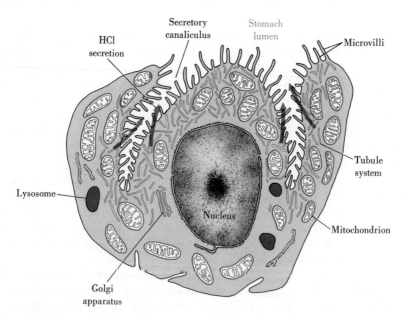

Fig. III-23 *Schematic diagram of a* parietal cell *from the mammalian* stomach. *(After S. Ito and R. C. Winchester.)*

ently uncertain. The two membrane systems are, however, extremely close to each other. In addition, it is clear that with the stimulation of hydrogen ion secretion the tubular system decreases in amount and the microvilli of the secretory canaliculi become longer. When ion secretion is decreased, or inhibited, the reverse occurs.

The tubule system differs from the scanty ER in several respects. There are thickness differences in the membranes delimiting the two systems, and they show different staining reactions for specific phosphatases found in ER (and in the Golgi apparatus). The tubule system membrane is similar to the plasma membrane but thicker than that of the ER (including the nuclear envelope). Golgi saccules, as in other cells (Figs. I-15 and I-16), show IDPase and TPPase activities, but no such activities are seen in the tubule system.

Marine birds (and some turtles) possess salt glands that function in salt excretion. The cells show numerous mitochondria and deeply infolded surfaces, presumably providing a large area for ion transport. Earlier we mentioned the folding of the basal portion of the kidney tubule cell (Section 3.5.6) which probably has the same significance.

For some important secretions, such as the steroid hormones, very little is known of the secretory mechanisms. The pertinent cells are rich in smooth ER (Section 2.4.4) and have a prominent Golgi apparatus.

3.6.2   *CNIDOBLASTS*   An interesting modified secretory mecha-
           IN HYDRA   nism has been described in the primitive an-
                    imal *Hydra*. The organism is made essen-
tially of two layers of epithelium. In the outer layer there are special cells, called *cnidoblasts* that produce small projectiles, *nematocysts*, that are shot out, to pierce and paralyze prey.

The cnidoblasts develop from undifferentiated precursor cells, the interstitial cells. The ER, sparse in the primitive cell, becomes extensively developed in the maturing cnidoblast. Ribosomes usually lie free in the cytoplasm in the primitive cell, but are arranged on the ER membranes during maturation (see Section 2.3.6). The Golgi apparatus also becomes highly developed as the cell begins to secrete the proteins that are stored in the *nematocyst*. The latter apparently begins as a Golgi vacuole, small at first and then enlarging greatly. Innumerable small vesicles develop from the much enlarged Golgi saccules and fuse with the nematocyst. As the nematocyst enlarges, its content becomes more electron dense (Fig. III-24). (A great many microtubules appear outside the nematocyst; they probably give rigidity to the area in which the secretory vacuole is rapidly enlarging.) When the nematocyst attains its maximum size, the ER and Golgi apparatus regress, breaking into vesi-

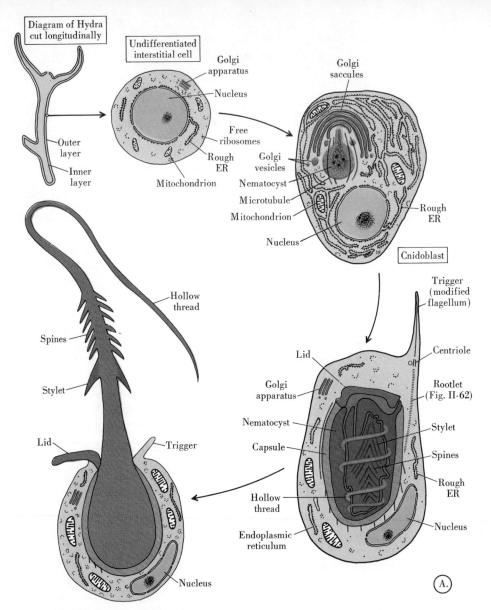

Fig. III-24 **A.** *Diagram illustrating the maturation of a nematocyst within a "cnidoblast". As the nematocyst increases in size there is a great increase in extent of rough ER and size of Golgi saccules and vacuoles. It is after the recession in development of ER and Golgi apparatus that the nematocyst undergoes the major part of its development of complex structure. (Based on studies of Hydra and other coelenterates by Slautterback, Westfall, Picken, Skaer, and Lentz.)* **B.** *Developing nematocyst (NC) in a cnidoblast of Hydra. Oriented microtubules surround the nematocyst; in this section most are sectioned transversely (T). As indicated in* **A,** *the nematocyst forms in association with the Golgi apparatus. Golgi saccules (G) and vesicles (V) are abundant. Much rough ER also is present (E). The arrow indicates a vesicle probably budding from the ER and contributing to the Golgi apparatus (or developing nematocyst; see Fig. II-22).* × 30,000. *(Courtesy of D. B. Slautterback.)*

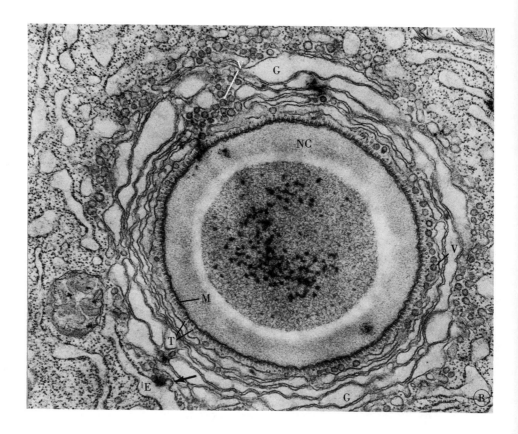

cles which progressively diminish in number. Upon triggering, the
nematocyst releases its contents to the extracellular environment (lower
left diagram in Fig. III-24A).

The striking development of the ER and Golgi apparatus, followed
by their virtual disappearance, is but one of the interesting features of
cnidoblasts. Another is the dramatic structural differentiation that occurs inside the nematocyst, *without apparent connections to other cell
organelles.* (Speculation on how this might occur is found in Section
4.1.6.) Thus far, little attention has been given to the manner by which
this differentiation occurs. Yet it seems likely to be a genetically determined process, involving a great many secretory proteins. In Hydra
there are four distinct types of nematocysts. In 100 species of coelenterates related to Hydra, 17 types of nematocyts have been described.
Each type is characterized by a structure that is characteristic for the
particular species. In one of the abundant nematocysts in Hydra, the
complex structure includes a capsule (which apparently contains a variety of phosphatases), a lid, a coiled thread, large stylets, and smaller
barbs of specific shapes (Fig. III-24).

3.6.3 *EXOCYTOSIS;* The release of secretions from cells is under
CONTROLS precise controls, but the mechanisms are
AND not well understood. In Hydra, a *cnidocil*
MEMBRANES protrudes from the cnidoblast surface; in
part this is a modified cilium. Nematocyst
release occurs when the cell is stimulated by the appropriate chemical
and mechanical stimuli, such as might result from the presence of the
small organisms used as food. The cnidocil is probably a mechanical
receptor, and additional chemical receptors are presumably built into
the plasma membrane. Release is also influenced by the primitive nervous system found in Hydra.

In higher organisms, nervous and hormonal stimulation strongly
influence the rates of release of different secretions, but how such extracellular influences can bring about appropriately controlled fusion of
membranes surrounding secretion granules with the plasma membranes
is not understood. Cells stimulated to secrete may show higher levels of
Ca^{++} ions due to changed plasma membrane permeability or to release
of Ca^{++} from internal "stores". Altered levels of molecules such as
cyclic AMP and cyclic GMP are also found. Secretion can be disturbed
by disruptive agents such as colchicine, which affects microtubules;
this could indicate that microtubules are involved in the motion of secretory structures to the cell surface. Microfilaments may also participate. But most available evidence is circumstantial, and most details of
the mechanisms are still unknown.

By conventional electron microscopy, exocytosis seems to involve
the sequential fusion of the three layers of the membrane delimiting a
secretion granule with the three layers of the plasma membrane.
Freeze-fracture techniques suggest that there may sometimes be special arrangements of membrane particles at the points of fusion. Circular arrays have been noted in the plasma membranes of the protozoan
Tetrahymena at the points where exocytic release of secretory bodies
(mucocysts) occurs.

Since actively secreting cells show only transient increases in their
surface areas, subsequent to exocytosis membrane must be withdrawn
from the cell surface. One way in which this can occur is through
endocytosis-like processes. Numerous vesicles and tubules appear in
the cytoplasm of actively secreting cells, and when tracers such as
horseradish peroxidase (p. 56), are present in the extracellular space
these structures are labeled, demonstrating their endocytic origin.

3.6.4 *ANTIBODIES* The immunological system of vertebrates
provides a means by which the organism
recognizes invading microorganisms and other foreign molecules or

cells, and initiates their sequestration and destruction. The most clearly understood phase of this complex system involves the production of antibodies, glycoproteins known as *immunoglobulins*. There are several related classes of immunoglobulins. Figure III-25 diagrams the structure of the IgG class, the best known of them. A molecule which evokes the production of an antibody that binds to it specifically is referred to as an *antigen*.

Immune phenomena involve a variety of different mechanisms. Thus rejection of tissue grafts or transplanted organs result from "cell-mediated" immune responses that are still incompletely understood. Blood cells called *T-lymphocytes* (because they originate in the thymus) are central participants in the recognition and destruction of the foreign tissue. The T-lymphocytes do not secrete antibodies, but they apparently do release chemical factors that induce the rapid entrance of phagocytes into the tissue from the blood. They also aid in the killing of the foreign tissue.

Generally responses to foreign organisms or macromolecules depend upon the production of antibodies that circulate in the blood stream. These form complexes with the corresponding antigens. The complexes interact with receptors on the cell surfaces of phagocytes, which induces binding and endocytosis of the complexes at rates much more rapid than is true for the antigen alone. Antigen-antibody complexes can also interact with a set of circulating blood proteins known collectively as *complement*. This initiates an ordered series of enzymatic and other reactions in the complement proteins, which has numerous biologic effects including attraction of phagocytes, inflammation, and the production of lytic factors that can kill foreign cells.

Circulating antibodies are secreted by plasma cells in the spleen, lymph nodes, and several other sites (Fig. II-26B). The plasma cells derive from B-lymphocytes (called B because they were initially analyzed in the bursa, an immunological organ of birds). As Figure II-26B indicates, plasma cells synthesize immunoglobulins in the rough ER. They lack distinctive secretion granules, and the pathways taken from the ER to the cell exterior are still unclear.

A fundamental question is how the specificity of the immunoglobulins is achieved. Figure III-25 illustrates some current views. The key point is that evolution has equipped organisms with the innate capacity to produce an enormous variety of antibodies. Presumably the range relates to the variety of important antigens which the species has encountered in its evolutionary history. As an organism matures, different clones of B-lymphocytes arise from stem cells that originate in the bone marrow. A clone consists of the progeny produced by division from a single initial cell. Through genetic mechanisms, different clones synthesize immunoglobulins with different amino acid sequences; each clone produces one fixed sequence. Antigens do not "instruct" the organism

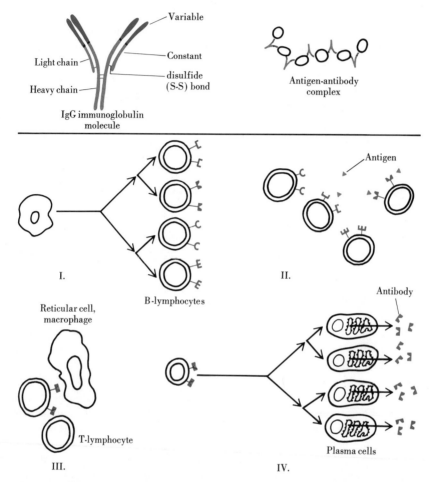

Fig. III-25   *Antibodies. Upper panel: IgG molecules are the predominant type of antibody found in the circulation. One IgG molecule consists of a pair of light chains (each with approximately 215 amino acids) and a pair of heavy chains (approximately 440 amino acids each). The amino acid sequences in the variable regions (red) differ among different IgG's, accounting for their abilities to bind to different antigens. The constant regions do not vary.*

Lower panel: Production of antibodies. I. B-lymphocytes arise by division of precursor cells whose origin in mammals is still uncertain. Different B-lymphocytes have different cell surface receptors, probably immunoglobulins. II. An antigen encounters B-lymphocytes in a wound or in spleen or lymph nodes. III. The antigen interacts with a B-lymphocyte carrying receptors than can bind the particular antigen. Other cells play roles in this interaction but the natures of these roles are not yet elucidated. IV. The lymphocyte, stimulated by the antigen binding, divides and differentiates into plasma cells. These plasma cells secrete antibodies of specificity similar to the surface receptors on the lymphocyte (these antibodies thus can complex with the antigen).

or its cells to make antibodies of a particular specificity. Rather, the specificity is already present in one or a few B-lymphocytes. Contact with the antigen stimulates the appropriate lymphocytes to divide and differentiate into plasma cells.

This analysis implies that the organism's antibody specificities are not limitless. A given organism can make thousands of different immunoglobulins, but within this set of molecules there may be overlaps in specificities and also gaps, which are expressed as poor immunologic responses to certain antigens.

The genetic mechanisms underlying antibody synthesis are still unsettled. The simplest explanation for the fact that different IgG molecules have the same amino acid arrangement in the *constant regions* of the molecules (Fig. III-25) is that the same DNA sequences are used by different clones of lymphocytes to code for these regions. As for the *variable regions* of the molecules, it has been argued that each of the progenitor cells giving rise to a clone of B-lymphocytes possesses DNA sequences for all, or many, of the possible variable regions. Perhaps as B-lymphocytes mature they somehow come to utilize only one of these sequences, linking it to a constant region DNA sequence. On the other hand, many investigators favor the proposition that at least some of the variable DNA regions do not preexist. Rather, they believe that variability in DNA sequences arises in some manner during B-lymphocyte development.

The detailed events by which antigens stimulate proliferation and differentiation of B-lymphocytes are yet to be ascertained. In various experimental situations the effective interaction of antigens and B-lymphocytes is found to depend upon the presence of other cells such as T-lymphocytes and macrophages (p. 141). In the spleen it is the antigen bound to the surfaces of "reticular cells," a large cell with extensive branched processes, that seems to stimulate the B-lymphocytes. The molecular mechanisms underlying such cell cooperation are being sought.

c h a p t e r **3.7**
NERVE CELLS

Neurons include the longest cells in the body, ranging to several feet in length. They are the coordinating elements of an elaborately interconnected network. The network consists of sensory receptor cells feeding into sensory neurons, enormous numbers of connecting and integrating neuronal circuits, and neurons leading to effector organs such as muscle. A given neuron can interact, more or less directly, with hundreds of other cells.

At the level of individual cells, the developmental processes responsible for the establishment of specific neuronal geometry and connections probably depend upon factors such as the times in development at which particular cells appear, grow, migrate and contact one another, cell surface "recognition" phenomena (Section 2.1.6), mechanical influences exerted upon a given cell by its neighbors, and the release by one cell of molecules that affect the direction of growth or migration of another (*chemotactic* interactions). A major task for modern biology is to determine how such factors operate and are integrated to produce a functioning nervous system.

Figure III-26 shows two of the major morphological types of neurons, *unipolar* and *multipolar*. Many unipolar neurons, such as vertebrate sensory neurons, are characterized by a single process that divides into two branches. One branch receives input from a sensory receptor, and the other connects to other neurons via *synapses*. Multipolar neurons like the motor neurons of the spinal cord have numerous receptor processes (*dendrites*), which receive impulses at synapses with other neurons, and a single transmitter process (the *axon*), which carries impulses from the cell body to effectors. Both types of neurons are fundamentally similar in other respects.

3.7.1   **PERIKARYON**   The neuronal cell body, or *perikaryon*, contains the cell nucleus and cytoplasmic organelles distributed around the nucleus in an arrangement that is roughly symmetrical (Fig. III-26). The Golgi apparatus is a well-developed network (Fig. I-16). Rough ER is extensive; it is concentrated in the so-called *Nissl substance*, patches (Fig. I-13) showing parallel rough cisternae plus many free ribosomes. Many lysosomes are present and tend to be concentrated in the Golgi zone. Mitochondria are numerous throughout the cell. In general the cytological characteristics suggest extensive macromolecular synthesis: large nucleolus, many ribosomes, extensive rough ER, and large Golgi apparatus. This is to be expected from the role of the perikaryon as a synthetic center that supplies macromolecules and other material to the rest of the neuron.

The Golgi apparatus and associated endoplasmic reticulum (GERL) appear to participate in forming the numerous lysosomes. As in other cells, autoradiographic studies indicate that some protein made in the rough ER passes into the Golgi apparatus of neurons (Fig. II-32). In special *neurosecretory* neurons, found in the pituitary gland and elsewhere, the Golgi apparatus packages hormone granules which travel down the axons and which are released at the axon endings to

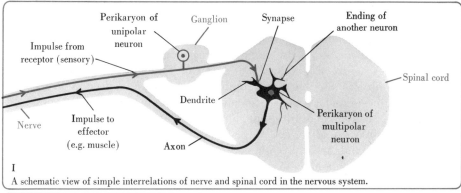

I

A schematic view of simple interrelations of nerve and spinal cord in the nervous system.

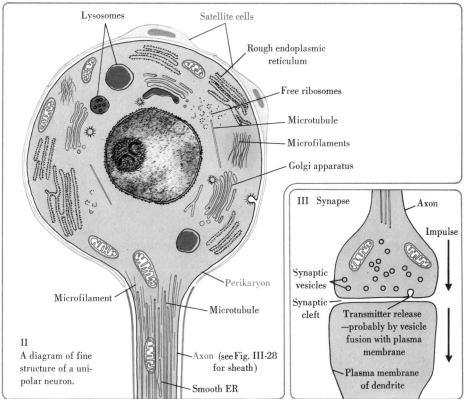

II

A diagram of fine structure of a unipolar neuron.

Fig. III-26   *Nerve cells.*

enter extracellular spaces or the blood stream. In most neurons the role of the Golgi apparatus is less clear, although at least some of the substances involved in the transmission of nerve impulses from one cell to another appear to be packaged in membrane-delimited structures that originate in it. As discussed below, transmission is based on phenomena akin to secretion which occur at axon endings.

The abundance of lysosomes is of uncertain significance. Some probably participate in the extensive turnover of membranes that accompanies neuronal function (Section 3.7.5). Many of the lysosomes of neurons are residual bodies. Neurons do not divide, and many live as long as the organism, slowly accumulating residual bodies. As mentioned in Section 2.8.5, with time, lipofuscin ("aging pigment") accumulates within residual bodies.

The abundance of RNA in neuron perikarya has led to the speculation that learning and memory might be based on the coding of information in nucleic acids. Unfortunately, experimental tests of this idea are difficult and the results are equivocal. In some learning experiments in which animals are trained to perform specific acts, changes in the composition and amount of RNA are observed. However, these could reflect general effects on neuronal metabolism in which RNA is involved rather than the specific synthesis of a "memory" RNA. Protein synthesis is thought to be involved in some phase of "memory storage." There also have been experiments in which some investigators believe that they have transferred "memory" by injecting into one organism RNA purified from another. The more widely held theory of learning considers that new experiences are "stored" in the form of connections between neurons that are either newly formed after a novel experience or in some sense newly activated. There is indirect evidence that seems to support this theory, but fuller evaluation will require an understanding of the controls of nerve growth and function. The complexity of nervous organization in organisms capable of observable learning has thus far hampered direct testing of this theory; it is difficult to trace the neuronal circuits that might be involved.

3.7.2   ***AXONS***   In contrast to the cell body, the axon contains neither Golgi apparatus nor rough ER. There are few ribosomes. On the other hand, microtubules and fine filaments are numerous; they are oriented along the long axis of the axon. There are some smooth-surfaced tubules and vesicles, probably including smooth ER. Mitochondria are frequent, lysosomes infrequent. The axon appears to be largely inactive in the synthesis of protein or other macromolecules.

Studies of living nerves indicate extensive cytoplasmic flow in axons. Although it occurs in both directions, a net flow down the axon from the perikaryon results. Thus it appears that the perikaryon is continually manufacturing molecules needed for the maintenance and functioning of the axons. These molecules are transported down the axon. Even mitochondria and other organelles are believed to pass into the axon from the perikaryon. Different components move at rates ranging from one to many millimeters per day, but the motive elements have not been identified. Probably the filaments, microtubules, and smooth ER are involved. Such continual manufacture and transport of axoplasm is an enormous metabolic task for the perikaryon, especially since the cell body represents a small fraction of the total cell volume.

Since neurons do not divide and no reservoir of undifferentiated "precursor" cells exists, if the perikaryon is destroyed, the neuron is not replaced. However, if the axons are cut, only the portion no longer attached to the perikaryon degenerates; the remainder is capable of regenerating. Under appropriate conditions, neurons of the peripheral nervous system (nerves and ganglia) can reestablish some of the original connections and so restore neuron function. Proper conditions for full regeneration and reestablishment of the connections of neurons of the mammalian central nervous system (brain and spinal cord) have not been found.

3.7.3   ***THE NERVE*** The most readily observable feature of the
IMPULSE passage of a nerve impulse along an axon
is a change in the potential difference that exists across the plasma membrane; this change (*action potential*) moves like a wave along the axon surface. Current theory holds that the underlying mechanism is based on the flow of ions accompanying a wave of permeability changes that passes down the axon membrane. In resting state, K^+ ion concentration is maintained high and Na^+ concentration low in the axon (as compared with the extracellular space) by an energy-requiring active transport mechanism (Section 2.1.3). A somewhat oversimple but useful view of the consequences of this asymmetric distribution of ions is as follows. It can be shown in model experimental systems and by theoretical treatment that when such asymmetries are established across a membrane, an electrical potential will result if some of the ions can diffuse through the membrane more rapidly than others. K^+ can pass across the axon membrane much more rapidly than can Na^+. (What is being considered here is the passive movement down concentration gradients that continually tends to restore equal concentrations on both sides of the membrane.) The tendency for K^+ to leave the axon more rapidly than Na^+ enters produces, at equilibrium, a net relative deficiency of positive charges on the inside of the membrane. In consequence the inside of the cell membrane is at an electrically negative potential as compared with the outside; this is referred to as the

resting potential. Impulse conduction by a given region of the axon is based on the following sequence (Fig. III-27):

(1) A localized permeability change permits the rapid influx of Na⁺. This influx leads to changes in the potential across the membrane; eventually the inside becomes positive with respect to the outside.

(2) A second set of permeability changes restricts the movement of Na⁺ but permits the rapid efflux of K⁺. This is associated with eventual restoration of the original negativity of the inside.

(3) Propagation of the impulse to adjacent axon regions occurs. This results from the fact that the local changes of (1) and (2) produce an axon region that differs in ion concentration and electrical potential from adjacent regions. A "flow of ionic current" takes place between

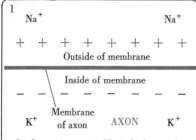

In the resting axon, Na is high outside and K high inside. The inside of the membrane is at a negative electrical potential compared to the outside.

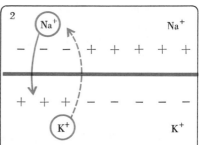

As the impulse reaches a given region the membrane permeability changes. Na enters and the potential difference reverses. Changes in permeability to K follow.

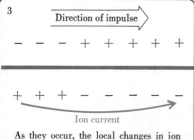

As they occur, the local changes in ion concentration and in potential lead to longitudinal currents of ions (or equivalent phenomena) which trigger permeability changes in adjacent regions of the membrane.

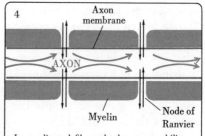

In myelinated fibers the key permeability changes take place at the nodes where the insulating myelin sheath is interrupted. Between nodes longitudinal current flow carries the impulse which thus "jumps" from node to node.

Fig. III-27   *Outline of nerve impulse propagation. Currents that flow outside the axon accompanying the currents inside are not shown.*

this region and the adjacent regions. The flow takes place both within the axon and outside it. It results in initiation of the cycle of permeability changes [(1) and (2)] in membrane regions adjacent to the local region under consideration. Thus the impulse moves along the membrane. (Exactly how a longitudinal ion flow can trigger the cycle is not known. However, the process probably depends on the fact that the flow decreases the potential difference across the membrane, and that decreases in potential difference can trigger nerve impulses.)

(4) The intracellular low Na–high K condition is restored, relatively slowly, by a "sodium pump" which produces active pumping out of Na^+ and reciprocal influx of K^+. This can be relatively slow since only a small percentage of the ions move in a single impulse. The very large initial asymmetry in Na and K distributions can permit the passage of many impulses before pumping becomes absolutely essential to further functioning. Normally the cycle of changes in permeability at a given axon region is completed in a few thousandths of a second. The impulse passes down the axons of different nerves at rates of tenths to tens of meters per second.

Only in the last step related to impulse propagation [Step (4) above] does the cell actively expend energy. In the living cell the mitochondria probably provide energy for the sodium pump, but the axon membrane itself (perhaps together with some closely associated material) seems to be solely responsible for conduction of impulses. This has been nicely demonstrated by using the giant axons, 500 μm in diameter, found in squid. Virtually all of the intracellular organelles and other cytoplasm can be removed from such axons and replaced by suitable solutions of ions in which the Na^+ concentration is kept low and the K^+ concentration high; the preparations still can conduct impulses quite efficiently. The nature of changes of the axon membrane that are responsible for passage of impulses has yet to be determined. Very probably, rapid realignment or changes in shape of membrane molecules produce the observed permeability changes. The identity of these molecules and the mechanisms whereby they change are being actively sought. One proposal is that membrane molecules contain charged regions which change their alignment in response to changes in electrical fields, such as those occurring when an impulse passes down the axon; another proposal is that specific chemical reactions take place.

3.7.4 **MYELIN** In vertebrate nerves two major types of axons are found, *myelinated* and *unmyelinated*. In both, *Schwann cells* (or *neuroglial* cells in the central nervous system) are closely associated with the axon. In the unmyelinated type, the axon occupies a pocket formed by the indentation of the Schwann cells (Fig. III-28). The Schwann cells of the myelinated axons wrap repeatedly around the axon, surrounding it with many layers of their plasma membranes. The multilayered membrane system constitutes myelin (Fig. III-28); it is part of the Schwann cell. The rest of the cytoplasm and the nucleus remain at the outer surface of the multiple layers of plasma membrane. Each Schwann cell contributes one segment of myelin to the sheath that covers most of the length of the axon. The segments of sheath contributed by two Schwann cells abut, but do not fuse, at *nodes of Ranvier*. The lipoprotein of which myelin is made is a good electrical insulator; unlike most membranes, there may be considerably more lipid than protein, by weight (Sections 2.1.2 and 2.1.6). Thus a myelinated fiber is surrounded by a layer of insulating material that is interrupted at the nodes. This modifies the conduction of the nerve impulse. Current tends to flow preferentially along paths of low resistance, and in myelinated axons the nodes define such a path. The changes in membrane permeability observable at the nodes are thought to be responsible for conduction (Fig. III-27). This kind of conduction, by "jumping" from node to node, is called *saltatory transmission* (saltation refers to jumping); each node may be thought of as amplifying the impulse and passing it down the axon to the next node. The permeability change at one node results in local concentration changes and sets off a flow of ionic current within the axon. This in turn triggers the permeability change at the next node. The "jumping" is responsible for the more rapid impulse conduction by myelinated nerves, as compared with unmyelinated nerves; the latter lack the extensive insulation that makes saltatory transmission possible.

Fig. III-28   *Myelin. A. Electron micrographs of a portion of sciatic nerve* ▶
(guinea pig) sectioned transversely. A myelinated axon (A) and several unmyelinated axons (UA) are shown. S indicates Schwann cell cytoplasm. A narrow space continuous with the extracellular space outside the Schwann cell (arrow) separates unmyelinated axons from the Schwann cells that surround them. The arrangement often resembles that shown in the lower left diagram. Myelin (M; enlarged at right) is of multiple membrane layers. Mitochondria in the axon are seen at MI. × 20,000; insert × 50,000. (Courtesy of H. Webster.) B. Schematic diagram of myelin formation by the establishment of a multilayered spiral-like pattern of Schwann cell plasma membrane. The cytoplasmic space and extracellular space originally separating membrane layers are obliterated so that the final pattern consists of layers where the original external surfaces of the membrane are closely apposed, alternating with layers where the original cytoplasmic surfaces are closely apposed. In mammals the myelin segment contributed by a single Schwann cell (this corresponds to the distance between successive nodes) may cover a length of several hundred micrometers of axon.

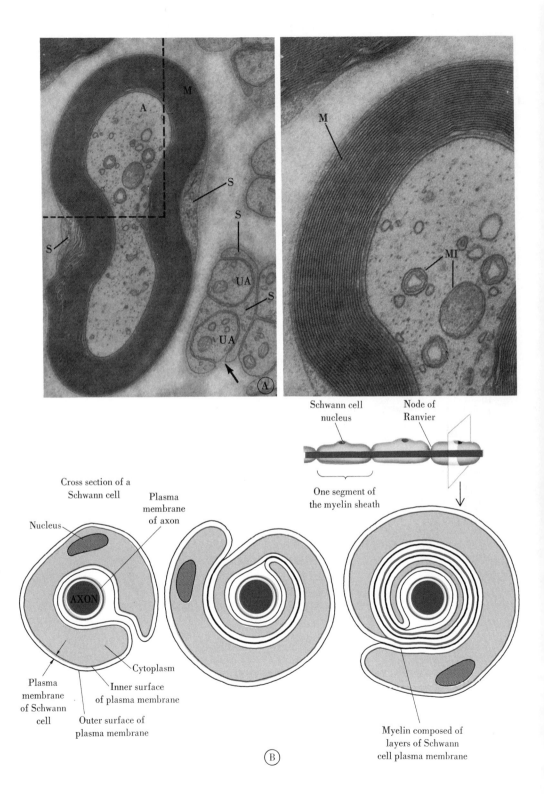

Schwann cell
nucleus

Node of
Ranvier

One segment of
the myelin sheath

Cross section of a
Schwann cell

Plasma
membrane
of axon

Nucleus

AXON

Plasma
membrane
of Schwann
cell

Inner surface
of plasma membrane

Cytoplasm

Outer surface of
plasma membrane

Myelin composed of
layers of Schwann
cell plasma membrane

A

B

3.7.5   ***SYNAPSES*** The impulse travels from one neuron to
                            another through synapses. Transmission
from cell to cell, in a few cases, appears to involve junctions of the
electrotonic junction type (Section 3.5.2). More commonly, however,
such direct transmission is precluded by the absence of the requisite
special junctions. Instead chemical transmission takes place. *Acetyl-*
choline and *norepinephrine* (a compound related closely to epinephrine
[p. 240]) are among the known chemical transmitters. Different classes
of neurons utilize different transmitters. Most of the known or suspected
ones are the size of amino acids or slightly larger. (Some are amino
acids, such as glycine and glutamic acid.) A synapse is diagrammed in
Figure III-26. The numerous vesicles are generally believed to hold
neurotransmitter substances. When the impulse reaches the synapse,
the vesicles apparently fuse with the membrane and release the transmitter into the extracellular "synaptic space." Involved in this is an
influx of Ca^{++} ions into the synapse. The transmitter interacts with
specific receptor sites on the plasma membrane of the "receiving" cell
(the *postsynaptic* plasma membrane). The effect varies depending on
the transmitter and receptor. Receptors on neurons or muscle cells that
respond to acetylcholine mediate initiation of an impulse. The initiation
mechanism is not known, but probably is based on changes in membrane permeability. Perhaps the interaction of transmitter and receptor
transiently opens a "pore" (Section 2.1.3). An enzyme, *acetylcholines-*
terase, is present in synapses of neurons transmitting by acetylcholine.
This enzyme can destroy acetylcholine very soon after it is released.
This gives an important measure of control to the system. If the acetylcholine remained intact in the synapse, it would continue to stimulate
impulses in the receiving cell. Thus, for example, one nerve impulse
might set off a long series of twiches in a muscle rather than the single
one that is usually obtained. Axon endings that release norepinephrine
also can absorb this compound; probably this is important in control and
conservation mechanisms.

While most investigators agree that the synaptic vesicles are central agents in transmitter release, there is still much to learn. For example, extensive synthesis of neurotransmitter molecules occurs in axon
terminals at synapses. Certain of the enzymes for such synthesis are
associated with the vesicles, and the norepinephrine vesicles seem also
to contain proteins and ions that can complex with the transmitter
molecule, forming aggregates that might be important in transmitter
storage. However, details of where and how transmitters are made and
stored are in dispute. Similarly, it is reasonably well established that the
macromolecules of synaptic vesicles are produced chiefly in the perikaryon and transported to the synapses, but the life history of the vesicles
is incompletely understood. Some seem to be formed in the perikaryon

in the Golgi zone, but there may be additional sites of vesicle formation such as the axonal smooth ER. Subsequent to fusion with the plasma membrane during neurotransmission, vesicle membrane is "retrieved" from the cell surface by endocytic-like processes (Section 3.6.3). It is widely presumed that the retrieved membrane can be reutilized to form functional synaptic vesicles; eventually the membrane is degraded, perhaps by transport back to the perikaryon and incorporation in lysosomes.

chapter **3.8**
SENSORY CELLS

A wide variety of sensory receptor cells occurs throughout the body. Each is specialized so that it responds to a given type of stimulus, initiating a nerve impulse. Some respond to mechanical pressure, others to specific classes of chemicals. One of the most interesting responds to light. This is the *retinal rod cell*, capable of mediating vision at low illumination levels. Another similar set of retinal cells, the *cones*, functions chiefly in high illumination and color vision. These photoreceptive cells line the retina, and they make synaptic contact with connecting cells that synapse with neurons of the optic nerve.

3.8.1   ***THE RETINAL***   The rod cell (Fig. III-29) consists of a cell
              ROD   body with a nucleus, numerous mitochondria and other organelles, connected by a cytoplasmic bridge or stalk to an *outer segment* specialized for light reception. At the base of the connection, a basal body with centriolelike structure is found, and within the cytoplasmic stalk, an arrangement of microtubules characteristic of cilia is present. From this appearance and from developmental studies, it has been concluded that the outer segment of the rod is a greatly modified derivative of a cilium. Other cilia of sensory type are known, but often the modifications are not as great as in the rod. Sensory cilia often lack the central pair of microtubules characteristic of most cilia and flagella. (See Section 3.6.2 for another example.) They are spoken of as 9 + 0 cilia, in contrast to the 9 + 2 cilia discussed earlier (Chapter 2.10).

The rod outer segment is a cylindrical body, delimited by a plasma membrane, containing several hundred flattened sacs (discs) stacked on top of one another (Fig. III-29). The membrane-bounded discs contain molecules of the visual pigment *rhodopsin*. The discs apparently origi-

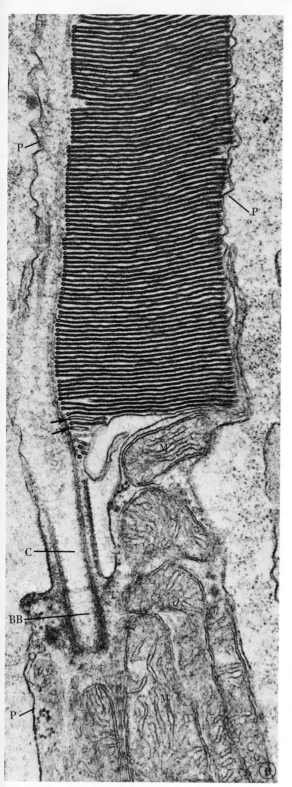

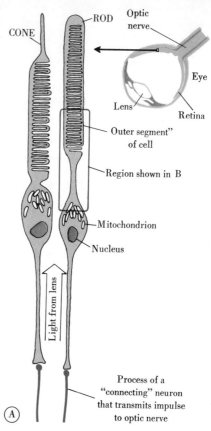

Fig. III-29 *Retinal receptor cells of vertebrates.* **A.** *Schematic diagram of a portion of the retina. The receptor cells have their light-sensitive portions directed toward the back of the eye. As indicated in the diagram, the stacked membrane systems in the cones remain connected to the plasma membrane.* **B.** *An electron micrograph of a portion of a rod cell like that outlined in* **A.** *The basal body (BB) and cilium (C) that connect the inner and outer segments of the cell are seen in longitudinal section. Many mitochondria (M) are present near this connection. P indicates the plasma membrane. The outer segment contains the stacked membrane systems (discs) in which the photo receptive pigments are located. These appear to form as sacs (arrows) which eventually flatten, obliterating the space within. (Courtesy of D. W. Fawcett.)*

nate as plasma membrane infoldings which then pinch off. (The chief morphological difference between cone cells and rods is the persistence of continuity of the stacked sacs with the cone cells' plasma membranes.) The rod discs gradually pass up the outer segment, reaching the tip in one to several weeks, depending on the species. At the tip the discs are shed from the rods and are phagocytosed and degraded by cells of the *pigment epithelium* that surround the tips of the photoreceptors.

The arrangement of the photoreactive material in stacked membranes provides a large surface where light can be absorbed. Studies with polarizing microscopes have shown that the rhodopsin molecules within the membrane exist in orientations that provide extraordinarily high efficiency of light absorption. When it absorbs light, the rhodopsin molecule splits into *retinene*, related to vitamin A, and *opsin*, a protein. The appearance of one or a few altered photosensitive molecules changes the bioelectrical behavior of the cell through mechanisms that are as yet unknown. The outcome is an alteration in the activity of the synaptic end of the cell. Present evidence tentatively suggests that photoreceptor synapses actively release neurotransmitters in the dark and that light may reduce the rate of such activity.

Rhodopsin accounts for approximately 50 percent of the mass of the disc membranes; most of the remainder is lipid. Thus the numerous particles (50Å diameter) seen in the membranes with freeze-fracture techniques almost certainly are rhodopsin molecules, or at least structures dependent upon the presence of rhodopsin. Studies now underway are aimed at determining whether the distribution or other features of the particles change during response to light, in ways that might help explain the functioning of the membranes.

3.8.2 **LIGHT RECEPTORS IN INVERTEBRATES** The structure of eyes differs widely among organisms. In contrast to the single light-sensitive retina of the vertebrate eye, many invertebrates (for example, crustaceans and insects) have *compound eyes* made of repeating photoreceptive units, each with its own set of cells. In these units, light reception centers at the *rhabdomeres* which are densely packed aggregates of microvilli protruding from the surfaces of the receptive cells. The microvilli, like the retinal rod sacs, are oriented with their long axes perpendicular to the surface from which light enters. They probably provide a large area of light-receptive material. One hypothesis suggests that light-induced changes in the molecules of the rhabdomere membrane lead directly to changes in permeability to inorganic ions. Since the microvilli are part

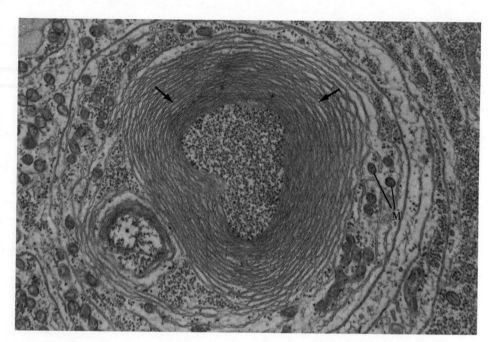

Fig. III-30   *Portion of a cell thought to be an abdominal light receptor in the crayfish. The extensive system of concentric membranes found in the cytoplasm is indicated by arrows. Mitochondria are seen at M. × 15,000. (Courtesy of A. J. D. DeLorenzo.)*

of the cell surface, this could initiate a nervelike impulse spreading across the surface and leading ultimately to transmission to connecting neurons.

Some invertebrates possess light-sensitive cells outside of their eyes. For example, crayfish have light receptors on their abdomens which apparently serve as protective devices against predators; the response to altered illumination leads to rapid movement of the organism. One of the cells that has been tentatively identified as an abdominal light receptor is shown in Figure III-30. The cell contains many membranes arranged in a whorl which presumably provide extensive membrane surfaces, comparable to those found in other light receptors and in sensory cells generally.

Many unicellular organisms possess special light-absorbing *eyespot* or *stigma* regions. These are probably parts of systems for orienting the organisms with respect to light. Such orientation may be especially important in photosynthetic forms. Sometimes the eyespots are discrete zones within chloroplasts (see Fig. II-46). In other cases they are separate structures. Often they are found near a basal body of a flagellum. The pigment in the eyespot regions may be in the form of large spherical bodies.

The role of light-absorbing pigments in eyespots seems to be quite different from that played by photosensitive pigments in receptors of higher organisms. For example, an interesting explanation has been put forth for the light responses of the flagellate *Euglena*. It is suggested that the small pigmented eyespot near the flagella is so positioned that when the organism is oriented at certain angles to the light, a shadow of the eyespot is cast on a special swelling located at the base of one flagellum. The swelling is presumed to be light sensitive and to respond to the alterations in light intensity (resulting from different shadow positions) by controlling flagellar motion and thus the movement of the organism.

chapter **3.9**
MUSCLE CELLS

Skeletal muscle is made of bundles of large multinucleate cell masses called *fibers* (see Fig. II-67). Both smooth and cardiac muscles are made of separate uninucleate cells; smooth muscle is composed of bundles or sheets of cells (Fig. III-19), cardiac muscle of a branched network of cells joined end to end.

The contraction of all three major types of muscles—skeletal, cardiac, and smooth—is dependent upon the presence of the proteins actin and myosin. As discussed in Section 2.11.2, it is the regular arrangement of actin- and myosin-containing filaments that is responsible for the *striations* of cardiac and skeletal muscle. The sliding of filaments past one another is the basis of contraction of these types of muscle.

Relatively little is known about the details of smooth muscle contraction. Longitudinally arranged filaments abound in the cytoplasm, but do not show the ordered arrangement seen in striated muscle. Scattered in the cytoplasm and near the plasma membrane are small electron-dense aggregates involving portions of filaments associated with amorphous material.

3.9.1   STRUCTURE AND FUNCTION OF THE SARCOPLAS-MIC RETICULUM   A skeletal muscle fiber (Figs. II-67 and III-31) is bounded by a *sarcolemma*, the plasma membrane with an overlying layer of extracellular material. Numerous peripherally located nuclei are present in each fiber. Golgi apparatus, rough ER, and ribosomes are scanty and are concen-

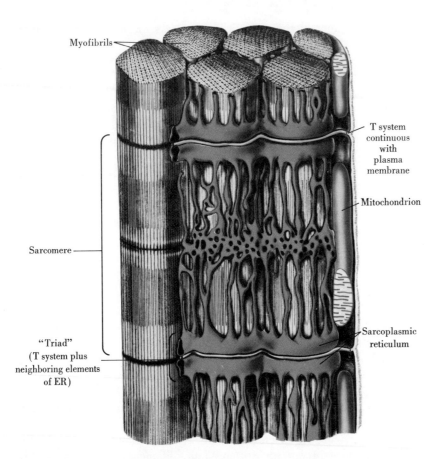

Myofibrils

T system
continuous
with
plasma
membrane

Mitochondrion

Sarcomere

Sarcoplasmic
reticulum

"Triad"
(T system plus
neighboring elements
of ER)

*Fig. III-31   The T system and sarcoplasmic reticulum of a frog muscle.
The conduction of an impulse by the plasma membrane at the fiber surface
depends on mechanisms comparable to those occurring at the axon membrane (Fig. III-27) and this probably is true within the T system as well. The
mechanism of coupling between T-system events and changes in the sarcoplasmic reticulum is incompletely understood. (Diagram modified from L.
D. Peachey and W. Bloom and D. W. Fawcett.)*

trated near the nuclei. Myofibrils occupy most of the cytoplasm. Closely
adjacent to the myofibrils are mitochondria; their close topographic
relation to the contractile material results in the efficient transfer of
ATP. Glycogen, in the form of particles 200–300 Å in diameter, is scattered through the cytoplasm, providing a storage supply of carbohydrates.

A membrane-bounded tubule is associated with each sarcomere.

The tubules constitute the *T system*. Marker molecules such as ferritin and peroxidase can be seen to pass rapidly from the extracellular medium into the T system. This confirms the finding by electron microscopy that the tubule system is continuous with the plasma membrane and is therefore open to extracellular fluids. The smooth ER of striated muscle is called the *sarcoplasmic reticulum*. The sarcoplasmic reticulum is closely associated with each T-system tubule and forms a network between tubules (Fig. III-31). Although the sarcoplasmic reticulum comes into close contact with the tubules, the reticulum and T system are not continuous with one another; extracellular peroxidase and other markers do not pass into the reticulum.

The elaborate arrangement of the T system and sarcoplasmic reticulum membranes is thought to coordinate the contraction of the fiber. Nerve impulses set off impulses in the plasma membrane of the muscle fiber that are probably similar in mechanism to the nerve impulses. A widely accepted hypothesis to explain coupling of the electrical impulse to the contraction of the muscle fibrils proposes that the impulse passes into the T system, and that this results in a release of calcium ions from the sarcoplasmic reticulum into the myofibrils. This, in turn, initiates contraction (Section 2.11.2). Following contraction, the calcium is reabsorbed by the sarcoplasmic reticulum. This theory is based in part on observations that injection or localized application of Ca^{++} to muscle fibers can cause contraction. Moreover, direct evidence for intracellular release of calcium associated with contraction has been obtained through the use of *aequorin*, a protein obtained from luminescent jellyfish that emits light (fluoresces) in the presence of calcium ions. When aequorin is injected into barnacle muscles and the muscles are stimulated, a burst of fluorescence is noted before contraction. Microsome fractions containing portions of the sarcoplasmic reticulum can concentrate Ca^{++} from the medium through active transport. ATPase activity and calcium binding proteins, such as *calsequestrin* probably related to this transport, are found in such microsome fractions and by cytochemistry. In the absence of something like the T system, it is difficult to see how an external plasma membrane change could produce virtually simultaneous response of all the sarcomeres of all the myofibrils throughout the mass of a muscle fiber many micrometers in diameter. If "trigger" substances had to diffuse in from the outer surface of the fiber, the outer fibrils would contract far in advance of the ones located at the center of the fiber. The T system penetrates throughout the fiber, so that even the sarcomeres of fibrils at the center can be affected directly by the impulse at the plasma membrane. Impulse conduction by the membrane is a much more rapid process than diffusion; molecules move by diffusion at rates measured in micrometers per second, while impulses are transmitted at rates of meters per second.

Muscles of different organisms often contain somewhat different arrangements of the T system and sarcoplasmic reticulum, but the basic function is probably the same in all striated muscles.

3.9.2 *COORDINATION OF CARDIAC AND SMOOTH MUSCLE* Sarcoplasmic reticulum and a T system are also found associated with the myofibrils of cardiac muscle. However, in this muscle type and in smooth muscle an additional problem in coordination results from the fact that both these muscle types consist of groups of separate cells, each relatively small in comparison to a skeletal muscle fiber. In both smooth and cardiac muscle, many gap junctions between cells are observed. It has been assumed that these junctions coordinate adjacent cells by providing a pathway for coordinating impulses (Section 3.5.2). An impulse triggered by a nerve ending on one cell could spread, via these junctions, to other cells.

c h a p t e r *3.10*
GAMETES

The details of cellular mechanisms related to sexual reproduction vary considerably among different organisms. However, in virtually all sexually reproducing eucaryotes, two features are constant. At some point in the life cycle, two special cell divisions occur; together the two divisions constitute *meiosis*. The meiotic divisions result in the halving of the number of chromosomes and, thus, halving of the amount of DNA per nucleus. At a subsequent stage two cells with the halved chromosome numbers, resulting from meiosis, fuse to form a zygote. The nuclei also fuse to produce a single nucleus combining the chromosomes from the two cells. The details of chromosome behavior and the genetic consequences of these processes will be considered in Chapter 4.3. For the moment it should be noted that zygote formation usually involves the fusion of cells and combination of chromosomes from two different parent organisms. This results in a zygote nucleus with a new genetic combination.

The cells that fuse and contribute the chromosomes of zygotes are generally referred to as gametes. Sometimes, as in many unicellular organisms and in some lower plants and animals, male and female gametes do not differ much in structure. In some cases fusion is partial and temporary. Thus during conjugation in *Paramecium*, two micronuclei are present in each cell and one from each cell passes into the other cell through a cytoplasmic bridge that forms as a transient structure. In most higher organisms male and female gametes are distinctly different, and gametes fuse completely (fertilization). Usually, though not invariably, the male gametes are smaller and motile; cilia or flagella are present on most male gametes of multicellular animals and lower plants, but some species have ameboid gametes. In flowering plants, pollen grains germinate to form a pollen tube that grows through the female structures of the flower and ultimately brings the male gametes into the ovary where the egg is located. In these plants "double fertilization" occurs. The egg cell is located in the embryo sac or female gametophyte (Section 4.3.1). One sperm cell fertilizes the egg cell to produce a zygote. A second sperm cell fertilizes another cell of the embryo sac; the products of this fusion give rise to nutritive "endosperm" tissue.

In this chapter we will discuss the gametes of higher animals which exemplify a high degree of gamete specialization.

3.10.1   ***NUCLEI***   The differences between sperm and egg are related to their different roles in zygote formation. Transmission of chromosomes from the male parent depends on specialized features of sperm cells. Egg cells carry the chromosomes from the female parent and also contain the reserve material used in the early development of the embryo. The gametes form from *gonial* cells (spermatogonia or oögonia). These divide to produce a population of *spermatocytes* or *oöcytes*. Each spermatocyte undergoes the two cell divisions of meiosis to produce four *spermatids*, which start off as more or less round cells but differentiate into mature elongate *sperm* (Fig. III-34). Oöcytes usually undergo extensive growth before or during meiosis; in many animals the meiotic divisions are not completed until after fertilization and usually only one of the four division products gives rise to a mature *ovum* (the others degenerate).

The *nuclei of mature sperm* are small and the chromatin is densely packed and entirely inactive in RNA synthesis. In some animals (for example, fish) the histones (Section 2.2.2) are replaced by *protamines*, very basic proteins unusually rich in the amino acid *arginine*. Often unusual lamellar or tubular patterns of chromatin distribution are seen (Fig. III-32). In other cases the chromatin is so densely packed that it appears to be a homogeneous mass. Some sperm nuclei are elongate; they may be more than ten times longer than wide. In such nuclei the chromosomes are often lined up as coiled fibers, with long axes running along the length of the head.

The *nuclei of growing oöcytes* and ova are quite large and the chromatin is dispersed within to such an extent that often it cannot readily be seen in the microscope, even after use of highly specific

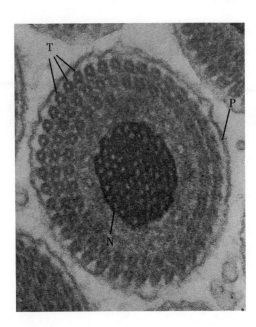

Fig. III-32   Cross section of the head of a sperm from the insect. Steatococcus. *The head has an elongate cylindrical shape. The many microtubules oriented longitudinally (T) are presumed to contribute to the establishment or maintenance of this shape. In the nucleus (N) the chromatin is arranged as dense material surrounding less dense tubular regions which appear in section as light circles. P indicates the plasma membrane. × 100,000. (Courtesy of M. Moses.)*

stains for DNA. The chromosomes may assume a special "lampbrush" configuration (Section 4.4.5), and sometimes a great many nucleoli are present. The large oöcyte nuclei engage in intensive RNA production, supporting their own cytoplasmic growth and providing RNA that is stored and used in later embryonic development; the special features of their nuclei are related to this RNA production.

At certain times the many nucleoli of amphibian oöcytes are seen to lie near the nuclear envelope and appear to contribute considerable amounts of material (probably ribosomes or ribosomal precursors) to the cytoplasm; ribosomes are abundant in the cytoplasm. Each nucleolus contains its own DNA, believed to derive from the nucleolar organizers on the chromosome (see Fig. IV-18). This has been explained by the proposal that a cell makes many extra copies of the DNA of the nucleolar organizer and that these are released into the nuclear sap where each functions independently to produce a nucleolus. In essence this leads to a great amplification in the number of sites specific for rRNA synthesis. It also results in the temporary presence of much "extra" DNA in the nucleus.

3.10.2   ***CYTOPLASM*** Sperm of higher animals are elongated, highly modified cells (Fig. III-34); widths of a few micrometers and lengths of 50–100 μm are not uncommon. Most of the length is accounted for by the flagellum. The *head* which contains the nucleus may make up only 5–10 percent of the length (Fig. III-34). The front of the head contains an *acrosome*. This large membrane-bounded structure is produced by the Golgi apparatus as the spermatid matures. Lysosomal enzymes have recently been shown to be present in the acrosome of some sperm, leading to the suggestion that acrosomes are highly modified lysosomes.

It is also during spermatid maturation that the flagella grow in association with the centrioles. As the flagella form, the centrioles become situated at the base of the nucleus. The flagella show the usual 9 + 2 tubule pattern, often accompanied by additional fibrous structures that run alongside the tubules (Fig. III-34). In the *midpiece*, the region just behind the head of the sperm, the mitochondria are aggregated near the flagellum. Quite often the mitochondria are in ordered arrays; sometimes they form a spiral ribbon that twists around the flagellum for some distance. Presumably this facilitates transmission of ATP used in sperm movement. The flagellum projects beyond the midpiece as the *tail*.

The entire sperm is covered by a plasma membrane, but aside from the acrosome, mitochondria, centrioles, and flagellum, little cytoplasm is present. Golgi apparatus, ER, ribosomes, and other cytoplasmic components are sloughed in a cytoplasmic bud that forms as the spermatid matures into a sperm and eventually separates from the sperm and disintegrates (in many species it is phagocytosed by the *Sertoli cells* of the testis).

In contrast to sperm, ova are often enormous cells with diameters ranging from roughly 100 μm in many mammals to over 1 mm in some invertebrates and amphibians; occasionally they are even larger, as in reptiles and birds. Their abundant cytoplasm accumulates during oöcyte growth. The cytoplasm contains the usual organelles, plus distinctive *yolk bodies* which may occupy much or most of the cytoplasmic volume (Fig. III-33). Yolk contains varying proportions of stored lipids, carbohydrates, and proteins, used later in development. Numerous ribosomes, some bound, many free (Section 2.3.6), are present in egg cytoplasm. These are used by cells of the early embryo. In some organisms new ribosomes are not synthesized by the embryo until relatively late in development. (Thus far this has been studied only in a few organisms.) The polysomes of mature but unfertilized eggs are inactive in protein synthesis. They are activated after fertilization.

The details of yolk formation in the oöcyte vary from species to species. Yolk may accumulate in Golgi vacuoles, endoplasmic reticulum (Fig. III-33), or occasionally in mitochondria. In many species considerable amounts of material are also taken up by pinocytosis and contribute to the yolk (Fig. III-33). Some of this material may come from follicle ("nurse") cells that surround many growing oöcytes. Much

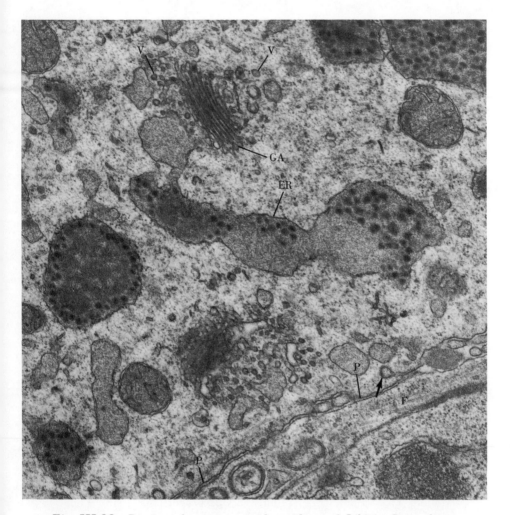

Fig. III-33   *Portion of an oöcyte in the spider crab* Libinia. *Part of one follicle cell (many surround the developing egg) is seen at F. Yolk accumulates as electron-dense material in cisternae of the ER. Dilated regions appear to separate from the ER as large yolk spheres. Vesicles, some showing "coats" (arrow; see Section 2.1.4) appear to form at the plasma membrane. These coated pinocytosis vesicles and other vesicles (V) associated with the Golgi apparatus (GA) are considered as probable sources of material added to the yolk. Presumably the vesicle membranes fuse with the ER-derived membrane of the yolk sphere.* × *50,000. (Courtesy of G. Hinsch und V. Cone.)*

material may come from the blood (for example, in amphibia the liver secretes large amounts of proteins destined for inclusion in yolk). The pinocytoic vesicles carrying extracellular material fuse with developing yolk bodies.

The cytoplasm just below the egg surface is often organized as a gel-like cortex. In many species special cortical granules or fluid-containing vacuoles accumulate there. The role of these granules in fertilization is discussed below.

The egg often shows asymmetric distribution of cytoplasmic components. Of particular interest are the special cytoplasmic regions in some eggs that contain substances involved in differentiation of specific cell types during development (Section 4.4.3).

A variety of extracellular coats, sometimes in multiple layers and usually rich in polysaccharides, surround the eggs of many species. In some cases a thin surface layer may be formed by the egg itself, but additional thick coats often are contributed by ducts down which the eggs pass after release from the ovary, or by follicle cells.

3.10.3 FERTILIZATION

Although the details of fertilization also vary from species to species, in many instances a general pattern is followed (Fig. III-34). The acrosome of the sperm probably plays an important role in sperm penetration through the exterior coats of the egg. On contact with the egg coats, the acrosome membrane fuses with the plasma membrane of the sperm, releasing the acrosome contents; this resembles the release of secretion from a secretory cell. Acrosomal enzymes of a class known as *hyaluronidases* probably break down polysaccharides in the egg coat; *proteases* that break down the proteins of the coats are also present. After the release of the acrosome contents, the acrosome membrane (which is now continuous with the *sperm* plasma membrane) forms one or more tubular projections that contact the *egg* plasma membrane and soon fuse with it. This establishes continuity between egg and sperm. The sperm nucleus thus can enter the egg without exposure to the extracellular environment. Microscopic observations on several species indicate that microfilaments are abundant in the aerosomal projections; elongation of the projections may involve rapid assembly of the (actin?) filaments.

As sperm and egg fuse, the egg undergoes changes. One rapid alteration, of unknown character, prevents fertilization by additional sperm. This barrier is soon reinforced in many eggs by the formation of a *fertilization membrane* that surrounds the egg. This is not the usual three-layered membrane, but rather a thick layer, probably of proteins and polysaccharides. The egg cortical granules contribute to the fertilization membrane. Fusion of their delimiting membranes with the plasma membrane releases the granule contents to the extracellular surface. Granule release occurs in a wave that proceeds around the egg from the point of sperm fusion. Fertilization also seems to evoke changes in the intracellular concentration and distribution of ions such

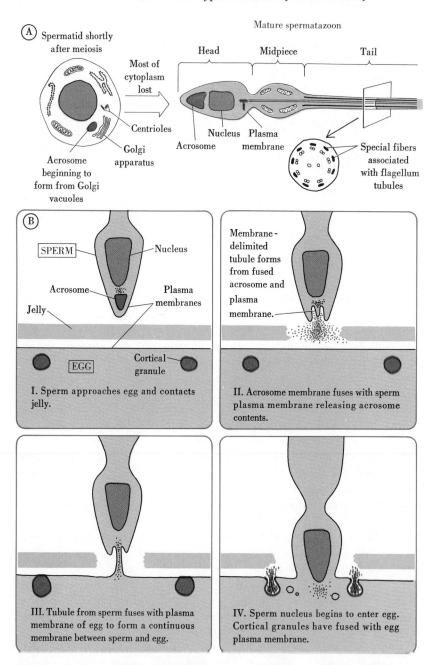

(A) Spermatid shortly after meiosis

Most of cytoplasm lost

Acrosome beginning to form from Golgi vacuoles

Golgi apparatus

Centrioles

Mature spermatazoon

Head   Midpiece   Tail

Nucleus Plasma
Acrosome membrane

Special fibers associated with flagellum tubules

(B)

SPERM Nucleus
Acrosome Plasma membranes
Jelly

Cortical granule
EGG

I. Sperm approaches egg and contacts jelly.

Membrane-delimited tubule forms from fused acrosome and plasma membrane.

II. Acrosome membrane fuses with sperm plasma membrane releasing acrosome contents.

III. Tubule from sperm fuses with plasma membrane of egg to form a continuous membrane between sperm and egg.

IV. Sperm nucleus begins to enter egg. Cortical granules have fused with egg plasma membrane.

Fig. III-34 **A.** *Schematic diagram comparing a* **spermatozoon** *(of a higher animal) with the* **spermatid** *from which it differentiates. The arrangement of mitochondria, the acrosome, centrioles, and other organelles varies in detail in different animals.* **B.** *Outline of fertilization based primarily on studies of invertebrates. (After L. Colwin and A. Colwin.)*

as Ca^{++} which might have important effects on subsequent events. These changes may in part be due to an altered permeability of the plasma membrane, and in part to release of ions from intracellular "stores," presently unidentified.

The role played by the sperm in bringing about these changes in the egg is still unclear. Even before the sperm nucleus enters the egg, a small amount of material from the region just behind the acrosome (Fig. III-34) passes into the egg. It is speculated that this, or perhaps material in the sperm plasma membrane or the acrosome membrane, normally initiates changes in the egg. In a few species the egg develops parthenogenetically (without sperm); eggs of other species can be induced to do so by pricking with a needle or by treatment with certain chemicals. Thus the effect of the sperm in activating the egg may be relatively nonspecific.

In many species not only the sperm nucleus but also the mitochondria, centrioles, and flagellum enter the egg. The sperm centrioles are believed to function in the division of fertilized eggs, though this is in dispute and probably is not the case in all species. The fate of the mitochondria is also not clear. In the eggs of some animals they have been seen to degenerate, as does the flagellum. The sperm nucleus and egg nucleus either fuse to form a single zygote nucleus or, in some cases, remain separate until development begins. In the latter instance the behavior of the maternal and paternal chromosomes is coordinated to achieve normal chromosome separation during the first division of the fertilized egg, and a single nucleus forms in each daughter cell.

c h a p t e r **3.11**
CELLS IN CULTURE

Early in the century it was shown that many embryonic cells and some cells removed from adult tissues of animals could grow and divide if adequate nutrients were added to the culture medium. About the middle of the century it was demonstrated that *single* cells of mammalian tissues, if properly fed, would divide to form *clones*. Addition of antibiotics to the medium prevented bacterial contamination and thus made large-scale cell culture for long periods of time a much simpler procedure. Human cells were first systematically used for cell culture in 1952.

Many cell types can be grown either on glass surfaces in single layers which are well suited for microscopic study of living cells (Fig. I-11), or suspended in liquid media to produce huge numbers of cells suited for many purposes. Cell lines from malignant tissue or cultured cells that become "transformed,"spontaneously or following addition to the cultures of tumor viruses or other agents (Section 3.12.3), appear to be immortal. That is, if the cell population is subdivided and transferred to fresh medium after some days (the interval in days depends upon the cell type and its rate of growth), they grow and divide indefinitely. The same is true of a few seemingly normal cells, that is, cells that do not form tumors if injected into the species of origin and that maintain their normal set of chromosomes in culture (over the long run most cultured cell lines do not do so—they become "aneuploid"; Section 4.5.1). However, most normal cells divide a limited number of times and then die; for example fibroblasts that are normal and maintain their *diploid* (p. 298) chromosome complement, will divide 40–50 times. Some investigators believe that study of this cell "senescence" will provide important information regarding the aging process in man and other organisms.

Growth of cells in culture requires salts, glucose or another sugar, amino acids, vitamins, lipid, and unknown growth factors that are provided usually by addition of fetal calf serum (or coconut milk for plant cells). The factors required for growth of specific cell types are being identified so that eventually fully defined artificial media may be used. Cultured cells can be stored apparently indefinitely in a deep-freeze at −70°C. When warmed rapidly, cells (not necessarily all of them) will resume growth and division.

Promising progress is being made in producing and isolating mutant strains of cultured cells. It is expected that, as with microorganisms, such mutants will be valuable for genetic and metabolic studies.

Cell cultures have been put to an increasing variety of experimental and practical uses. For example, the experiment illustrated in Figure II-19 was done with HeLa cells, one of the most widely used of mammalian cultured lines that originated from a human cervical cancer. This experiment required use of a large cell population, accessible to labeled precursors and to drugs, and uniform enough that representative samples could be conveniently taken at intervals. Doing such experiments with, for example, the livers of intact animals involves waiting until injected materials reach the liver through the blood stream. The materials are greatly diluted in the blood, and large proportions enter organs other than the liver. There may be serious problems in obtaining successive samples. If, as is customary, several animals are used for different time periods, it often is impossible to be sure that the animals are sufficiently similar to one another in nutrition, genetics, and other properties to permit reliable comparison. If the experiments require exposures to drugs, hormones, toxins, or similar compounds, the direct effects on the liver sometimes are difficult to distinguish from secondary effects in which the agent has acted upon another organ (such as an endocrine gland) which in turn has acted upon the liver. Many examples of the use of cultured cells for experiments are found through this book.

Cell and tissue cultures have been useful in the production of some vaccines, for example, to grow attenuated strains of viruses that will evoke immune responses, but not infections, in humans. More recently the ability to grow large populations of cells from a small sample has been put to important medical use, for example, in prenatal diagnosis of possible biochemical defects (Chapter 5.2).

Most cell lines that are carried in continous ("immortal") culture are relatively undifferentiated in structure and function. But there are numerous exceptions, and they often prove very interesting. For example, about ten years ago a clonal line was established from a mouse neuroblastoma (an embryo nerve cell tumor) whose cells can form long cytoplasmic processes resembling those of neurons in some aspects of structure and of electrical and enzyme activities. Molecular biologists, biochemists, and neurobiologists are vigorously studying this line for clues to nerve growth and function. Even when such lines have not been established, much information can be obtained from so-called *primary cultures*. These are tissues removed from the animal (embryonic or fetal material is generally best) and grown for a limited period as dispersed cells or in some cases (such as fetal liver tissue) as small pieces of tissue. These preparations often maintain many of the special structural and biochemical features of the corresponding cells and tissues of the intact animal.

Cell cultures are being extensively used to study the interactions of cells in populations. When mixtures of several cell types are grown in a common culture dish, like cells often associate with one another (Section 2.1.6), indicating an ability of cells to "recognize" one another. In many cases when cells of a given type are grown on glass or plastic surfaces, the cells move and proliferate only until their number and distribution reaches the point where most of the glass surface is covered by a single layer of cells (a *confluent monolayer*). Such inhibition has historically been referred to as "contact" inhibition; the effects on growth and division are now often called "density-dependent" inhibition. If the cells are separated subsequent to forming a confluent monolayer, they resume active movement and proliferation, indicating that no permanent change has occurred. There are many hypotheses but little general agreement as to factors that are responsible for such behavior. Many investigators feel that direct cell surface interactions may be involved. Others stress the possibility that the cells influence their immediate extracellular environment in some crucial fashion, perhaps by producing inhibitory substances or by depleting the growth medium of molecules needed for continued proliferation. The latter viewpoint has received some tentative support from experiments indicating that the final density attainable by a growing cell population can be increased by repeatedly replacing the growth medium with fresh medium or even by careful choice and maintenance of the pH of the medium. Of great interest is the observation that many cultured cancer cells are less sensitive to density-dependent inhibition of movement and growth than are normal cells (Fig. III-36).

With tissues taken from early embryos one can study some of the cell interactions involved in development. Work is underway on certain fairly general effects (such as the clustering of cells in groups as a prelude to differentiation) and also on specific interactions of different cell types—sometimes the presence of one cell type or its products in a culture is required for the differentiation of a second cell type. Single uninucleate embryonic muscle cells will divide repeatedly; then they will cease dividing and fuse to form multinucleate skeletal muscle fibers. This is important evidence that numerous nuclei in one skeletal muscle fiber result from fusion of separate cells, rather than from repeated nuclear division without division of cytoplasm.

Cell fusion techniques applied to cultures have great potential. When certain viruses or glycoproteins from their envelopes are added to the medium, fusion of cells with each other occurs on a large scale. The most popular tool is *Sendai virus* made noninfective by exposure to ultraviolet light. Two cells adhering to the same virus particles eventually fuse with each other. Thus human cells (HeLa) can be fused with a mouse tumor line that was initiated by Paul Ehrlich early in the century. If the DNA of the HeLa cells is prelabeled with ^{3}H-thymidine and the Ehrlich tumor cells are unlabeled, the fused cells can be shown to contain both types of nuclei, labeled and unlabeled. Fused cells containing separate nuclei of different cell types are called *heterokaryons*. Section 4.4.3 discusses the activation of the ordinarily inactive nuclei of red blood cells of birds that occurs in red blood cell–HeLa heterokaryons. Such experiments are providing insights into nuclear–cytoplasmic interactions.

If in fused cells the nuclei also fuse, the product is a *somatic hybrid*. In many cases such cells can divide and be perpetuated in culture as more or less stable types. Hybrids have been produced by fusion of the cultured pigment-producing cells from hamsters with the nonpigmented cells of mice; the hybrids do not synthesize pigment (melanin). This suggests that some mouse genes can cause the synthesis of molecules able to repress hamster genes which, when unrepressed, lead to melanin synthesis.

Somatic hybrids have greatly facilitated the genetic mapping of human chromosomes. As mouse cell–human cell hybrids are continued in culture, many human chromosomes are eliminated. Such elimination is partially random, so that different hybrid clones come to contain different combinations of human and mouse chromosomes. By determining which proteins are synthesized by clones possessing or lacking particular chromosomes one can determine which chromosomes carry genes for which proteins.

The drug cytochalasin B (p. 177) causes a small percentage of the cells in a culture to extrude their nuclei into the medium. This has led to the development of a simple technique for obtaining large numbers of enucleated cells. Cells are cultured as monolayers on coverslips. If before the cells become confluent the coverslips are placed face down in the medium containing cytochalasin B and are gently centrifuged, the nuclei are extruded and go to the bottom of the centrifuge tube. The coverslip is left with cells that almost all are enucleated; these are still viable for a while if returned to fresh medium without the drug. The metabolic activities of the cells without nuclei can be studied and the preparation can be used for important experiments. For example, enucleated cells were made with mutant lines of mouse fibroblasts that were *resistant* to the drug chloramphenicol, which ordinarily affects mitochondrial protein synthesis (Section 2.6.3). They were then fused with intact (nucleated) mouse fibroblasts of the "wild type" (*sensitive* to chloramphenicol). Some of the products gave rise to cell lines that were resistant to the drug. Since the resistant cells used for the fusions lacked nuclei, the likely explanation is that in this experiment chloramphenicol resistance is due to a mutation inherited through the cytoplasm, probably through mitochondrial DNA (Section 4.3.6).

Masses of unspecialized *plant cells* (callus) can be grown in culture, and they can be used for developmental and other studies. Structures resembling roots or buds can be induced to form in cultured callus tissue by treatments with appropriate levels of cytokinins and plant hormones that promote cell division and affect growth and differentiation. Cytokinins are best known for their stimulation of cell division. Plant hormones, such as *auxin* (indole acetic acid) and *gibberellin* (gibberellic acid), can affect cell growth and elongation. But these various agents may have multiple effects and they interact with one another. To cite one example, with some cells addition of cytokinin affects the *kinds* of peroxidase "isoenzymes" produced. (Isoenzymes are distinguishable forms of a given type of enzyme that differ in details of molecular structure.) Auxin addition will then modulate the *levels* of isoenzymes synthesized, increasing some and decreasing others. It is expected that such studies in cultured cells will elucidate molecular events in the living plant.

Cells from mature carrots have been grown in culture for prolonged periods; small clusters of such cells can be made to complete new plants. Apparently a single cultured cell can give rise to an entire plant. The usual interpretation of such experiments is that many plant cells are not irreversibly fixed in differentiation. Normally the cells of the mature carrot from which the culture was begun would have restricted functions; however, after growth and division in culture they can produce all the cell types of a plant. The ability to grow entire plants from cultured cells may have practical importance. Plant scientists are attempting to use the full range of cell culture techniques such as mutant selection and fusions of mutated cells to create cell lines that will generate plants with desirable agricultural properties.

If a tobacco plant is treated with an extract of a specific bacterium, within three to four days a tumor, called a crown gall, forms (the responsible factor has not been characterized). As with normal plant cells, single tumor cells can be isolated in culture, and they will grow to form multicellular masses. When these tumor masses are grafted to normal tobacco plants, they appear to form normal tissues. Such reversal of tumor characteristics in cells is being analyzed for possible implications for reversal of malignancy in animals as in plants. In some cases, simply growing the tumor cells in culture will yield normal plantlets again.

c h a p t e r *3.12*
CANCER CELLS

This part ends with the consideration of a most important form of cell abnormality. The basis (or bases) or cancer remains unknown. But optimism regarding its conquest is generated by current progress toward understanding the genetic machinery of cells. In a manner not understood, permanent changes in the heredity of cells can upset the mechanisms that restrict their division and that keep them in their normal places. When the cells divide to form abnormal large masses, they are called *tumors*. *Benign* tumors remain as discrete masses. *Malignant* tumors (cancers) generally *metastasize*–cells spread to distant parts of the body through the blood or lymph, and there they take root and grow into new cancer masses.

The chief sources of cancer cells for morphological and biochemical studies have been the transplantable tumors of mammals. In transplanting tumors, a small piece or a suspension of tumor cells is injected under the skin or into the leg muscle or body cavity. Before the tumor has grown to a size that kills the animal (from five days to several months, depending upon the growth rate of the tumor), a bit is again transplanted. Some transplantable tumors are obtained in *ascitic* form. These grow in the body cavity where they elicit the production of fluid (called *ascites*); the cell-laden ascites can be removed from the body cavity as a convenient source of malignant cells. For the study of molecular biology of cancer, growth of malignant cells *in culture* has proved of great value.

Ultimately all studies are directed toward understanding the nature and cause of the changed heredity. Many, but not all, cancer researchers believe that despite their diversity [animal and human cancers vary enormously in etiology (developmental history), histology, and biology], the fundamental nature of the transformation to the malignant state is essentially the same in many, and perhaps all, cancers.

3.12.1   ***TRANSPLANT-***   Cancers of a wide variety of organs have
ABLE   been studied, but none more extensively
HEPATOMAS   than tumors of the liver, or *hepatomas*.
This is because the liver is so well known biochemically and morphologically, and because a broad spectrum of hepatomas, varying in growth rate, are available for investigation.

Initially much of the work on the biochemistry of cancer was done with rapidly growing tumors of the liver and other organs. However, many of the findings, both morphological and biochemical, seem chiefly to be secondary features, manifestations of the rapid rate of division of the cells; they shed little light on the nature of the underlying transformation from normalcy to malignancy. Thus attention is focused on the more slowly growing hepatomas. It is thought that analysis of differences between these tumors and the normal liver may more readily lead to understanding the crucial nature of the transformation to malignancy. Perhaps the most firmly established feature of malignant cells had been their maintaining a high rate of glycolysis (Section 1.3.1), even in the presence of oxygen (aerobic glycolysis). Analysis of the slowly growing hepatomas, however, showed some to have low rates of aerobic glycolysis, similar to that of normal liver.

Both rapidly growing and slowly growing hepatomas possess organelles like those of normal cells, although the number and arrangement of the organelles may differ somewhat from normal. The rapidly growing *Novikoff hepatoma* differs from the slowly growing *Morris hepatomas* in possessing much less endoplasmic reticulum, having many more "free" polyribosomes, smaller mitochondria, and a smaller Golgi apparatus (Fig. III-35) and lacking nucleoid-containing peroxisomes (perhaps it was derived initially from nonhepatocyte cells of the liver, cells that possess only peroxisomes without nucleoids). In general the cytology of the slow-growing Morris hepatomas is like but not quite identical to that of normal hepatocytes. For example, differences from normal hepatocytes are evident in the morphology and enzyme activities

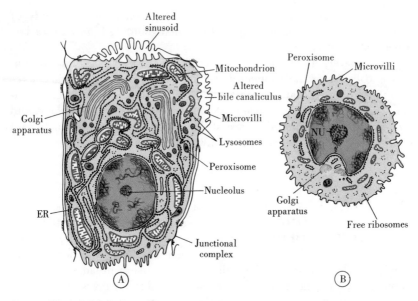

Fig. III-35   *Diagram of the cells of two transplatable hepatomas of rat.* **A.** *A slowly-growing (differentiated) Morris hepatoma.* **B.** *A rapidly growing (undifferentiated) ascites Novikoff hepatoma. Comparison with Figure I-2 will confirm that the differentiated hepatoma cells resemble hepatocytes in many respects far more closely than do the undifferentiated hepatoma cells. Yet even the differentiated type differs from hepatocytes in cell-surface features, decreased amounts of ER and in the size and nature of the lysosomes.*

of the plasma membrane of Morris hepatoma cells. The microvilli are less regularly arranged and there is a different distribution of some phosphatases among different regions of the cell surface. It is possible that this reflects other, more subtle, important differences of the surfaces of normal and malignant cells. Such surface differences could result in abnormal cell-to-cell interactions. Some investigators believe that cancer cells generally differ from normal ones in their surface interactions with one another.

A number of other differences have been detected between normal and malignant hepatocytes, but their significance for malignancy is a matter of speculation. The fact that tumor cells may retain many of the specialized features of the tissues from which they derive suggests that the transformation to malignancy may be based on fairly subtle changes. One theory suggests that abnormalities in polysomes or in their attachment to the ER may underlie the observed failure of certain enzymes (for example, an enzyme involved in the metabolism of the amino acid tryptophan) to increase in hepatomas when substrate (tryptophan) is added to the diet; such induction of enzyme increase does occur in normal livers. This has led to speculation that all cancers, whatever their diverse nature, may show a defect in the control of enzyme synthesis and turnover. Some investigators maintain that in malignant cells nucleolar RNAs are abnormal in composition, others that differences exist in the nonhistone chromosomal proteins that combine with DNA. But no suggestion seems more likely, at least for many cancers, than the possibility of alterations in the DNA itself, including the incorporation of foreign DNA of viral origin. Observations on *carcinogenesis*, the process by which normal cells are converted into cancer cells and, ultimately, into cancers, both in animals and in cultured cells, point to this suggestion.

3.12.2 CARCINOGENESIS IN ANIMALS

Agents that produce cancers are called *carcinogens*. A great many chemical carcinogens are known in animals and man. There are also physical carcinogens, such as X-ray and ultraviolet radiation, and biological carcinogens, mostly viruses. Numerous *oncogenic* (carcinogenic) viruses are known in animals (Fig. III-3). They include DNA viruses, such as the polyoma virus, and RNA viruses, such as the mouse leukemia virus. There is a widespread belief that leukemia, and possibly other human malignancies, will prove to be virus induced. Viruses are believed to induce some tumors in plants.

Some chemical carcinogens or products of their metabolism are known to combine with specific proteins of the cell. An increasing number have been shown, at least in the test tube, to affect nucleic acids, especially DNA. Similarly, the effects of X rays and of the oncogenic viruses that have been studied most closely can best be explained in terms of their action on DNA. A widely discussed theory derived from studies on cultured cells (Sections 3.12.3) suggests that the DNA of some oncogenic viruses becomes incorporated into the malignant cell's chromosome, as is known to occur with some bacteriophages (Section 3.1.2). A change in the cell's genetic constitution, or *genome*, may well be the common pathway to malignancy in many instances, however different the initiating factor or "cause" may be in each particular case.

Such genetic changes are apt to be much more subtle than the abnormalities in numbers (aneuploidy, Section 4.5.1) and shapes of chromosomes that are seen in many, but not all, cancer cells. The chromosomal abnormalities are likely to be an effect rather than a cause of the malignant transformation; they may reflect cell division "errors" engendered by a very high rate of division. In one type of chronic leukemia in man, particular chromosomes change consistently (absence or reduction in size of one or two of the chromosomes); in this case there *might* be a causal connection between the chromosomal abnormality and the malignancy.

The transplantable rat hepatomas considered above were induced by chemical carcinogens. Prior to the appearance of tumors, cell clusters called *hepatic nodules* develop. The nodules are being carefully studied for clues to the early stages in the development of tumors. Do nodules form as clones from single altered cells?

3.12.3 CARCINOGENESIS IN CULTURE; CULTURED CANCER CELLS   Important ideas concerning the subtle changes in the cell genome come from studies on cultured cells. Most work on carcinogenesis in cultured cells centers on oncogenic viruses. These are used because some viruses are exceedingly simple, and because the large fund of information on the molecular biology of viral infection aids in the interpretation of findings.

Polyoma virus is one of the simplest known viruses. It produces cancers in a variety of tissues when inoculated into newborn mice, rats, or hamsters. When embryonic fibroblasts are infected with virus, some of the cells are transformed. Their morphology then resembles that of rapidly dividing cells, and when transplanted into an animal, the cells multiply and produce cancers. Like cells from naturally occurring cancers, transformed cells continue growing in culture after normal cells cease to do so—their growth is not as markedly inhibited by increased density (Chapter 3.11). Unlike many normal cells, most malignant cells grow over each other and form clumps of unoriented cells several layers thick (Fig. III-36).

How do the cell surfaces of cancer cells differ from normal? Plant proteins known as *lectins*, including one from jack beans called *Concanavalin A*, bind to specific sugars in cell surface glycoproteins of many cell types. Apparently as a result of differences in such binding, lectins can cause clumping (agglutination) of transformed cells (also of fetal cells) at lower concentrations than are needed to agglutinate normal cells. The features of cell surface organization underlying this are being sought; studies are underway on the synthesis and turnover of glycoproteins and other components of the surface.

Observations on cultures suggest that some transformed cells may release unusually high levels of certain proteolytic enzymes into the growth medium. Moreover, treatment of normal cells with proteases can induce changes in their growth and cell surfaces resembling some of the

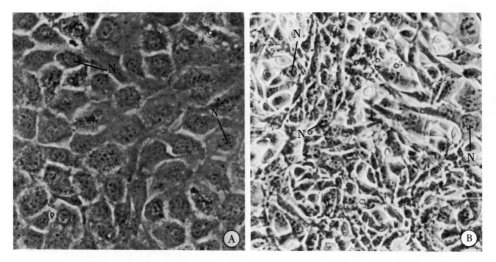

Fig. III-36 *Cells of a strain of cultured mouse fibroblasts photographed through a phase-contrast microscope. Nuclei of a few of the cells are indicated by (N). The cells in A are normal cultured cells that have formed a single layer (monolayer) of flattened cells on the glass surface. The cells in B have been infected with polyoma virus and thereby "transformed." They no longer form a monolayer but overlap in multiple layers, clumps, and irregular arrays. (Courtesy of R. Dulbecco.)*

changes seen in transformation. As yet the significance of such findings is uncertain.

One important observation on polyoma and some other viruses is that, once the cell has transformed, the virus is no longer detectable as a separate infective particle. However, the presence of viral DNA in the host cell (probably in its nucleus) is indicated by several facts. RNA that hybridizes (p. 65) with viral DNA and some virus-specific proteins have been identified in transformed cells. When fusions are induced (Chapter 3.11) between certain transformed cells, the virus "reappears," that is, separate viruses are again detectable.

With polyoma and other viruses as few as 4–8 viral genes appear to be involved in a cell's transformation to malignancy. These are such relatively simple systems that they are expected to contribute much to a rapid progress in understanding the basic nature of cell transformation to malignancy.

In 1970 an RNA-directed DNA polymerase was discovered in a virus that causes a chicken tumor (*Rous sarcoma*) and in another virus that causes mouse leukemia (*Rauscher leukemia*). The enzyme is generally referred to as "reverse transcriptase" because the information flows from RNA to DNA, reversing the usual direction. Evidently the enzyme generates DNA copies of the viral RNA, and these transform the cell to a malignant state. Such RNA viruses containing reverse transcriptase have now been found in a variety of animals, including the wooly monkey and gibbon ape. Very recently such a virus has been recovered from a human patient with "acute myelogenous leukemia." If and when viruses are found in other cancers, then immunological therapeutic approaches may be feasible, including perhaps the development of vaccines against these particular viruses. However, it should be noted that in mouse leukemia, the most studied in this regard, carcinogenesis is not simply a question of the presence of the virus. Genes of the mice influence the incidence of leukemia even in the presence of virus. Mice have been inbred so that in one strain 80–90 percent of the animals develop leukemia, in another inbred strain fewer than 5 percent do so, and still other strains show intermediate incidences.

FURTHER READING

Aidley, D. J., *The Physiology of Excitable Cells*. Cambridge: Cambridge Univ. Press, 1971, 468 pp. An advanced but introductory account of nerve function.

Albersteim, P. "The Walls of Growing Plants." *Scientific American*, **232**: no. 4, p. 80, April 1975.

Austin, C. F. *The Ultrastructure of Fertilization*. New York: Holt, Rinehart and Winston, 1968, 196 pp.

Axelrod, J., "Neurotransmitter." *Scientific American*, **230**: no. 6, p. 58, June 1974.

Berg, H. "How Bacteria Swim." *Scientific American*, **232**: no. 2, p. 36, August 1975.

Bjorkman, O., and J. Berry. "High Efficiency Photosynthesis." *Scientific American*, **229**: no. 4, p. 80, Oct. 1973.

Cold Spring Harbor Symposium on Quantitative Biology 37: Mechanism of Muscle Contraction. Cold Spring Harbor, N.Y.: Cold Spring Harbor Press, 1973. A collection of research articles.

Bogorad, L. "Phycobiliproteins and Complementary Chromatic Adaptation." *Ann. Rev. Plant Physiol.*, **26**: 369, 1975.

Cooper, M. D., and A. R. Lawton. "The Development of the Immune System." *Scientific American*, **231**: no. 5, p. 58, Nov. 1974.

Dodge, J. D., *The Fine Structure of Algae Cells*. New York: Academic Press, 1973, 261 pp.

Dulbecco, R. "The Induction of Cancer by Viruses." *Scientific American*, **216**: no. 4, p. 28, Apr 1967.

Ephrussi, B., and M. C. Weiss. "Hybrid Somatic Cells." *Scientific American*, **220**: no. 4, p. 26, Apr. 1969.

Hall, J. L., J. T. Flowers, and R. M. Roberts. *Plant Cell Structure and Metabolism*. London: Longmans, 1974, 426 pp.

Hall, R. H. "Cytokinins as a Probe of Developmental Processes." *Ann. Rev. Plant Physiol.*, **24**: 415, 1973.

Jensen, W. A. "Fertilization in Flowering Plants." *Bioscience*, **23**: 21, Jan. 1973.

Laetch, W. M. "The C$_4$ Syndrome: A Structural Analysis." *Ann. Rev. Plant Physiol.*, **25**: 27, 1974.

Ledbetter, M. C., and K. R. Porter. *Introduction to the Fine Structure of Plant Cells*. New York: Springer-Verlag, 1970, 188 pp. An atlas of micrographs with explanatory material.

Lentz, T. L. *Cell Fine Structure*. Philadelphia: Saunders, 1971, 437 pp. An atlas of schematic drawings of many cell types of mammals with explanatory material.

Mandelstam, J., and K. McQuillan. *Biochemistry of Bacterial Growth*, 2nd ed. New York: Wiley, 1973, 582 pp. A discussion of bacterial structure and biochemistry.

Maramorosch, K., Ed. *Mycoplasma and Mycoplasma-like Agents of Human, Viral and Plant Diseases*. Annals of the N.Y. Academy Science, **225**, 1973, 532 pp. An extensive collection of research articles.

Northcote, D. H. "Chemistry of the Plant Cell Wall" *Ann. Rev. Plant Physiol.*, **23**: 113, 1972.

Pollack, R. *Readings in Mammalian Cell Culture*. Cold Spring Harbor, N.Y.: Cold Spring Harbor Press, 1973, 735 pp. An excellent compilation of research articles with concise summaries by Pollack.

Raschke, K. "Stomatal Action." *Ann. Rev. Plant Physiol.*, **26**: 309, 1975.

Ray, P. M. *The Living Plant*, 2nd ed. New York: Holt, Rinehart and Winston, 1972.

Robards, A. W. "Plasmodesmata." *Ann. Rev. Plant Physiol.*, **26**: 13, 1975.

Ross, R. "Wound Healing" *Scientific American*, **220**: no. 6, p. 40, June 1969.

Ruddle, F. H., and R. S. Kucherlapati. "Hybrid Cells and Human Genes." *Scientific American*, **231**: no. 1, p. 36, July 1974.

Sharon, N. "The Bacterial Cell Wall." *Scientific American*, **220**: no. 5, p. 92, May 1969.

Smith, D. S. *Insect Cells: Their Structure and Function*. Edinburgh: Oliver and Boyd, 1968, 372 pp. A collection of electron micrographs with explanatory material.

Spector, D. H., and D. Baltimore. "The Molecular Biology of Poliovirus." *Scientific American*, **232**: no. 5, p. 25, May 1975.

Stanier, R. Y., M. Doudosoff, and E. A. Adelberg. *The Microbial World*, 3rd ed. Englewood Cliffs, N.J.: Prentice Hall, 1972, 873 pp.

Stevens, C. F. *Neurophysiology–A Primer*. New York: Wiley, 1968, 182 pp. A brief introduction.

Troughton, J., and L. A. Donaldson. *Probing Plant Structure*. New York: McGraw-Hill, 1972.

Workshop on Viral-Cell Interactions (*Advances in Biosciences*, **11**, 1973). Oxford: Pergamon Press, 248 pp. A collection of research articles on the interactions of viruses and cells.

Textbooks of Histology: Many fine ones are available including Bloom, W., and D. W. Fawcett, *A Textbook of Histology*, 10th ed. Philadelphia: Saunders, 1975, 1033 pp.; Greep, R. O., and Weiss, L., *Histology*, 3rd ed. New York: McGraw-Hill, 1973, 1044 pp.; Rhodin, J. A., *Histology*. New York: Oxford Univ. Press, 1974, 803 pp.

part 4

DUPLICATION AND DIVERGENCE: CONSTANCY AND CHANGE

Duplication of even the simplest procaryote involves far more than replication of DNA. Cell growth and division are integrated processes based on coordinated activities of virtually all cell components. Mechanisms operate for duplication of all cell structures and for separation of the cell into two daughters, each with its share (in general, approximately equal) of the structures and macromolecules of the parent.

In bacteria such as *E. coli*, and presumably in other procaryotes, only a single chromosome is present per nuclear region. In some bacteria the mesosomes appear to function in separating the two daughter chromosomes produced by DNA replication (Section 3.2.4).

In eucaryotic cells there is much more DNA than in procaryotic cells, and it is "packaged" in several or in many chromosomes. The chromosomes, unlike those of bacteria, contain much protein and other non-DNA material. In almost all eucaryotic cells the chromosomes are duplicated and separated into daughter cells by the process of *mitosis*, which involves the formation of a temporary intracellular structure, the *spindle*.

Mechanisms for synthesizing nucleic acids, proteins, and other macromolecules are now reasonably well analyzed. Organelle duplication has only recently begun to be understood. As has been previously indicated, mitochondria, plastids, and possibly centrioles make direct contributions to their own duplication. Other structures, such as flagella and ribosomes, may not do so. Much attention is now being devoted to analysis of the ways in which organelles are actually assembled from macromolecular building blocks. Centrally involved in such organellogenesis is the process of *self-assembly*, the spontaneous organization of molecules into specific multimolecular configurations. Some macromolecules possess built-in features that automatically produce a correct association with other macromolecules, so that a given three-dimensional structure results from properties inherent in the molecules of which it is composed. No special, external, "structure-specifying" system is needed to impose three-dimensional ordering on groups of macromolecules in a manner analogous to the way in which mRNA controls the amino acid sequence of proteins. If such structure-specifying systems are involved at one level or another, they themselves must ultimately derive their structure from a self-duplicating system (presumably containing nucleic acids) that essentially specifies its own structure. Otherwise one must postulate an endless chain of systems specifying the structure of other systems.

The evidence for the operation of self-assembly is strongest for relatively simple structures, such as some viruses and protein fibers. These systems have been dissociated into their component macromolecules or subunits and then reconstructed by mixing the components *in vitro* (in the test tube). Even the simplest *in vitro* self-assembly system requires careful control of the composition of the medium in which the subunits are suspended. Ions such as calcium (Ca^{++}) or magnesium (Mg^{++}) can have dramatic effects on structures; for example, the two subunits of isolated ribosomes separate if the magnesium concentration in the suspension medium is reduced below a certain level. The two positive charges of calcium or magnesium can bind two separate negative charges and, in so doing, hold separate molecules or regions of molecules in fixed relations to one another. Probably many other effects of charged ions and molecules also operate in the control of structure. For example, two positively charged molecules will tend to remain separate due to the fact that similar charges repel one another. But if each complexes with negatively charged ions, the net charge of the complex will be zero and the molecules can more closely approach each other. For assembly of structures *in vivo* (in the living cell), mechanisms must operate to control the synthesis of the necessary macromolecules at the proper time, to collect appropriate amounts of components at the proper point, to control levels of ions, acidity, ATP, and other factors, and probably to ensure the presence of enzymes required to speed assembly.

The degree to which self-assembly operates in the formation of complex cell organelles, particularly those delimited by membranes, is not known. Experiments have been outlined previously in which mitochondrial and bacterial enzyme systems have been reconstructed from their components (Sections 2.6.5 and 3.2.4). Experiments like these suggest that parts of organelles may self-assemble. However, it is unlikely that all the molecules of a chloroplast or mitochondrion, mixed in correct proportions in a medium of proper composition, would produce a *complete* organelle. Mitochondria and plastids grow and divide; the relationship between organelle growth and structure assembly has only begun to be explored, but it seems likely that as an organelle grows, the previously existing structure may aid in orienting newly added molecules. In the next chapter we will outline experiments dealing with the formation of reasonably complex structures. The experiments provide hope that the duplication of organelles will soon be understood in molecular terms.

Duplication leads to similar cells, but cells also undergo change. Change occurs on three time scales: the very rapid physiological changes that have been discussed at several points in the preceding parts, cell differentiation during embryonic development, and the slower processes of evolutionary change.

The cell diversification that occurs in embryonic development results from complex patterns of interaction between nucleus and cytoplasm and between different groups of cells. When an egg is fertilized, a precise program of events is set in motion, a program that depends on factors many of which are only beginning to be appreciated. An important part of the story is the formation of different messenger RNAs in different cells as different genes are activated or freed from the repression that keeps them inactive. Interesting cytological changes are associated with this gene activation.

It is self-evident that basic cellular processes, such as photosynthesis or mitosis, have an evolutionary history. At some point in evolution they arose from simpler processes. Their history is known only in vaguest outline, but some research and much speculation have been devoted to its elucidation.

Evolutionary changes are based on the natural selection of variant organisms resulting from mutations of DNA. Mutated DNA is often expressed as altered proteins which engender changes in cellular metabolism. A change in any major step of metabolism tends to affect many distantly connected metabolic processes due to the interweaving of pathways. Variation is greatly increased by sexual reproduction which generates new combinations of genes. In most eucaryotic organisms, sexual reproduction depends on *meiosis*, a special pair of cell divisions.

The evolution of cells involves the evolution of chromosomes, as evidenced by differences in the number and shape of chromosomes in different species. Cytoplasmic factors also are involved; some cytoplasmic organelles possess genetic information that is expressed partly independently of the nuclear genes. Mutations of cytoplasmic genetic factors can occur and can result in permanent alterations of cell characteristics.

c h a p t e r **4.1**
MACROMOLECULES AND MICROSCOPIC STRUCTURE

4.1.1   ***SELF-ASSEMBLY***   If a protozoon or a tissue culture cell is cut
             OF MEMBRANE   or torn with a microneedle, rapid healing
             STRUCTURE   occurs, unless the opening created is large
                                    enough to permit extensive loss of cell con-
tents. The simplest explanation for healing is that the plasma membrane can grow and repair itself rapidly, using components already present in the cytoplasm.

When artificial mixtures of proteins and lipids, or mixtures of components isolated from cellular membranes, are placed in proper solvents, they often form arrays resembling three-layered membranes (Section 2.1.1). The phospholipids play a central role in this; they can spontaneously assume an ordered arrangement (Fig. IV-1) in which hydrophilic regions (Section 2.1.2) associate with the water and hydrophobic regions associate with one another. This creates layered structures in which each layer is two lipid molecules thick. When different proportions of various other lipids (for example, cholesterol) are added to the phospholipids and proteins, membranes of different thicknesses form. Variation of lipid or protein content may be a partial explanation of the different thicknesses of membranes of different organelles.

The distribution of hydrophilic and hydrophobic regions in the various proteins that participate in a membrane would be expected to influence strongly their interactions with the lipids, and hence the way in which they are oriented and bound to the membrane. Extensive hydrophobic regions would permit a protein molecule to penetrate deep into the lipid layer, whereas proteins with a predominantly hydrophilic exterior would assume a more peripheral location.

These observations, though not conclusive, raise the possibility that the basic structure of cellular membranes may form by self-assembly (or by relatively simple addition of molecules to preexisting membranes). How enzymes become arranged with specific geometry on the membrane remains unknown, and we cannot yet account for many of the functional differences among different membranes. Different enzymes might assemble in complex groups before they associate with lipids and become incorporated in membranes. One important feature of natural membranes, difficult to duplicate in test-tube assembly, is the asymmetric arrangement of components such as proteins (Figs. II-3B and II-42). It seems likely that cellular membranes generally form by growth or alteration of previously existing membranes (Sections 2.4.4, 2.5.4, 4.1.6). The preexisting organization might contribute to asymmetry. For example, a cell could add an individual protein molecule to the inner surface of its plasma membrane more readily than to the outer, since the membrane itself impedes access to its outer surface. For the ER and its derivatives, and perhaps for mitochondria and chloroplasts, ribosomes attached to the membranes may be a major source of proteins of the integral type (p. 48), such as those that span the membrane—perhaps some of these proteins are inserted into the membranes as they are synthesized. The proteins added to membranes by bound ribosomes might enter in asymmetric fashion; this is a plausible conjecture based on the unidirectional passage of secretory proteins from ribosomes into the ER cavity (Section 2.4.1).

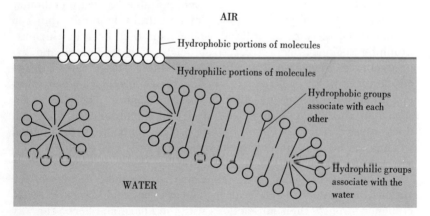

Fig. IV-1   *Polar lipid molecules associate spontaneously in ordered arrays when placed in water and at air–water interfaces.*

4.1.2   SELF-ASSEMBLY OF PROTEIN STRUCTURES   Proteins have specific three-dimensional conformations (Fig. IV-2) which are central to enzyme activity and other properties. The folding of the chain into a specific three-dimensional structure will be strongly influenced by the *primary structure* (the amino acid sequence). For example, if sulfhydryl (–SH) groups on the amino acid cysteine are present at two points on the chain, they may react to form a disulfide bond (S–S). This bond will hold the two portions of the chain in a fixed relation to one another (Fig. IV-2). Similarly, interactions between other amino acids may form *noncovalent* stabilizing bonds such as the ionic bonds resulting from attraction between the oppositely charged NH_3^+ and COO^- groups, hydrogen bonds (Section 2.2.1), and complex interactions (hydrophobic bonds) that promote associations between hydrophobic groups (Sections 2.1.2 and 4.1.1). Most of these bonds are easily broken and reestablished, in contrast, for example, to the bonds that hold amino acids together within the polypeptide chain (Fig. II-15). The latter are *covalent* bonds based on sharing of electrons by two atoms. Generally, drastic chemical treatment or enzymatic digestion is needed to break covalent bonds, whereas noncovalent bonds are responsive to mild treatments such as changes in the ion concentration of the medium.

Experiments have suggested that some proteins of given primary structures may fold spontaneously into the correct three-dimensional conformation. Enzymes have been unfolded (*denatured*), with consequent loss of function, by exposure to agents such as *urea* which breaks hydrogen and hydrophobic bonds. When the denaturing agents are removed, some enzymes are capable of returning spontaneously to a conformation indistinguishable, in terms of enzymatic activity, from the original. For other proteins, conditions for such reversal of denaturation have not yet been found, and it is possible that they fold by more complex processes. Perhaps, during protein synthesis, the completed portion of a polypeptide detaches from the ribosomes and begins to fold, while the rest of the chain is being completed. Sequential folding in this way, with the part synthesized earliest folding partly independently of portions synthesized later, could lead to complex three-dimensional patterns. When the protein is denatured in the test tube and allowed to re-form its structure, the original distinction between early and late synthesized ends does not exist, and it may fold in a different manner. The importance of timing is even more dramatic with proteins such as insulin. It is difficult to get the two chains of insulin to come together properly if the chains are separated in the test tube. Apparently the actual folding of the insulin molecule occurs when the two chains are still parts of the larger proinsulin molecule (Section 2.4.1). Without the small portions nicked out by the protease that converts proinsulin to insulin, the correct three-dimensional structure will not form.

Polypeptide with SH groups and
other groups at specific points.

Polypeptide folds into a three-dimensional
pattern stabilized by S-S and other bonds.

Two folded polypeptides associate in a
specific pattern based on the distribution of bonding groups at their surfaces.

Formation of a structure
with self-limited size.

Subunits of a given shape (more precisely,
with a given three-dimensional distribution
of binding groups) can associate to form
specific more complex and larger structures.

Continued growth by oriented
addition of subunits.

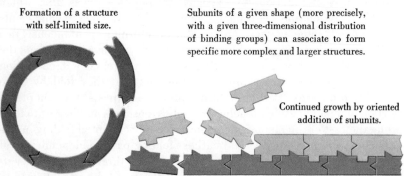

Fig. IV-2   *The self-assembly of macromolecules to form complex shapes
and arrays.*

Many proteins have subunits joined together in specific manners.
For example, hemoglobin consists of four subunits, each a folded polypeptide chain. If the hemoglobin subunits are dissociated from one
another, then mixed in a test tube, they readily reconstitute the origi-

nal tetrapartite structure. Apparently the proper amino acid groups are present at the surfaces of the folded polypeptide chains to promote association of subunits in specific patterns (Fig. IV-2).

These observations indicate that the information encoded in nucleic acids, by specifying the amino sequences of polypeptide chains, can contribute directly to the determination of three-dimensional protein structures.

4.1.3 FILAMENTS AND FIBERS The muscle proteins actin and myosin exist in the striated muscle cell as distinct filaments containing linked molecules of actin (thin filaments) or myosin (thick filaments) (Section 2.11.2). In the test tube the molecules can be dissociated from one another and then reassociated to reconstitute filaments. The reconstituted filaments from myosin resemble the thick filaments of striated muscle; those from actin resemble the thin filaments. However, there are differences between reconstituted filaments and *native filaments* (those found in muscle). For example, in most vertebrate muscles the thin filaments are uniformly about 1 μm in length; reconstituted filaments are of variable length, extending up to 10 μm or more. Reconstituted filaments consist of two helically wound chains in which each chain is composed of apparently identical globular subunits about 50 Å in diameter; each subunit is an actin molecule. Similar chains are present in the native filaments (see Fig. II-67), but there are several hundred subunits in native filaments and a thousand or more in reconstituted filaments.

Collagen, the major protein of connective tissue, appears in the electron microscope as thick fibers with distinct crossbandings reflecting orderly arrays of constituent molecules (Fig. IV-3). Tropocollagen, the collagen subunit, consists of three intertwined polypeptide chains. Tropocollagen can be extracted from connective tissues by cold salt solutions. When the solutions are warmed, collagen fibers form spontaneously and show crossbanding identical to that seen in connective tissues.

In the organism, fibroblasts apparently secrete *procollagen* into the external milieu where it is enzymatically cleaved to tropocollagen (Section 3.6.1). Collagen fibers can then form by self-assembly. Subsequently the fibers are stabilized by enzymatically produced covalent bonds between the tropocollagen molecules. (A similar sequence of events starting with the circulating protein *fibrinogen* is involved in the formation of the fibrin fibers in blood clots.)

Collagen fibers can also be dissociated into tropocollagen by dilute acids. When the acid solution is neutralized, fibers reconstitute. However, if ATP or certain negatively charged macromolecules are added to the acid solution, the reconstituted fibers have crossbands spaced 2800 Å apart, rather than 700 Å as in native collagen. This altered crossbanding indicates a different arrangement of tropocollagen molecules.

These observations on self-assembly systems lead to interesting conclusions. The mechanisms that control the sizes of some self-assembling structures must be sought. For actin it is suspected that other proteins in the muscle cell cooperate in formation of thin filaments to terminate the addition of subunits when the proper length is reached. The width of collagen fibers in connective tissue varies from place to place in the body. The differences may be due to variations in the rate of release of tropocollagen resulting in different rates of fiber formation, or to variations in the concentration of ions or other substances that interact with collagen.

Also, it appears that the same set of subunits may aggregate in different patterns. Conditions needed to make tropocollagen form fibers that do not have the 700-Å spacing are very rarely found in the body. However, collagen fibers themselves are the subunits of larger aggregates, and these aggregates are of varying form. In tendons, collagen fibers are grouped in thick bundles, while in some other connective tissues they form large flat layers. Probably interaction of collagen with other connective tissue components influences the form of the aggregates.

4.1.4   MICROTUBULES;   Bacterial flagella (Section 3.2.4) can be dis-
BACTERIAL   persed into separate molecules of the pro-
FLAGELLA   tein flagellin. In the presence of short pieces
broken from intact flagella, the flagellin molecules rapidly reconstitute long fibers similar to flagella. The fragments serve as "seeds" for the growth of the fibers; flagellin molecules attach to them in oriented fashion.

The flagella are relatively straight in some bacterial strains and wavier in others. If long fragments are used to "seed" growth, then fragments from "wavy" flagella combined with dispersed flagellin molecules from "'straight" flagella produce reconstituted fibers that are wavy. If both fragments and flagellin are from "straight" flagella, reconstitution results in straight fibers. These experiments suggest that the flagella can grow by adding additional subunits in oriented fashion. Most likely, binding sites are present that control the addition of subunits (Fig. IV-2).

Section 2.11.1 outlined evidence that microtubules can form in the test tube by self-assembly, and Section 2.10.4 described the growth of

Fig. IV-3   *Collagen fibers shadowed with evaporated metal atoms to bring out their three-dimensional appearance. The prominent pattern of crossbanding is evident. × 50,000. (Courtesy of J. Gross.)*

microtubules in flagella by addition of subunits to preexisting tubules. Microtubule formation appears to be initiated at "organizing" or "nucleating" sites. Basal bodies, satellite bodies near centrioles, and other structures are thought to function in this capacity. Once started, tubule elongation can continue by sequential addition of subunits. How elongation is terminated once the proper length has been reached is not known. There is no obvious intrinsic limit to the lengths tubules can attain, and both short cilia and long flagella are made of the same basic units. Another point requiring elucidation is how special sites, such as those to which dynein (Section 2.10.3) binds, come to be present along certain tubules.

4.1.5   **VIRUSES**   Those *spherical* viruses with simple polyhedral shapes are thought to form by the assembly of protein subunits constructed in such a way that they assume automatically the proper three-dimensional array. For simple closed surfaces such as polyhedra or spheres it is not difficult to conceive of subunit shapes that will associate in a structure of fixed size and shape (Fig. IV-2). Note that in such a case, the control of size may not be a problem; no sites are left for additional subunits. This is a *self-limiting* structure.

The cylindrical tobacco mosaic virus can be dissociated into RNA and separated protein subunits. The proteins can reassemble alone to form cylinders of two alternative types, only one of which has the subunits arranged as in the native virus (Fig. IV-4). If RNA and protein

Ⓐ

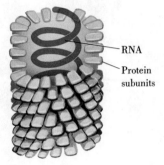

Intact virus — consists of RNA surrounded by helical array of protein subunits. The RNA and protein subunits can be separated and their behavior studied in the test tube.

RNA

Protein subunits

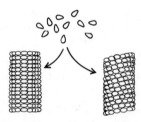

Isolated protein subunits can aggregate in two arrays, helical and nonhelical. The helical array is similar to the intact virus.

Isolated protein subunits mixed with isolated RNA reconstitutes the virus with the protein subunits arranged as in the native virus.

Ⓑ

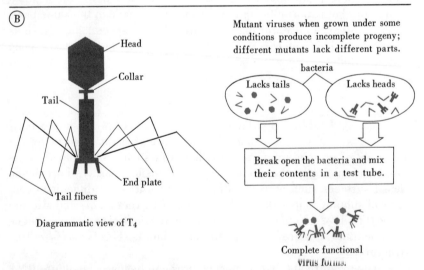

Mutant viruses when grown under some conditions produce incomplete progeny; different mutants lack different parts.

Head

Collar

Tail

End plate

Tail fibers

Diagrammatic view of T₄

bacteria

Lacks tails

Lacks heads

Break open the bacteria and mix their contents in a test tube.

Complete functional virus forms.

Fig. IV-4 *Viral reassembly in the test tube (see Fig. III-1 for electron micrographs of the viruses). **A.** Experiments with tobacco mosaic virus. **B.** Experiments with T₄ bacteriophage. (From the work of Frankel-Conrat, Kellenberger, Wood, Edgar, and others.)*

subunits are mixed under the proper conditions, viruses are reconstituted. Both nucleic acid and protein subunits have the same arrangements as in the native viruses.

Different mutants of the bacterial virus called T_4, which has a complex morphology, make incomplete viruses when they replicate in bacteria. Some fail to make heads; others cannot make tails or parts of tails. If test-tube mixtures of incomplete viruses are made so that all the parts are present, complete viruses may form (Fig. IV-4). However, there are restrictions. In the systems so far studied, for example, tail filaments will not attach to tails unless a head is attached (at the other end).

Two of these findings are of particular interest. In the tobacco mosaic virus experiments the RNA not only carries the genetic information for the structure of individual protein subunits, but the nucleic acid also appears to act as a core or scaffold that interacts with the protein subunits in promoting association into one of several alternative configurations. In addition, the RNA is of fixed length and probably determines the length of the assembled structure. In a sense the RNA may be operating as a system that is both self-duplicating and can also specify the three-dimensional ordering of other molecules (as discussed in the introduction to this part). However, the final disposition in the virus of the RNA and proteins can probably best be thought of as reflecting mutal interaction in which the RNA "helps" align the protein molecules, and the interaction with the proteins contributes to determining the helical arrangement of the RNA.

The observations on T_4 indicate that the formation of complex structures may depend on specific *sequences* of assembly. In normal virus formation, heads, tails, and tail filaments apparently all assemble from their components as separate structures. Completed heads and tails associate spontaneously; once this has occurred, tail filaments are attached. Still unsolved is how the DNA, which is many micrometers long, becomes packaged within a head with a diameter of only a fraction of a micrometer. For a number of viruses the available evidence suggests that prior to the formation of a mature virus, the DNA is in the form of very long molecules, each composed of multiple copies of the viral sequences. The DNA for an individual virus is enzymatically cleaved from such a molecule prior to or during its entry into the virus.

For various viral strains, proteins coded for by the viral nucleic acid are synthesized within the cell and function in the assembly of the virus, but do not become incorporated in the finished structure. Certain of these are enzymes. For example, in T_4 an enzyme may function in the attachment of filaments to the tail. Since they can catalyze the formation of specific bonds between specific groups, the participation of enzymes in assembly processes adds a level of complexity and potential

control. Other, apparently nonenzymatic, proteins function as assembly
"scaffolds." They interact with other viral proteins to promote a specific
three-dimensional arrangement, but once this arrangement is complete,
the scaffold proteins dissociate from it. Few of the molecular details of
such events are clear.

Bacteriophage T_4 is the most complex assembly system studied
thus far. About 40 different genes control the sequences of T_4 head, tail,
and tail filament proteins and of other proteins that affect the structure
of the virus. By comparison with a mitochondrion or chloroplast, this is
a fairly simple system.

4.1.6   ***CELL***   Leaving aside microtubules (Section 4.1.4),
   ORGANELLES   most progress in analysis of organelle assembly has been made with ribosomes.
When ribosomal proteins and RNAs are mixed, functional ribosome
subunits can form (Section 2.3.2). From reconstitution experiments in
which various proteins are omitted it appears that certain of the proteins
must be associated with the RNAs if others are to associate. Probably
specific protein–RNA complexes form a basic core structure to which
other proteins can then attach.

Sections 2.4.4 and 2.5.4 summarized evidence relating to the formation of the membranes of the ER, Golgi apparatus and plasma membrane. The *tentative* conclusions are (1) the pertinent processes involve, at least to some extent, the passage of membrane from one cell
compartment to another, such as from the Golgi apparatus to the cell
surface; (2) this may be coupled with molecule-by-molecule growth
maintenance, and replacement mechanisms which might contribute to
the transformation of one type of membrane into another.

Evidence is also fragmentary for organelles such as mitochondria
or chloroplasts. Still to be elucidated are the pathways by which the
molecules made outside these organelles are brought together with
those made within (Sections 2.6.3 and 2.7.4). We have outlined reconstitution experiments (Section 2.6.5) and work on membrane formation
(Section 2.7.4), indicating that the membranes might grow by addition of
molecules or of previously assembled multienzyme complexes. But we
know almost nothing of the controls governing the arrangements of the
membranes in the organelles. Why are the mitochondrial cristae tubular
in some cell types and flat in others? What processes generate the
complex arrays of grana and stroma thylakoids in plastids? There are a
few observations that may eventually provide experimental "handles;"
for example, abnormalities have been observed in the stacking of
chloroplast membranes in mutants of Chlamydomonas (Section 2.7.3).

Earlier (Section 3.6.2) it was shown that the nematocysts of Hydra have a complicated structure which arises inside a vacuole originating from the Golgi saccules. A wide variety of proteins and other substances is probably present inside the vacuole. From these, the capsule, lid, coiled thread, stylets, and barbs are fashioned. All these have a morphology that is specific for a given species; therefore they must be under genetic control. The structures develop at a time when the endoplasmic reticulum and Golgi apparatus are rapidly regressing, suggesting that protein synthesis has been turned off and that little, if anything, is being added to the vacuole.

An interesting possibility is that nematocyst structures within the vacuole form, like the T_4 virus, by the separate assembly of individual parts, with the presence of binding sites on each part for the attachment to other parts and by the sequential addition of parts to form a whole structure. Perhaps enzymes present in the vacuole are involved. This kind of assembly process has not been demonstrated for nematocysts, and the possibility cannot be excluded that more complex processes are involved. However, it would be interesting to attempt isolation of individual nematocysts at a time when protein synthesis by the cnidoblast is turned off and when visible differentiation of nematocyst structures has not yet progressed. If the notion of T_4-like assembly within a vacuole containing all necessary components has any merit, differentiation of structure might well proceed in isolated nematocysts. The next experimental steps would be the extraction of nematocyst proteins and attempts to reconstitute the complex structures in the test tube.

c h a p t e r 4.2
CELL DIVISION IN EUCARYOTES

4.2.1   **CHROMOSOME CONSTANCY AND MITOSIS**   That genes are arranged in linear order on chromosomes was one of the great discoveries of this century. It opened the route to a series of studies that have led in recent years to the analysis of heredity in terms of DNA base sequences. Some major conclusions emerging from these studies are relevant to the discussion here.

(1) In dividing and nondividing cells the nuclear DNA is located in the chromosomes. When division is not occurring, the chromosomes are usually uncoiled, and the DNA participates in controlling cell structure and function. During division the chromosomes coil into compact structures whose behavior accounts for transmission of DNA to daughter

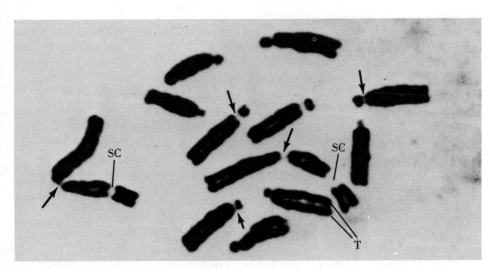

Fig. IV-5 *The twelve chromosomes (six pairs) of the bean* Vicia *during mitotic metaphase (see Fig. IV-8). Each consists of two chromatids (T) and each has a kinetochore (arrows) seen as a depression or lightly stained region of the chromosome. The chromosomes differ in length and position of the kinetochore. Two of the chromosomes, members of one pair, show a second "gap" (lightly stained region) along their lengths, the secondary constrictions(SC). These are the nucleolar organizer sites, the places where the nucleolus is attached when it forms (see Fig. II-18). × 2000. (Courtesy of J. E. Trosko and S. Wolff.)*

cells. Chromosomes of a dividing plant cell are shown in Figure IV-5. Note that the chromosomes are of different lengths and that recognizable specialized regions are visible. *Kineotochores* (also called *centromeres* or *primary constrictions*) are present in all chromosomes; they function in chromosome movement during cell division. *Secondary constrictions* are present only on a few chromosomes; they include the sites of nucleolar organizers. Each chromosome is divided longitudinally into two identical *chromatids*; this is characteristic of cells soon to enter division.

(2) Figure IV-6 shows the human *karyotype* (chromosome complement of man). The chromosomes of one cell have been arranged to demonstrate the fact that they occur in pairs of similar (*homologous*) chromosomes, referred to as *homologous pairs*. Cells in which the chromosomes are present in such pairs are referred to as *diploid*. The pairs may differ considerably from one another in sizes of chromosomes and in position of kinetochores. Although unicellular organisms and many lower plants differ from this pattern (see Section 4.3.1), most cells of usual multicellular animals or of higher plants contain identical dip-

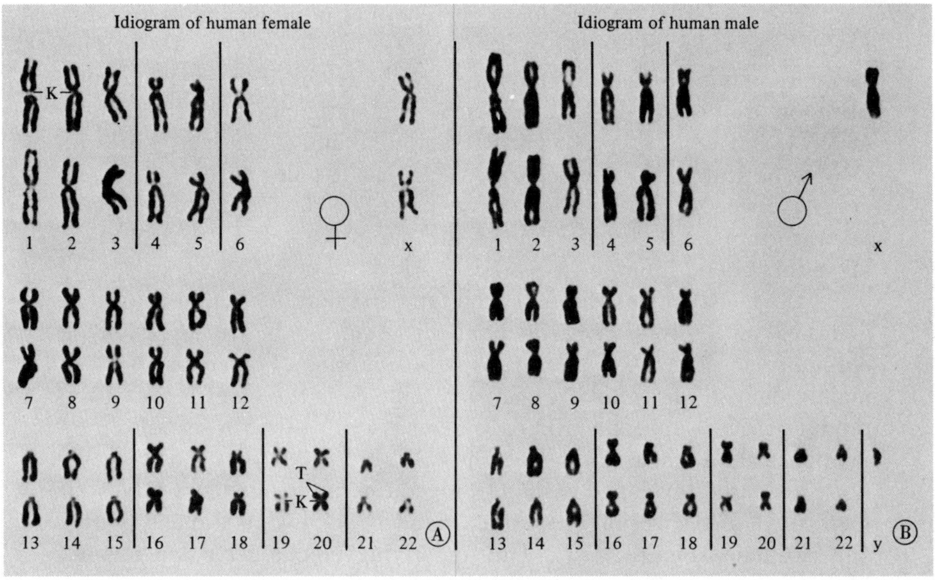

Idiogram of human female   Idiogram of human male

Fig. IV-6   Metaphase chromosomes (Fig. IV-8) from human cells. The cells have been spread on a slide, stained, and photographed. The chromosomes have been cut from the photographs and arranged as an "idiogram" showing the 22 pairs of "autosomes" and the two "sex chromosomes" (in females, two X chromosomes, in males, one X and a smaller Y chromosome). K indicates kinetochores, and T, chromatids. × 2000. (Courtesy of T. Puck and J. H. Tijo.)

loid sets of chromosomes. The major exceptions are the mature sex cells which are *haploid*; only one chromosome of each pair is present. Also, especially in plants, some cells may contain twice (or greater) the normal total number of chromosomes with corresponding numbers of extra identical copies of each chromosome (polyploidy; Section 4.5.1).

(3) Most organisms of a given species have the same karyotype. Different species differ in chromosome number, which varies from two to several hundred, and in chromosome morphology (lengths, positions of kinetochores, and so forth).

(4) A particular chromosome in a given species carries a particular set of genes controlling the characteristics of that species, and nonhomologous chromosomes carry different genes. (Differences between homologous chromosomes will be discussed in Section 4.3.1.) A gene may be thought of as a DNA segment, usually with the information necessary to specify, via mRNA, the amino acid sequence of a particular polypeptide chain or protein molecule. (However, some genes produce RNAs which are not translated into protein, for example, the tRNAs.)

The constancy of chromosome type and number among the different cells within an organism is dependent on mitosis (Fig. IV-8). The two chromatids of each chromosome are the result of chromosome duplication prior to division. Chromosome duplication includes replication of DNA and results in two chromatids containing identical copies of DNA. The chromatids are separated to opposite poles of the spindle; in consequence, both daughter cells possess one daughter chromosome derived from each chromosome of the parent cell. Both thus contain identical genetic information.

The constancy of chromosomal type and number within a species is based on reproductive mechanisms resulting in transmission of chromosomes from parent organism to offspring. For some species, mainly those of unicellular organisms, mitosis is the chief mode of reproduction. The special chromosome behavior involved in sexual reproduction will be discussed in Section 4.3.1.

Recently developed staining procedures utilizing colored dyes or fluorescent dyes, result in characteristic banded appearances of the chromosomes (Figure IV-7). The underlying mechanisms are incompletely understood. Banding may partly reflect variations along the chromosome in the frequency of different bases in DNA, and in features of the chromosomal proteins. The techniques have proved very valuable for chromosome identification. The banding pattern for a given chromosome in the karyotype is constant from cell to cell, whereas chromosomes that may differ little in size or kinetochore position, frequently are readily distinguishable on the basis of their banding (compare Figs. IV-6 and IV-7).

4.2.2   **THE CELL**   The life cycle of dividing cells is usually
            LIFE CYCLE   divided into (1) *interphase* and (2) the mi-
                            totic division stages *prophase*, *metaphase*,
anaphase, and *telophase* (Fig. IV-8). Interphase was originally given this name because it is the stage between successive mitotic divisions and usually shows few dramatic chromosomal changes that are readily recognizable in the microscope. It was once regarded as a resting stage. However, interphase is now known to be the period of intense cellular metabolic activity; cells usually spend most of their time in interphase.

Cytochemical and autoradiographic evidence indicates that DNA synthesis (replication) in preparation for division occurs during the interphase prior to division. In interphase a period of DNA synthesis (S), a pre-S period (G_1), and a post-S predivision period (G_2) (Fig. IV-9) are recognized; these are distinguishable in experiments like the following.

If tissue culture cells are exposed very briefly to tritiated

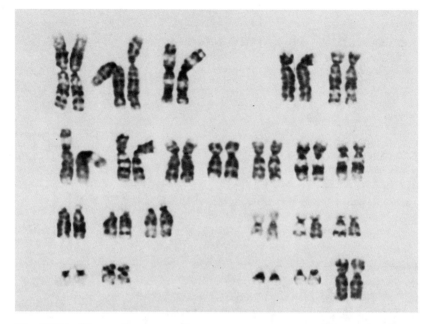

Fig. IV-7 *Human karyotype from a preparation stained by a method that produces a banded appearance of the chromosomes. The cells used were from a female; the two X-chromosomes are at the lower right corner. The method used employs a staining mixture known as Giemsa's mixture and thus the banding is sometimes referred to as "G-banding." (Courtesy of T. C. Hsu.)*

thymidine (Section 1.2.4) and then grown in nonradioactive medium, only those cells actively synthesizing DNA during the brief exposure will incorporate label. If the population is studied by autoradiography at intervals thereafter, it is observed that the first cells to enter division will show no label in their chromosomes during the division stages. These were in G_2 at the time label was given. They had already completed DNA synthesis and so incorporated no radioactive thymidine. After a time the division figures will begin to show label as the cells that were synthesizing DNA when label was given pass through G_2 and go on to divide. The time between exposure to tritiated thymidine and the *first* appearance of labeled cells in division is a measure of G_2, since it represents the time taken by cells almost finished synthesizing DNA to proceed to division. Similar approaches are used to determine the duration of the other interphase stages. For example, under some circumstances the percentage of cells found by autoradiography to incorporate ^{3}H-thymidine during a very brief exposure gives a rough measure of the length of the S phase as compared to the other cycle stages. This

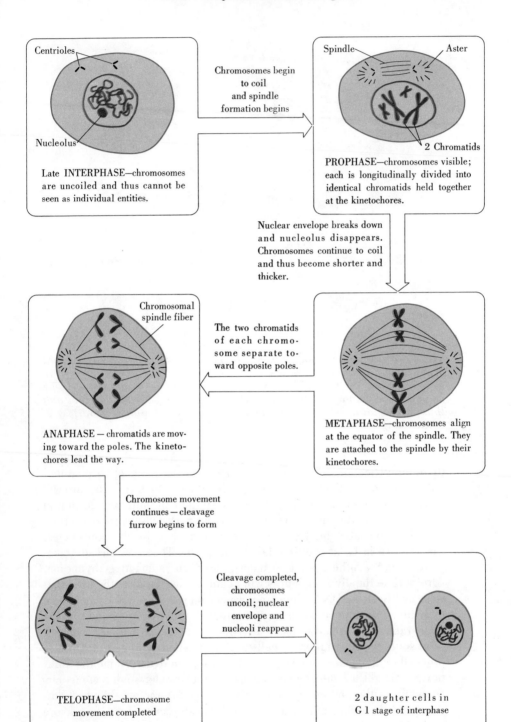

Centrioles

Nucleolus

Late INTERPHASE—chromosomes are uncoiled and thus cannot be seen as individual entities.

Chromosomes begin to coil and spindle formation begins

Spindle   Aster

2 Chromatids

PROPHASE—chromosomes visible; each is longitudinally divided into identical chromatids held together at the kinetochores.

Nuclear envelope breaks down and nucleolus disappears. Chromosomes continue to coil and thus become shorter and thicker.

Chromosomal spindle fiber

The two chromatids of each chromosome separate toward opposite poles.

ANAPHASE — chromatids are moving toward the poles. The kinetochores lead the way.

METAPHASE—chromosomes align at the equator of the spindle. They are attached to the spindle by their kinetochores.

Chromosome movement continues — cleavage furrow begins to form

Cleavage completed, chromosomes uncoil; nuclear envelope and nucleoli reappear

TELOPHASE—chromosome movement completed

2 daughter cells in G 1 stage of interphase

◄ **Fig. IV-8**   *Mitosis in an animal cell. The behavior of chromosomes is similar in virtually all eucaryotes, plants, and animals. Some of the other organelles vary in behavior as outlined in the text. See Figure IV-23 for one of the special features of plant divisions. Once the two chromatids have separated after metaphase, it is a matter of terminological convenience as to when they should be referred to as daughter chromosomes.*

is so because the probability of any given cell being in S when the label is presented depends directly on the percentage of the total cycle represented by S. (Note that these techniques assume that the population being studied is reasonably homogeneous in terms of the duration of the different stages of the cell cycle, and that the cells are distributed at random among the stages at any given moment—some are in mitosis, some in G_2, some in S, and some in G_1.)

Not only DNA synthesis, but also many other cellular activities, show regular variations during the different stages of the cycle. For example, RNA synthesis and protein synthesis decline sharply during mitosis; by metaphase very little RNA is being made. The synthesis of new histones for the chromosomes occurs chiefly during S.

	Interphase	Mitosis (M)
Sea urchin egg cleavage		
at 2 cell stage	0.25-0.5 hours	0.5 hours
at 200 cell stage	2	0.5
Plant root meristems	16-30	1-3
Cultured animal cells	12-24	0.5-2
Mouse intestinal epithelium	12-24	1

	G_1	S	G_2
Mouse intestinal epithelium	9	7	1-5
Cultured mouse fibroblasts	6	8	5
Bean root meristems	9-12	6-8	4-8
Ameba proteus	undetectable probably no G_1	3-6	30

Fig. IV-9   *Approximate rates of division and durations of cell cycle stages for some rapidly dividing cells. For a given cell type G_1 is the most variable period. Nondividing cells are sometimes thought of as remaining in a modified G_1 called G_0. (Data from the original observations of Howard and Pelc, from Mazia, and from others.)*

4.2.3   ***RATES AND*** Rates of division vary considerably among
            CONTROLS OF cell types (Fig. IV-9). The rates are de-
                DIVISION pendent on conditions of growth. For
                        example, *Ameba proteus* at 23°C divides
every 36–40 hours, but if the temperature is lowered to 17°C, this rate
slows to one division every 48–55 hours. Few eucaryotic cells divide as
rapidly as bacterial cells; a bacterial population may double in cell
number and mass every 15–20 minutes. Early cleavages of developing
eggs are relatively rapid, probably because no growth occurs between
divisions; the egg is separated into smaller and smaller cells. Such
cleavages may occur at rates of one per hour to several per hour. The
fastest are found in some insect eggs where only the nuclei divide in the
earliest stages of development (see Section 4.4.3 and Fig. V-2). Cells of
plant root tips or mammalian cells in culture double in mass and divide
every 12–24 hours. Most other cells divide much less frequently, and
some highly specialized cells, such as mature neurons, do not divide at
all. See pages 272–274 for growth and division of cultured cells.

A number of organs maintain populations of "unspecialized" cells
which divide relatively rapidly. Some of the division products differentiate; others remain as a reservoir for future divisions. In the intestinal
epithelium there is a continual loss of cells by sloughing from the tips of
the villi and a compensatory replacement by migration up the villi of
cells originating by division in the crypts (Fig. III-19). In the bone marrow "stem cell" populations divide and provide replacements for the
aged blood cells that are continually being destroyed (Section 2.8.2). In
plants, tissues known as *meristems* are present in growing regions of
roots, shoots, and other organs (Section 2.4.1). Some of the progeny of
mitoses in a meristem differentiate into the specialized cell types of the
organ.

If 70 percent of a rat's liver is removed, the remaining cells, which
normally divide very slowly, begin to divide at rates exceeding cancers
of rat liver. When the original mass is regained, cell division slows to the
original rate.

When embryonic blood-forming tissue, muscle cells, or cartilage
cells are placed in tissue culture under suitable conditions, they often
show an initial stage of rapid proliferation followed by a period when
multicellular aggregates form (the muscle cells fuse; Chapter 3.11). In
these aggregates cell division slows or stops. It is only then that large
amounts of specialized tissue products are synthesized: hemoglobin,
muscle proteins, or cartilage matrix. Interestingly, if the compound
bromouracil deoxyriboside (BUdR) is included in the growth medium,
the cells continue to divide but may not synthesize their specialized
products. BUdR is incorporated in replicating DNA in place of the
nucleotide thymidine, but beyond this, its mode of action is incom-

pletely understood. The fact that it seems to inhibit synthesis of specialized products but not of macromolecules needed for cell survival is of interest to developmental biologists concerned with the mechanisms leading to cell specialization.

It is sometimes said that the more complex or specialized a cell becomes in its morphology and metabolism, the more unlikely it is to divide, or that cells preparing to divide do not make specialized products. To some extent this may hold for various cell types (such as neurons or blood cells) in multicellular organisms. But there are no hard and fast rules. Many protozoa are quite complex and have specialized morphologies, but still divide rapidly. A more useful conclusion from the observations just considered and the studies on amebae, cancer cells, and cell cultures discussed in Section 3.3.2 and Chapters 3.11 and 3.12 is that cell growth, division, specialization, and organ size are under complex controls; these produce different behavior for different cell types, and for a given cell type under different conditions. The controls result, for example, in the fact that many cells which undergo repeated divisions almost precisely double their mass during the interphase preceding each division. As mentioned above, however, others (such as the cells in early divisions of developing embryos) may not grow at all between divisions. Perhaps the most important thing to be borne in mind is that *specialized* is a vague word. It is often used to designate cells that produce a unique product, such as hemoglobin, or those that have a special architecture, such as neurons. However, while every cell has divided at some stage in its history, it is not absurd to think, for example, of the plant meristems as groups of cells specialized for continued rapid division.

It is not yet clear whether the controls governing division rates result from activation or triggering processes (such as the accumulation of threshold amounts of some component) or whether division is initiated by the reversal of some inhibitory situation. Research with plants has revealed the intervention of hormones that stimulate division (auxins, cytokinins). For cell populations undergoing repeated mitoses, many investigators believe that the "clocks" governing the timing of cell cycle stages depend on cyclic alterations in intracellular levels of DNA-synthesis enzymes, cyclic AMP, ATP, or some unknown regulatory molecules. Thus work on the transcriptional, translational, and degradative events that produce specific patterns of increase and decrease of specific synthetic pathways during the cell cycle may eventually provide clues to cell-division timing mechanisms. Studies on cell proliferation associated with the repair of wounds in the skin or the lens of the eye, and on the regeneration of damaged organs such as the liver, have stressed the involvement of tissue-specific mitotic inhibitors. Inhibitory proteins called *chalones* have been tentatively identified in some tissues. The theory is that wounding somehow reverses the inhibition. (Alternative explanations rely on hypothetical mitotic activators supposedly released in damaged tissues.) As implied in our discussions of density-dependent growth inhibition (Chapter 3.11) and oncogenesis (Section 3.12.3), some pertinent agents may act at the cell surface, perhaps via cyclic AMP or other "second messengers" (Section 2.1.6). Thus lymphocyte divisions follow the binding to the cell surface of antigens (Section 3.6.4) or of certain lectins (Section 3.12.3) such as phytohemagglutinin, a protein isolated from beans.

Experiments may show that looking for *the* inhibitor or *the* activator will not be a fruitful approach in studying the control of division. Since many organelles and diverse activities are involved in division, several possible control points probably exist in a given cell type, and somewhat different controls might operate in different cell types.

Division of cells may be reversibly prevented by conditions such as exposure to abrupt temperature changes, to certain metabolic inhibitors interfering with nucleic acid metabolism, or to colchicine (Section 2.11.1) which disrupts the mitotic spindle. These treatments prevent cells from passing through one or another critical stage in the cycle necessary for initiation or completion of mitosis. If a population of cells is treated for a time period long enough to permit most cells to reach the same stage and to stop there, and if the conditions preventing division are then reversed, then synchronous division often occurs. Over 90 percent of the cells may divide at the same time. Normally, dividing cell populations are asynchronous; different cells divide at different times (early cleavages of a developing embryo being among the few exceptions). The availability of methods for artificial synchronization is an important asset for biochemical studies of division stages, since these techniques provide homogeneous populations for analysis. It should be possible to investigate the accumulation and utilization during the division cycle of components such as ATP, specific enzymes, or spindle proteins and thus, perhaps, to identify key control points.

Another promising approach is the study of the interaction of nucleus and cytoplasm by the direct experimental alteration of the cytoplasmic environment surrounding a given nucleus. In several different situations [adult toad brain nuclei transplanted into toad eggs; nuclei transplanted from amebae in G_2 to amebae in S; and hybrid cells such as the ones formed by fusing HeLa cells with chicken red blood cells (Chapter 3.11)] DNA synthesis is induced in nuclei that ordinarily do not synthesize DNA, by exposure to the cytoplasm of cells preparing for division. Chromosome coiling can be induced in S phase nuclei by fusion with cells in mitosis. Efforts are underway to identify the specific activating factors that are involved. Are enzymes or other proteins transferred from the cytoplasm to the nucleus? Is there some critical cytoplasmic concentration of specific metabolites in dividing cells, or are inhibitors absent? Hopefully such questions will be answerable in specific molecular terms, and insight will be gained into the remarkable coordination of nuclear and cytoplasmic events during division.

Mutations disrupting cell division are difficult to study because ordinarily they prevent the obtaining of large populations of affected cells. However, analysis of cell division is now being aided by "temperature-sensitive" mutants blocked at one stage or another of the cell cycle. Such mutants are essentially normal when grown at one temperature, but abnormalities appear as the growth temperature is changed (usually raised). Probably these effects depend on the fact that the three-dimensional structure of proteins can be strongly affected by temperature, so that raising the temperature can magnify a minor defect. With bacteria, study of such mutants has led to the conclusions that specific proteins of still uncertain function must be synthesized to initiate DNA synthesis. The mechanism regulating the actual division of the bacterial cell normally seems tied closely to, or even triggered by, the completion of chromosome duplication, but the ties are disrupted in some mutants so that offspring lacking chromosomes may be formed. For eucaryotes, studies have begun with yeast and with algae. Yeast normally duplicate by mitosis, coupled with the budding of a small daughter cell. Mutants have been isolated in which chromosome duplication does not occur, but many of the cytoplasmic events of bud formation take place nonetheless. There are other mutants in which the budding is selectively inhibited. It is inferred that the sequence of events yielding budding are partially independent of those involved in nuclear replication. Efforts are underway to identify the factors that coordinate nuclear and cytoplasmic sequences in the normal cell.

4.2.4   ***CHROMOSOME***   As we use the terms in this book a *molecule*
        DUPLICATION   or *duplex* of DNA refers to a *double helix*,
                                      composed of *two polynucleotide strands.*
DNA duplicates by a "semiconservative" mechanism (Figs. II-9 and IV-10). The double helix separates into its two polynucleotide strands. Each strand remains intact and acquires a new complementary partner that is formed by the sequential alignment of nucleotides along the old strand by base pairing and the enzymatic linking of the nucleotides into a polynucleotide chain. Figures V-1 and V-2 are micrographs showing duplicating DNA. Important gaps remain in our understanding of the enzymatic details of the process. What is known comes chiefly from work on bacteria and viruses. Much progress is to be expected in the

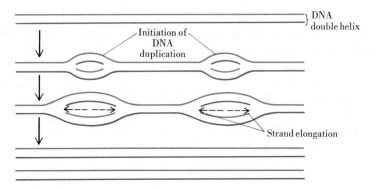

Fig. IV-10 *Duplication of DNA. Two adjacent replicating units are shown (See Fig. V-2 for electron micrographs of this process.)*

near future. Although the matter is still in dispute, work on viruses has led some investigators to believe that the initial segment of a new forming DNA strand is a short sequence of RNA which later is degraded. Elongation of a forming DNA strand depends on the interplay of DNA polymerases, other enzymes and proteins that unwind the "old" double helix. In several eucaryotes, as well as procaryotes, it has been shown that replication is bidirectional; each newly forming DNA strand elongates at both ends (Figs. IV-10 and V-1). For eucaryotes the rate of elongation of a given growing end is on the order of 1μm per minute (1 μm of DNA contains 3000 base pairs); procaryote DNA growth is roughly 10 times faster. Elongation is not simply a matter of adding nucleotides one at a time to the growing strand. Rather, at one, or perhaps both ends of the strand, the nucleotides soon to be added are first linked together into a short chain and this chain then is enzymatically joined to the growing strand. [The two ends of a polynucleotide differ in features related to the polarity of the sugar-phosphate backbone (Fig. II-9) and there are corresponding differences in details of their elongation.]

In *E. coli* and other bacteria the chromosome is a single circular molecule of DNA. Replication begins at a specific point of the circle and proceeds in both directions from that point. In higher organisms the chromosomes are larger, more complex, and more numerous. They consist of DNA, histone and nonhistone proteins. However, in some respects the duplication of chromosomes parallels the replication of DNA, as in bacteria.

When cells of higher organisms are exposed for a few minutes to tritiated thymidine and then grown in a nonradioactive medium and studied by autoradiography, not all the chromosomes of a given cell are invariably labeled. Those that are labeled show radioactivity at several discrete, and often quite distant, points along their length. These findings indicate (1) that different chromosomes may duplicate at different times during interphase (only those that are duplicating during the brief exposure to label become radioactive); and (2) that several regions replicate simultaneously on a given chromosome (if only one replication point per chromosome were operating, only one site would be labeled). Of interest is the fact that the sequence of DNA replication in different chromosomes and chromosome portions appears to be the same in successive divisions of a given cell type.

The conclusion from such observations is that in its replication, the DNA of a eucaryote nucleus behaves as if composed of several thousand replicating units (replicons), each on the order of 10–50 μm long (see Figs. IV-10, V-1, and V-2). The mechanisms coordinating the initiation of replication in each unit with that in the others are presently unknown. The DNA made in each unit eventually is joined enzymatically to that made in adjacent units.

Once the DNA is replicated, how is it apportioned between daughter chromosomes? An experiment designed to answer this question is outlined in Figure IV-11. The key finding is that the distribution of label in chromosomes after the second duplication is different from that at the first duplication following label administration. This is interpreted as indicating that, as in bacteria, the DNA of a duplicating chromosome is distributed between the products of duplication (the chromatids in Fig. IV-11) as if it were in the form of one long DNA double helix. That is, in chromosome duplication two longitudinal subunits separate from one another, each forms a new partner, and the two chromatids of a duplicated chromosome consist each of one old and one newly built subunit. Presumably the subunits are equivalent to single strands in a DNA double helix. The words "equivalent to" are important. This experiment by itself does not establish unequivocally that there actually is only one long DNA duplex per chromosome before duplication and per chromatid after duplication. However, as summarized in the next section, additional more direct evidence does tend to support such a straightforward conclusion.

The findings so far discussed concern only the DNA of the chromosome. New chromosomal proteins are also made during interphase, but few details of the processes by which they become associated with the DNA are understood. When chromosomal proteins in rapidly dividing cell populations are labeled through incorporation of radioactive amino acids, some radioactivity in the chromosomal proteins is still often detectable several generations later. This suggests that at least some chromosomal proteins can be conserved and distributed among daughter chromosomes. The extent to which such protein conservations hold for different classes of chromosomal proteins is not

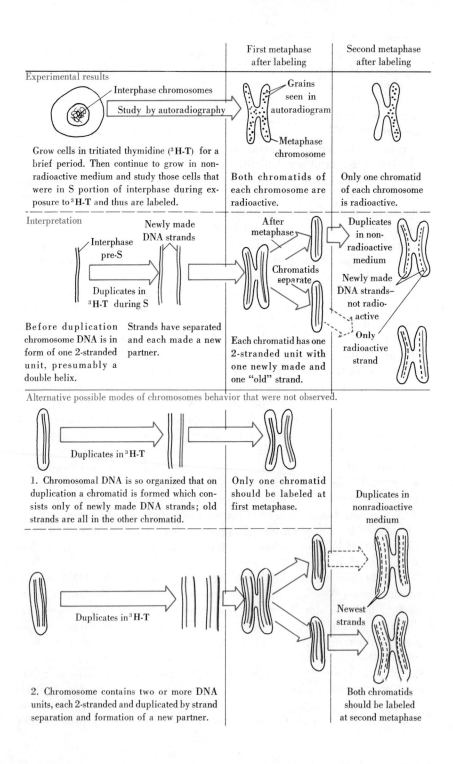

	First metaphase after labeling	Second metaphase after labeling

Experimental results

Interphase chromosomes

Study by autoradiography

Grains seen in autoradiogram

Metaphase chromosome

Grow cells in tritiated thymidine (^{3}H-T) for a brief period. Then continue to grow in non-radioactive medium and study those cells that were in S portion of interphase during exposure to ^{3}H-T and thus are labeled.

Both chromatids of each chromosome are radioactive.

Only one chromatid of each chromosome is radioactive.

Interpretation

Newly made DNA strands

Interphase pre-S

Duplicates in ^{3}H-T during S

After metaphase

Chromatids separate

Duplicates in non-radioactive medium

Newly made DNA strands— not radio-active

Only radioactive strand

Before duplication chromosome DNA is in form of one 2-stranded unit, presumably a double helix.

Strands have separated and each made a new partner.

Each chromatid has one 2-stranded unit with one newly made and one "old" strand.

Alternative possible modes of chromosomes behavior that were not observed.

Duplicates in ^{3}H-T

1. Chromosomal DNA is so organized that on duplication a chromatid is formed which consists only of newly made DNA strands; old strands are all in the other chromatid.

Only one chromatid should be labeled at first metaphase.

Duplicates in nonradioactive medium

Duplicates in ^{3}H-T

Newest strands

2. Chromosome contains two or more DNA units, each 2-stranded and duplicated by strand separation and formation of a new partner.

Both chromatids should be labeled at second metaphase

known, nor is it clear how long the most persistent of the proteins survive. In nondividing cells, at least some of the chromosomal proteins appear to turn over every few days, as do most other cell constituents.

The total nuclear protein content doubles during interphase. The doubling of chromosomal histones takes place during the S period. In cultured cells such as HeLa cells, studies with radioactive amino acid incorporation and cell fractionation at successive intervals indicates that the actual synthesis of the histones takes place in the cytoplasm with the proteins then migrating into the nucleus. Polysomes carrying histone mRNAs become detectable as this synthesis begins. Synthesis of nonhistone proteins has been less thoroughly analyzed; their increase occurs throughout interphase.

The morphology of chromosome duplication is difficult to study because the interphase chromosomes are uncoiled, and their structure is poorly understood. Indirect evidence based on the patterns of breaks induced by X rays in interphase chromosomes suggests that in some sense the chromosome begins to separate into chromatids early in S. As mentioned earlier, fusing cells in interphase with cells in mitosis will induce a premature coiling of the interphase chromatin. This may eventually aid in visualizing the structural rearrangements occurring in interphase. It is found, for example, that prematurely condensed S-phase chromatin appears fragmented.

4.2.5   ***CHROMOSOME***   A chromosome that measures, at mitotic
             STRUCTURE   metaphase, 5–10 μm in length and a mi-
                                 crometer or less in width may contain an
amount of DNA that would be several millimeters to several centimeters in length if stretched out in one straight double helix. "Packaging" of this DNA in chromosomes involves extensive coiling and folding. This is readily evident through light microscopy; chromosomes during cell division are seen to consist of coiled fibers (Fig. IV-12).

On occasion, by light microscopy longitudinal subdivisions are seen in chromatids. This suggests the presence within the chromatid of two or more subunits, seemingly much larger and more complex than

◄ *Fig. IV-11   J. H. Taylor's experiment on chromosome duplication. Taylor used the drug colchicine to suppress cell division (Section 4.2.8) without preventing chromosome duplication. Chromatids separate, but in the absence of a spindle, they remain in the same cell. This provides a convenient method for determining the number of times the chromosome set being studied has duplicated during the experiment. Each duplication results in doubling of the number of chromosomes per cell. Red strands are radioactive. See also Figure V-3.*

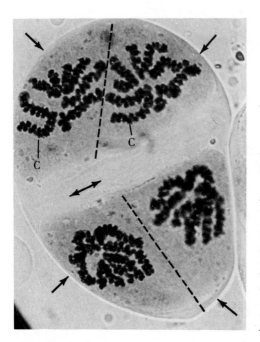

Fig. IV-12   *A microsporo-cyte of the plant* Trillium. *It has been fixed at the end of anaphase (second meiotic division) as the chromosomes (C) begin to uncoil. The two meiotic divisions (Fig. IV-24) produce four cells that give rise to male gamete nuclei. In the present case the first division took place along the plane of the double-headed arrow; the second would have been along the dotted lines. The single-headed arrows indicate cell borders.* × *1000. (Courtesy of A. Sparrow.)*

the two strands of a DNA helix. The nature of such subunits is not clear, but it is conceivable that the observed subdivisions are artifacts of one sort or another. However, such observations and some other indirect evidence have contributed to a long debate between those who believe that the chromosome is "uninemic" and those who believe it is "multinemic." The suffix *-nemic* denotes "threadlike" and in this context refers to a hypothetical basic unit of chromosome structure. A uninemic chromosome would contain one longitudinal structural unit that presumably is equivalent to a DNA duplex. A multinemic chromosome might contain several such units lying side by side.

The "current status" of the debate is as follows: Indirect evidence already outlined (Section 4.2.4) suggests the presence of a single functional unit equivalent to a DNA molecule in a preduplication chromosome or in a postduplication chromatid. It is not yet possible for electron microscopy to definitely associate such a functional unit with a structural unit and thus settle the "uninemic" versus "multinemic" issue at the morphological level. When carefully isolated whole (unsectioned) chromosomes are examined in the electron microscope, they appear to be masses of long fibers, as large as 100–300 Å in diameter (Fig. IV-14). No delimiting membrane is present. The fibers often show looping, as would be expected if there were actually one very long fiber folded into a compact chromosome. But technical difficulties in examining thick structures in the electron microscope and in minimizing the

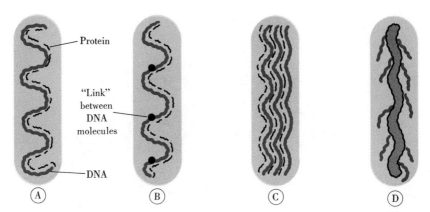

Fig. IV-13 *Schematic sketches of several hypothetical models of chromosome structure. **A.** One elongate coiled DNA molecule (double helix) with protein and other material closely associated. **B.** Several DNA molecules linked together end to end by unknown materials. **C.** Several to many DNA molecules arranged longitudinally in parallel strands or coiled as in a rope. **D.** A non-DNA backbone (for example, of protein) to which DNA molecules are attached at intervals. In all cases, the material in the actual chromosomes would be far more extensively folded and coiled.*

breakage of fibers and other alterations during preparation have hampered these studies. The presence of a few distinct fibers rather than one cannot be unequivocally ruled out, though most think this unlikely.

Sections of chromosomes show numerous fine fibers (*fibrils*), grouped in a variety of bundles. In diameter they are as small as 20–30 Å, the dimensions of a DNA molecule or DNA–protein complex (Fig. IV-19). Only a short length of a given fibril is included in a section so that it is impossible to tell whether the chromosome is of one or a few coiled fibers, of many independent fibrils, or of a complex, interconnected network. Attempts at serial-section reconstruction (Section 1.2.1) have been of limited value owing to the small dimensions of the fibrils and to their high degree of coiling.

Many efforts have been made to dissect chromosome structure by chemical and enzymatic procedures. Proteins may be removed from isolated chromosomes by digestion with proteolytic enzymes. When this is done, fibers normally 100–200 Å thick are replaced by much thinner fibers. Some feel that this may indicate that the 100–200 Å fiber is a bundle of several DNA molecules, bound together by protein. But proponents of the "uninemic" view are convinced that the 100–200 Å fiber is seen to contain only a single DNA molecule when protein is removed. Normally this DNA molecule is extensively coiled and cannot be seen because of the presence of much chromosome protein.

Some chromosomes are fragmented into shorter pieces when

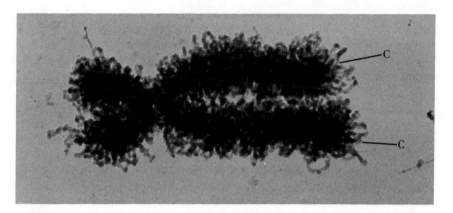

Fig. IV-14 *A human chromosome isolated at the metaphase stage of mitosis and viewed without sectioning. The two chromatids are readily visible (C); they are attached to one another at the kinetochore. Since very few ends of fibers are seen, it has been argued that each chromatid consists of one long looped and coiled fiber. Approx. × 50,000. (Courtesy of E. J. DuPraw.)*

treated with the enzyme DNase (Section 4.4.5) or when cells are grown in solutions of abnormal metabolites that interfere with DNA replication. This seems to indicate that the structural continuity of some if not all chromosomes depends on DNA; if DNA polynucleotide chains are interrupted, chromosome structure is interrupted.

Many models of chromosome structure have been proposed to explain the variety of observations (Fig. IV-13), but no one is universally accepted. One model pictures a chromosome as consisting of a continuous non-DNA backbone along which DNA molecules are attached at intervals. Experiments just cited indicate that this arrangement is unlikely, since it predicts that breaking DNA molecules might make the chromosome thinner but should not interrupt chromosome continuity. Some models propose that the chromosome is actually made of one elongate DNA molecule or that it consists of several molecules linked end to end, so that they behave as one. Because of their simplicity and because they agree with much of the available cytological and genetic information, these models are widely favored. Their validity is difficult to prove since it is difficult with available means to study DNA molecules that are very long. Long DNAs tend to fragment into shorter pieces during homogenization, centrifugation, and other isolation steps. However, an ingenious effort may have succeeded in circumventing this problem. Cells of several species have been carefully broken open (lysed) in a solution of detergents, proteolytic enzymes, and other components that open the cell very gently and simultaneously inhibit the cell's DNases that would otherwise degrade the DNA. The lysis is carried out within the chamber of a device used to measure *viscosity* properties of solutions. With such a device one measures rates of flow and related features of a solution and this provides information about the size and geometry of the dissolved molecules. Under the conditions used, cell lysates were found to contain individual DNA threads sufficiently long to account for all the DNA of a chromosome. Convincing evidence for end-to-end linking of chromosomal DNA molecules by non-DNA materials is not at hand, and therefore the tentative conclusion is that each thread is simply an elongate DNA duplex. A few skeptics still believe in more complex arrangements of the DNA.

The coiling and folding of DNA into a compact structure presumably depends largely on its association with proteins. How are the chromosomal proteins arranged? For the histones, a model now being debated suggests that they associate with the DNA to form a kind of repeating subunit along the chromosome. Nuclei of many species, animal and plant, contain five distinct type of histones, varying in molecular weights from 10,000 to 20,000 and in the relative proportions of the basic amino acids arginine and lysine. One histone type is relatively lysine-rich, two are arginine-rich, and two intermediate. How the lysine-rich histone is associated with DNA is uncertain. For the others it is proposed that two molecules of each type complex with each other and with a coiled portion of DNA producing a structure, containing 200 DNA base pairs and measuring roughly 100 Å in diameter. Such complexes would be repetitively arranged along the chromosome, being linked by DNA stretches poorer in protein. This specific model may require revision, but there is some intriguing evidence for some sort of repetitive substructure along the chromosome fiber. Under suitable conditions, chromatin fibers isolated from interphase nuclei, and "chromatin" reconstituted from purified DNA and histones, can appear as in Figure IV-15. Further, when isolated chromatin is gently and briefly digested with DNase, the fragments produced tend to contain 200 base pairs of DNA or a simple multiple thereof. This would be expected if the fibers were made of alternating regions of lesser and greater susceptibility to DNase, perhaps the "subunits" and the "links," respectively. Some call the subunits "nucleosomes." Evidently the chromosome resembles a coiled string of beads.

The architecture of the eucaryote chromosome is one of the outstanding unsolved problems of cell structure. When it is finally understood, it may bring quick resolution to problems such as the nature and control of chromosome coiling mechanisms, the arrangement of replication and transcription enzymes, and the structure and functioning of specialized chromosome regions such as the kinetochore. Although kinetochores can often be readily recognized by light microscopy (Fig. IV-5), it is only recently that the electron microscope has shown that their structure differs characteristically from the rest of the chromo-

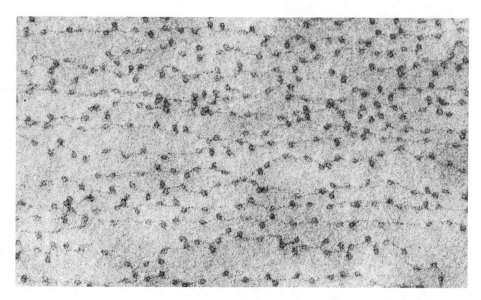

Fig. IV-15   *Chromatin prepared from a chicken red blood cell nucleus by a procedure involving swelling of the nuclei, fixation in formaldehyde, and negative staining. Note the numerous small globules (100 Å diameter) connected to each other by a fine fiber. × 170,000. (Courtesy of D. E. Olins and A. L. Olins.)*

some (Fig. IV-19). Before long it may be possible to describe such differences in molecular terms. Sections 4.4.4 and 4.4.5 will describe the structures of some special types of chromosomes.

4.2.6   **DNA AND CHROMOSOMES: WORK IN PROGRESS**   Given the likelihood that a chromosome contains a single longitudinal unit equivalent to a DNA double helix, how is the genetic information arranged in that molecule? The conclusion from classical genetic analysis and from the study of special types of chromosomes is that genes are arranged in linear order along chromosomes (Sections 4.2.1 and 4.4.4). Genes are usually detected by mutation—their presence is revealed by the discovery of inheritable changes in the organism's appearance, chemistry, or other characteristics. From the number of genes thus detected, plus estimates of the numbers not readily observed in this way, and from the work on some special chromosomes (Section 4.4.4) it seems that well-studied organisms, such as fruitflies or mice, have on the order of 10,000 genes. Relatively simple calculations like those outlined in Section 4.1.3 indicate that most eucaryotes have far more DNA than is needed to account for such gene numbers. Several suggestions for the significance of the "excess" DNA are being investigated. Perhaps the estimates of gene numbers should be revised (figures as high as 100,000 or more are being discussed for mammals). Repetition of sequences, as discussed for nucleolar organizers, accounts for some of the "excess." Some might be important in evolutionary terms—as "debris" left over in the evolution of genes, or as a kind of pool from which new genes arise (Section 4.5.1). The DNA might play "structural" roles, such as involvement in chromosome folding or pairing during meiosis (Section 4.3.1). Or some DNA, might have regulatory functions such as turning genes on or off. Recent progress in analyzing chromosomal DNA encourages optimism that much of the fuzziness in the following paragraphs will soon be eliminated.

Figure IV-16 illustrates studies on DNA "reannealing." The strands are separated by gentle heating, and upon cooling, strands with complementary base sequences will form duplexes by base pairing, as in RNA-DNA hybridization (Fig. II-10). The rate at which reannealing occurs depends upon the speed with which appropriate strands "find" one another during the random "collisions" occurring in solution. If, before the strands are separated, the DNA is fragmented into pieces a few hundred base pairs long, reannealing may have a complex time-course (Fig. IV-16). This is observed with virtually all eucaryotes studied, whereas viruses and procaryotes tend to show the simpler reannealing pattern illustrated in the same figure. The observations are interpreted to mean that a given nucleus contains several populations of DNA sequences that reanneal at vastly different rates. The most reasonable explanation is that the populations differ in frequency within the nucleus (Fig. IV-16). A proportion of the DNA usually ranging from roughly 40 to 75 percent for different organisms is "unique-sequence" DNA, that is, the sequences of base pairs are present once or a very few times per haploid chromosome complement ("per genome"). This is what one might expect for most genes in the traditional sense, that is, those coding for most proteins. Confirming this, a few specific mRNAs have been shown to hybridize with unique-sequence DNAs. The remainder of the DNA is "repetitive." It consists of various sets of DNA sequences. Some of these are present in hundreds or thousands of similar or identical copies per genome ("middle repetitive DNA). Other sequences are very highly repetitive; there are hundreds of thousands, or millions, of copies of such DNA sequences per genome.

From the few cases studied it is suspected that many of the very highly repetitive DNAs are of simple base sequences. For example, one of these DNAs in the fruitfly *Drosophila virilis* has the sequence ACAAACT in repeating stretches ("tandem arrays") up to hundreds of base pairs long. (ACAAACT is the sequence on one DNA strand; the partner strand has the complementary sequence). Other highly repeti-

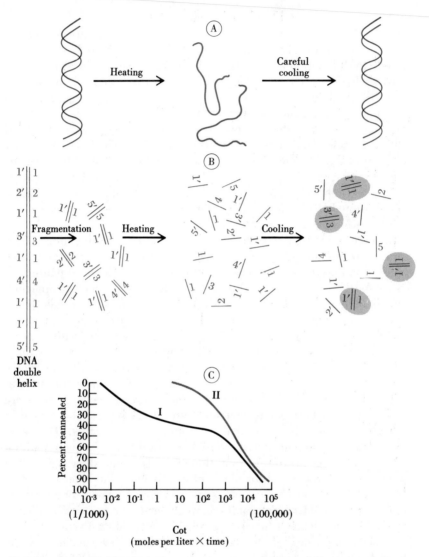

Fig. IV-16 DNA reannealing. Reannealing is also called reassociation or renaturation.

A. The reestablishment of a double helix by two DNA strands separated by heating.

B. Reannealing of DNA, containing repetitive and nonrepetitive sequences, that has been fragmented by ultrasonic vibrations or other means. The annealing has reached an intermediate stage; at a later time, the remaining single-stranded DNA will also return to the double-helical form (C). 1 and 1', 2 and 2', and so on, indicate complementary sequences present on the two DNA strands. The double-stranded 1-1' fragments form more rapidly during reannealing because there are several 1' fragments with which a given 1 fragment can reassociate; it need not "find" its original partner. For the nonrepetitive sequences (2, 3, 4, 5) there is but one partner.

tive DNAs of *D. virilis* differ from this, and more complex sequences are found in various other organisms. Such DNAs are unlikely to code for proteins. When the RNA present in a given cell is tested by hybridization with the DNA of that cell, there are no RNAs that will hybridize with these simple-sequence DNAs. In other words, the sequences appear not to be transcribed by the cell.

DNA density increases with increasing proportions of G-C base pairs. Thus DNA fragments especially rich or poor in G and C can be separated from other fragments by centrifugation on a density gradient (Fig. I-23). For example, G-C poor subpopulations may form distinct "satellite" bands separate from the main DNA band. By adding required precursors and polymerases such DNAs can be transcribed in the test tube, producing highly radioactive complementary RNAs. An *in situ* hybridization experiment (p. 82) utilizing RNAs made from a highly repetitive satellite DNA is shown in Figure IV-17. The results indicate that in the cell such DNA is localized in the heterochromatin near the kinetochores of most or all the chromosomes. This striking finding has been made for a number of species. Its significance remains to be determined. There has been speculation that it might reflect a special structural role of the DNA, or even some kind of participation in attachment of the chromosome to the spindle.

The nucleolar organizer DNA (Section 2.3.4) is of the middle repetitive type. One way of studying its organization is the use of the extra nucleoli produced in the amphibian oöcyte nucleus during maturation (Section 3.10.1). Each nucleolus contains a DNA molecule, probably circular, that is composed of many copies of the rRNA sequences. This DNA can be observed directly by isolating the nucleoli and spreading their contents on an electron microscope grid (Fig. IV-18). It consists of a cluster of tandemly-arrayed segments, each coding for a large rRNA precursor molecule and separated from one another by "spacer" segments that are not transcribed into RNA. Apparently the transcribed

---

C. Reannealing time course for eucaryote DNA. Curve I is the type of pattern generally obtained. If few or no repetitive sequences were present, a curve like II would be obtained. Each point on the curves represents the percent reannealing for a given "Cot" value. Cot is t (the time elapsed since reannealing started) multiplied by Co (the total concentration of DNA in the solution used for the experiment). The use of Cot facilitates combining and comparing data obtained using different Co's. Reannealing is faster at higher DNA concentrations. Thus for convenience and accuracy in constructing curves like the present ones, the relatively slow reannealing of the unique sequences (right-hand portion of curve I) is often studied with high Co's while the very rapid reannealing of the highly repetitive sequences (left-hand portion of the curve) is studied at low Co's.

(Based primarily on work by R. J. Britten and D. E. Koehne.)

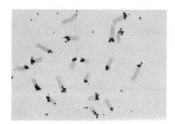

Fig. IV-17   In situ *hybridization preparation (see Fig. II-21) of mouse chromosomes, fixed at mitotic metaphase. After denaturation of their DNA, the chromosomes were incubated with highly radioactive RNA. (The latter was prepared by adding radioactive RNA precursors plus RNA polymerases to a highly repetitive "satellite" DNA fraction isolated by centrifugation from mouse nuclei. Thus the RNA is complementary to the highly repetitive DNA.) Clusters of grains are found close to the kinetochore present near the end of each chromosome, indicating that it is here that the highly repetitive DNA is located. × 750. (From M.L. Pardue, Science, **168**: 1356-1358; June 12, 1970. Copyright 1970 by the American Academy for the Advancement of Science.)*

region also contains "spacer" DNA, in the sense that some stretches of the RNA produced are degraded when the precursor is modified to generate the rRNAs (Section 2.3.4). The DNAs coding for 5S RNA (Section 2.3.2), for histones, and for tRNAs also are of the middle repetitive type and occur in clusters of many copies. But most of the middle repetitive DNAs are of unknown function. There seem to be many sets of such sequences, but the hundreds or thousands of copies in a given set often are not clustered; rather they are interspersed among unique-sequence DNAs. One model suggests that many of the unique-sequence DNAs are associated with a representative of one or another of the middle-repetitive DNAs. Evidence for this includes the fact that it is possible to prepare DNA fragments from *Xenopus* each of which contains a 1000-base-pair unique sequence linked to a 300 base-pair sequence of the middle repetitive type. There are indications that some hnRNA molecules (Section 2.3.5) as well as certain mRNA molecules (for example, in the slime mold) contain sequences coded for by DNAs of the intermediate repetitive type, in addition to sequences from unique DNAs. Clarification of the roles and organization of the middle-repetitive DNAs should be forthcoming before too long. Favored hypotheses are that they regulate the activities of unique-sequence DNAs or produce RNA segments important for intranuclear processing of RNAs.

Evolutionary considerations related to the repetitive DNAs will be discussed in Sections 4.5.1. It should be noted that the problem posed at the outset of this section still is not completely resolved. There is much more unique-sequence DNA than is accounted for by the traditional estimates of gene number.

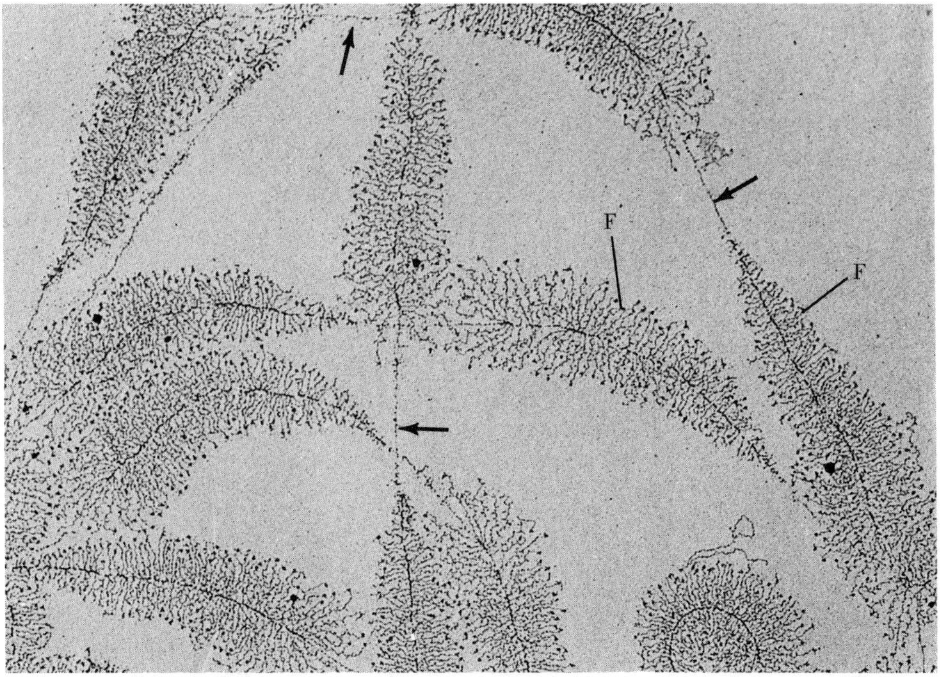

Fig. IV-18 *A part of the material isolated from the fibrillar core of one of the extra nucleoli of a salamander* (Triturus) *oöcyte (Section 3.10.1). Arrows point to the elongate DNA-containing fiber identifiable as such by its disruption in DNAse-treated preparations. Lateral filaments (F) protrude from the fiber and are grouped in regions 2–3 micrometers in length with each region showing about 100 filaments. 2 to 3 micrometers of DNA contain enough information to specify one of the large (about 40 S) rRNA precursor molecules, and each group of filaments is thought to represent many such RNA molecules being synthesized simultaneously by one gene. The filaments are known to be RNA from autoradiographic demonstration of the incorporation of appropriate precursors. The smaller filaments of a group are considered to be at an earlier stage in synthesis. Each group of filaments is separated from the next by "spacer" DNA that apparently is not transcribed into RNA (Section 4.2.6). × 20,000. (Courtesy of O.L. Miller, Jr., and B.R. Beatty.)*

4.2.7 SPINDLES, KINETOCHORES, AND CENTRIOLES

The precision of chromosome behavior in mitosis is related to the *spindle*. This highly organized, fairly rigid, gel-like region of cytoplasm has a fibrous appearance in the light microscope. Three types of spindle fibers seem to be distinguishable by light microscopy: *continuous* ones that pass from pole to pole, *chromosomal* fibers that attach to the chromosome kinetochores, and *interzonal* fibers that are present between two groups of chromatids as they separate at anaphase (Figs. IV-8 and IV-23). The interzonal fibers probably include the continuous fibers that persist after the chromatids separate under the influence of the chromosomal fibers.

In animal cells, spindles generally form in association with the centrioles (Fig. IV-8). At each pole an aster is often present. In the light microscope asters appear as roughly spherical arrays of fibers oriented radially around the centrioles and merging with the spindle. Centrioles and asters have not been seen in many plant cells.

Observations with the polarizing microscope resolved a long debate among light microscopists as to the reality of spindle fibers. The presence of oriented molecules, in a fiberlike organization, was shown in the spindle of living cells (Fig. IV-20). In the electron microscope spindles (and asters) show numerous microtubules (Fig. IV-19), in addition to ribosomes and vesicles (some of which probably derive from the endoplasmic reticulum). The fibers of light microscopy are shown to contain groups of microtubules. Some of the tubules terminate at the chromosomes; a few may be continuous from pole to pole. There are other types such as those spanning much but not all of the distance between the poles or reaching part way from chromosomes to poles.

The *mitotic apparatus* (spindles, centrioles, asters, and chromosomes) can be isolated as a unit from developing eggs (Fig. IV-21). The apparatus retains much of its original morphology, and chromosomes remain attached to it. However, it has not yet been possible to induce normal movement of chromosomes on isolated spindles, although progress in this direction is being made (Section 4.2.8). Much material is lost from the spindle during isolation, and there may be other alterations rendering it nonfunctional. As expected, some of the proteins isolated from the spindle resemble the tubulins of other microtubule-rich structures. But microtubules as such may account for only 10–25 percent of the spindle's mass. Little is known of the remainder. Spindle proteins are present in the cell in large amounts long before the spindle forms, and spindle formation probably involves primarily assembly of previously made subunits. Under appropriate ionic and temperature conditions, asters will assemble in homogenates (Fig. I-23) of fertilized clam eggs; centriole-related organizing centers seem to play a role in this. In similar homogenates, preexisting spindles are found to enlarge when tubulin-rich solutions are added. In the cell the mitotic apparatus can undergo reversible dissolution relatively easily (Section 4.2.8).

In most cells with centrioles, the spindle forms between the centriole pairs as they separate (in prophase). Centriole duplication occurs well in advance and may begin as early as telophase of the prior division. In interphase, as DNA synthesis progresses, each centriole of the initial pair is accompanied by a new daughter procentriole that will

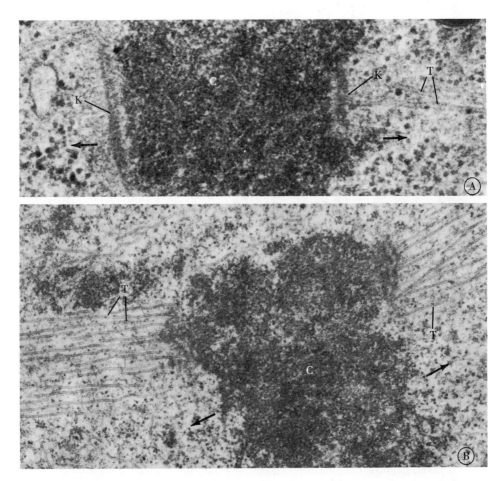

Fig. IV-19 *Chromosomes during mitosis as seen in sectioned material. A. Portion of a chromosome (C) from a cultured fibroblast (Chinese hamster). Kinetochore regions (K) are seen as special bands of dense material pointing toward the spindle poles (directions indicated by arrows). The structure of the remainder of the chromosome is difficult to interpret; it includes fine fibers sectioned at various angles with only a small portion of a given fiber region included in any given thin section. × 80,000. B. Portion of a chromosome in a cultured cell from the "rat-kangaroo" or Tasmanian Wallaby (Potorus). Spindle microtubules are indicated by T. (Courtesy of B. Brinkley and E. Stubblefield.)*

mature into a centriole oriented at right angles to the parent (Section 2.10.5). Generally the nuclear envelope breaks down in prophase as the chromosomes become associated with the spindle.

However, this pattern is not universal. Many plant cells have no

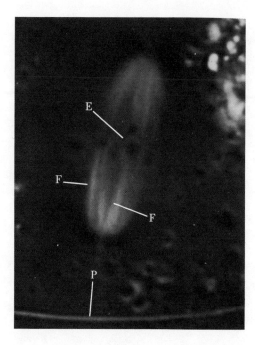

Fig. IV-20 The spindle of a living *egg of the worm* Chaetopterus, *photographed through a polarizing microscope (Section 1.2.2). The equator where the metaphase chromosomes (not visible in this photograph) are located, is indicated by* E. *Clear indication of the fibrous organization of the spindle may be seen at* F. *The plasma membrane is present at* P. × *1500. (Courtesy of S. Inoue.)*

obvious structures corresponding to centrioles. In some protozoa and other cells, the spindle forms within the nucleus, and the nuclear envelope does not break down. In other cases, even though the envelope remains, the chromosomes attach to an extranuclear spindle in a manner not understood. In a few fungi in which no centrioles are present, an intranuclear spindle forms between special thickened regions of the nuclear envelope. In some protozoa, even if the nucleus is removed from the cell, a spindle of continuous fibers forms between the centrioles. On the other hand, in some insect and plant cells, spindle fibers are seen to form initially in association with the kinetochores of the chromosomes.

If only a single cell type were studied, one or another structure would be assigned a primary role in controlling spindle formation. Usually it is the centrioles or kinetochores that are most obviously involved. However, until the formation of the spindle is understood in molecular terms, and we have more information on microtubule organizing centers (Section 2.11.1), it will be difficult to reconcile the variety of observations on spindle formation with the apparent uniformity of its function in cell division. Recent studies, including some on the test-tube reassembly of microtubules, suggest that both kinetochores and centrioles (or structures closely associated with centrioles) can "nucleate" the growth of microtubules from tubulin (see Section 2.11.1).

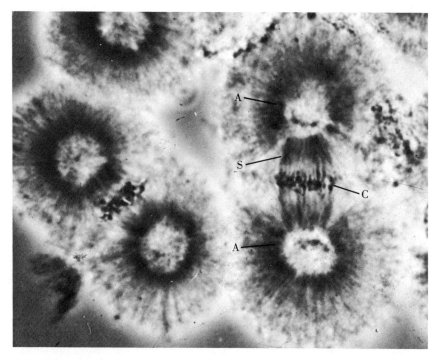

Fig. IV-21 *Phase-contrast micrograph of the mitotic apparatus isolated from sea urchin eggs. The spindle (S), asters (A), and attached chromosomes (C) are visible. (Courtesy of D. Mazia.)*

Among further questions, to which answers are still being sought, are the following. (1) How do the chromosomes attach to the spindle so that the two chromatids separate one to each pole, rather than both moving to the same pole? (2) How is the precise alignment of chromosomes at metaphase brought about? Is the notion that the alignment results from an equal "pull" from both poles valid, or are much subtler mechanisms also involved? (3) What role, if any, do asters play in cells where they are present? (4) Do special associations exist, during interphase, among chromatin strands, centrioles, and the nuclear envelope? (For example, some microscopists claim that portions of chromosomes may be attached to the envelope.) Do such associations account for the behavior of chromosomes during duplication and early stages of division? (For example, in some cells the chromosomes move within the prophase nucleus along a path paralleling the movement of the centrioles outside the nucleus.)

The participation of membranes in controlling the separation of chromosomes in cell division at some stages of cellular evolution is suggested by work on procaryotes (Section 3.2.4) and by observations on the unusual primitive eucaryotes, dinoflagellates. In dinoflagellates the chromosomes remain inside the nuclear envelope throughout division. They show attachments to the envelope that may be involved in their movement. In the cytoplasm adjacent to these attachments bands of microtubules are found, as if the tubules might play some structural or guiding role, or take part in the movement. But some investigators assert that the tubules themselves do not attach directly to the chromosomes; this matter is still under investigation.

4.2.8   ***CHROMOSOME*** In separating at anaphase, chromatids
   MOVEMENT move at rates of about one or two micrometers per minute (sometimes they are
faster). The movement often involves two components. In most cells, separating chromatids approach the spindle poles. In many cells the spindle as a whole *also* elongates, the ends move apart, and the chromatids separate without actually getting closer to the poles (Fig. IV–22).

When the kinetochore is near the middle of a chromatid, a characteristic V shape is seen, with the kinetochore in the lead as the chromosome moves in anaphase. This probably results from the effects of spindle fibers which pull on the kinetochores or exert some other influence leading to kinetochore motion with the rest of the chromosomes pulled behind. Chromosome fragments without kinetochores can be produced by treating cells with radiation or certain chemicals. Such fragments do not attach to spindle fibers and do not move.

Many mechanisms have been proposed to explain chromosome motion. The simplest idea to account for the poleward motion of chromosomes is that the spindle fibers contract. If the fibers were like muscles or like stretched rubber bands, they should thicken as the chromosomes move toward the pole. Although the spindle fibers shorten (sometimes to only 25 percent or less of their original length), no such thickening of the fibers and no appropriate changes in spindle microtubule thickness have yet been observed.

An alternative theory has been proposed, based upon observations indicating that there is continued movement of components within the spindle and that the spindle can readily undergo reversible disorganization.

(1) Chromosomes can be detached from the spindle by fine glass needles. They are able to reattach and move in normal fashion during division.

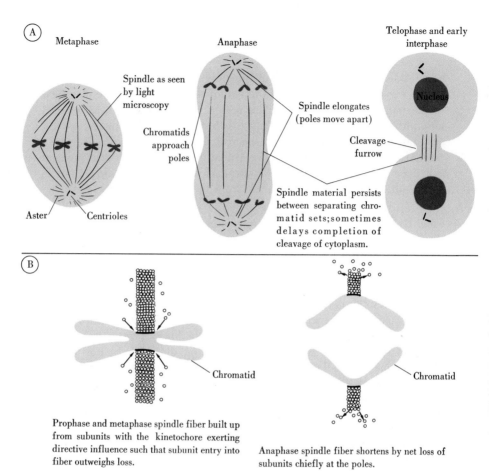

Metaphase Anaphase Telophase and early interphase

Spindle as seen by light microscopy

Spindle elongates (poles move apart)

Chromatids approach poles

Cleavage furrow

Nucleus

Aster Centrioles

Spindle material persists between separating chromatid sets; sometimes delays completion of cleavage of cytoplasm.

B

Chromatid

Chromatid

Prophase and metaphase spindle fiber built up from subunits with the kinetochore exerting directive influence such that subunit entry into fiber outweighs loss.

Anaphase spindle fiber shortens by net loss of subunits chiefly at the poles.

Fig. IV-22 *Chromosome movement during cell division. **A**. A diagram of a hypothetical animal cell showing the separation of sister chromatids. The relative contributions to chromosome movements of spindle elongation and of movement toward the poles vary in different cell types. **B**. A diagram of the theory developed by Inoue that during cell division spindle fibers grow and shorten by gain and loss of subunits. The diagrammed fibers are hypothetical structures presumed to be identical to the microtubule bundles and associated material found in cells. One key element in the theory being actively investigated is the nature of the coordinated operation of kinetochores and centrioles in controlling the assembly and disassembly of the fibers.*

(2) Exposure of cells to cold or to high pressure dissolves the spindle, and the spindle reappears upon return to normal conditions.

(3) Treatment of cells with heavy water, D_2O (where D is deuterium, a heavy isotope of hydrogen), "freezes" the spindle and prevents mitosis; probably the presence of D_2O strengthens interactions among hydrophobic groups and other relatively weak bonds holding the spindle structure together. The spindle structure thus is made more stable than is normally the case.

(4) Localized exposure to ultraviolet light will abolish the organization of the irradiated portion of the spindle. The effect can be observed in the polarizing microscope. The region of a spindle disorganized at metaphase or anaphase moves toward the pole and disappears, even if chromosome motion has not yet begun; this leaves the spindle with a normal organization. Other observations (Section 4.2.10) also suggest that even at metaphase, when the chromosomes are not moving, there is some sort of movement of material from the equator toward the pole.

(5) The drug *colchicine* dissolves the spindle. Like those of many other treatments (D_2O, altered temperature, or pressure), the effects of this drug on the spindle parallels its effects on other microtubule-rich structures (Section 2.11.1).

The main point of the theory that emerges from these considerations (Fig. IV-22) is that a spindle fiber attached to a chromosome is in dynamic equilibrium with subunits that are continually entering and leaving it. During prophase, as the spindle fiber forms, the gain of subunits outweighs the loss, perhaps owing to the influence of the chromosome kinetochores. During anaphase the situation is reversed, and the loss of subunits at the spindle pole outweighs the gain; as a result, the fiber shortens without thickening. In some manner, as the fiber loses subunits, the attached chromosome moves toward the pole (Fig. IV-22).

This still leaves open the question of what generates the motion. Does the shortening of the microtubules exert a force? Do the microtubules interact with one another in a fashion comparable to the sliding of muscle filaments or of the microtubules of cilia? Or is there interaction with other spindle components, yet to be characterized? (Are actin filaments or myosin involved? Recent work suggests that actin may be present in the spindle.) Conceivably the microtubules do not actually generate the poleward chromosome movement, but serve instead to orient motion produced by some other devices. Studies now underway on the changes in microtubule number, length, and distribution during mitosis should help clarify the situation.

Spindle elongation is usually thought of in terms of the pushing apart of the two poles. Most investigators propose that as part of elongation the spindle grows by addition of material at its center, or, perhaps more likely, at the ends of the microtubules of continuous fibers. Some maintain that elongation is a consequence of the sliding or other interactions of microtubules.

When dividing mammalian cells are partially disrupted in a solution rich in tubulins, chromosome movement continues. Efforts are now underway to determine how much of the cell's integrity can be destroyed without abolishing division, and what supplements are needed to sustain chromosome motion. Studies on partially disrupted cells have often proved useful for work on aspects of cell motility. For example, muscle fibers can be gently fragmented and soaked in glycerol which removes small molecules but leaves the organization of the fiber sufficiently intact to permit contraction. The disruption of the plasma membrane facilitates introduction of ATP and other reagents.

4.2.9   ***CYTOPLASMIC DIVISION***   In general, cells at telophase divide along the plane of the spindle equator (where the metaphase chromosomes had been located), and two cells of equal size are produced. The two daughter cells generally receive similar shares of cytoplasmic organelles, although there are some dramatic exceptions, such as the divisions of a maturing animal oöcyte into a large future ovum and small polar bodies destined to degenerate. In some algae, where a single plastid that occupies much of the cytoplasm is present, the plastid divides in synchrony with the rest of the cell. In cells with large numbers of mitochondria or plastids, these organelles are often distributed apparently at random in the cytoplasm, so that roughly equal numbers of organelles are contained in the two daughter cells. Sometimes equal distribution results from a special arrangement of organelles. Thus lysosomes often tend to cluster near the poles. In the dividing spermatocytes of some insects, the mitochondria group around the spindle to form an aggregate of elongate mitochondria; this is cut in half as the cell divides.

In animal cells a cleavage furrow moves in perpendicularly to the long axis of the spindle. Many hypotheses for the mechanism of furrowing have been proposed. Early investigators ascribed the process to surface tension forces like those that maintain droplets of liquids in more or less rounded forms. Such forces are generated by the orientation and interaction of molecules at surfaces, and furrowing could be produced by local changes in such surface organization. However, most present investigators emphasize special "contractile" systems in their proposals to explain furrowing. A ring of microfilaments is often located just below the surface of the furrow and might be responsible for generating the furrowing forces.

If the mitotic apparatus is experimentally shifted during prophase or metaphase in a fertilized egg cell or tissue culture cell, the plane of the cleavage furrow is shifted accordingly. If the mitotic apparatus is shifted after metaphase, or even removed, it has no effect upon the cleavage furrow. Some interaction between the spindle and the rest of the cell appears to "set" the cleavage plane; once this has occurred, the spindle is no longer required. One view suggests that attachments sometimes found between asters and the cell surface may contribute to setting the cleavage plane.

In plants cytoplasmic division occurs by the formation of a *cell plate*. Vesicles, many derived from the Golgi apparatus, accumulate in the middle of the cell, flatten, and fuse to establish the new cell boundaries (Fig. IV-23). This begins at the *phragmoplast*, a region of vesicles that accumulate near microtubules which persist at the spindle equator for a time after the chromosomes have separated (Fig. IV-23).

Interesting observations have been made on the division of the "guard mother cell" that produces the guard cells bordering the stomata of onions. The mother cell is elongate, and the initial movement of the chromosomes is toward opposite ends of the cell (the orientation actually is somewhat oblique). Late in division, however, the spindle and phragmoplast rotate in the cell so that the final separation into daughters occurs along a plane parallel to the mother cell's long axis, rather than the usual transverse plane (Fig. IV-23). Thus two elongate guard cells result (Fig. III-13). Many comparable cases are known in which specially oriented cell division contributes to the geometry of developing tissues in plants and animals. The underlying mechanisms are being sought.

4.2.10　　　　　*NUCLEAR*　Disappearance of the nuclear envelope in
　　　　ENVELOPE AND　prophase results from fragmentation of the
　　　　　NUCLEOLUS　envelope into numerous sacs and vesicles.
In telophase the new nuclear envelope forms by fusion of sacs and vesicles that accumulate around the chromosomes. It is not yet known whether the fragments of the "old" envelope generally persist and contribute to formation of the new one. Presumably for cell types in which the envelope does not disappear (Section 4.2.7) such persistence is extensive.

The nucleolus, as an organized entity, generally disappears in division. However, in at least some cell types nucleolar components probably are preserved during division and participate in forming a new nucleolus. Microscope and autoradiographic studies suggest that during division some material from the nucleolus passes to the chromosomes, to the cytoplasm, or to both, and that some of this returns to the new nucleolus formed after division. In a few cell types nucleoli persist as intact bodies throughout the division cycle; sometimes they pass into daughter cells with the chromosomes to which they are attached. In some plant cells portions of nucleoli, associated with the chromosomes until metaphase, can move to the spindle poles before chromosome movement begins. Apparently the nucleoli are carried passively by

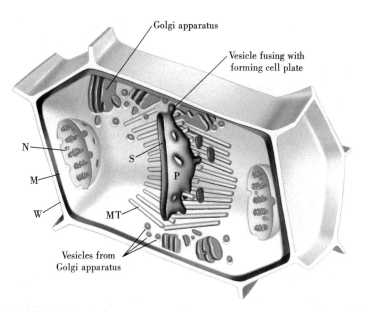

Golgi apparatus

Vesicle fusing with forming cell plate

N

S

M

P

W

MT

Vesicles from Golgi apparatus

Fig. IV-23 *Schematic representation of a cell of a higher plant as seen at telophase in mitosis. The cell wall is indicated by (W), the plasma membrane by (M), and one of the two daughter nuclei by (N). In the* phragmoplast *region, a region of membranes and microtubules (MT), a cell plate forms (P) and grows until it separates the cytoplasm into two daughter cells. The cell plate develops as a membrane-delimited structure enclosing a space (S) in which a new cell wall will form. The Golgi apparatus contributes many vesicles to the phragmoplast membrane. The vesicle membranes apparently are incorporated into the membrane of the cell plate, and the vesicle contents enter the forming cell wall. (Modified after M. Ledbetter.)*

some sort of poleward flow of material within the spindle. Thus they are transported independently of the chromosomes.

c h a p t e r **4.3**
GENETIC BASIS OF CELL DIVERSITY

Thus far, discussion has included mechanisms that tend to keep structures and cells *constant*. The present chapter considers the origins of *diversity* through *genetic mechanisms*, diversity that is the basis of evolution. It is usually not the genes themselves, but rather the genetically determined characteristics of organisms, that are exposed to the environment and thus to natural selection [although there are some special cases of fairly direct interaction of gene and environment, such as enzyme induction (Section 3.2.5)]. Those organisms with characteristics that are advantageous in the particular environment produce more offspring than organisms with fewer advantageous characteristics. They contribute a larger number of genes to the next generation of a population. With time, populations undergo evolutionary change.

4.3.1   **MEIOSIS**   Mutations in DNA result in altered base sequences which ultimately are translated into proteins with altered amino acid sequences. Sometimes the change of just one amino acid in a protein has profound effects. For example, in *sickle cell anemia*, the human inheritable disease, a single amino acid in the hemoglobin molecule is replaced by another. This results in drastically altered structure and function of red blood cells. (The cells are fragile, and under certain conditions they tend to assume a shriveled sickle shape, in contrast to the normal disk shape.)

Mutations give rise to *alleles*, which are alternate forms of a gene. (For example, one allele of an eye color gene may produce blue eyes, whereas brown eyes are produced by another of the several alleles of the same gene.) For the most part, inheritable differences among individuals of a given species are attributable to the fact that different individuals carry different alleles.

All sexually reproducing eucaryotic organisms (that is, many unicellular organisms and almost all multicellular organisms) go through a *zygote* stage or its equivalent. Usually a single nucleus is formed containing a mixture of alleles from both parents. As outlined in Chapter 3.10, the zygote nucleus may result from the fusion of nuclei carried by distinctive gametes, as in most higher organisms. In many unicellular organisms and in some lower plants, nuclei contributing to the zygote are transferred between two cells without the formation of obviously specialized gametes by processes such as partial and temporary fusion (*conjugation*) of ciliated protozoans.

Meiosis is complementary to zygote formation in the life cycle of organisms, and it accomplishes the segregation of alleles. This results from the fact that the two members of each homologous pair of chromosomes (Section 4.2.1) are separated by meiosis into different cells. Usually the members of a homologous pair both carry the same genes arranged in the same sequence along their lengths; however, the particular alleles present on each homologue may differ. Meiosis involves two rounds of cell division, as outlined in Figure IV-24. The first meiotic division is preceded by DNA replication during interphase. The division is different from mitosis in two key respects. (1) In the first meiotic prophase, homologues pair gene for gene; they come together and align so that the site (*locus*) of a given gene on one homologue lies next to the same gene locus on the other homologue (Figs. II-18, IV-26, and IV-27); no such pairing occurs in mitotic prophase. (2) The chromosomes remain paired until metaphase. Then one *chromosome* of each pair goes to each pole. Unlike mitosis, the two *chromatids* of each chromosome remain together. Thus the number of chromosomes per nucleus is halved. The *second* meiotic division resembles mitosis in that no pairing occurs and chromatids separate from one another. However, the second division is not preceded by DNA replication or chromosome duplication; the chromatids that separate are the ones already present during the first division. The net result of the two divisions is four haploid cells (Figs. IV-12 and IV-25), each having half the DNA and half the number of chromosomes (one of each pair) of the parent diploid cell.

The subsequent behavior of the meiotic products varies considerably in different organisms. Many algae and protozoa are normally haploid; the meiotic products each divide by mitosis to establish a new cell line. The members of a given line produced in this way are genetically identical, except for mutations occurring subsequent to the meiotic divisions that produced the first cell of the line. Lines differ from one another in alleles, since they have resulted from products of meiosis that differ in alleles (see Fig. IV-24 for an explanation of such differences). Sexual reproduction is an occasional process; it occurs, for example, in *Paramecia* under conditions of low food supply.

In multicellular lower plants, such as the seaweeds (multicellular algae), the haploid cells resulting from meiosis are released as spores. These divide by mitosis to form a gamete-producing plant that is haploid and multicellular (*gametophyte*). Gametes are produced by differentiation of some of the gametophyte cells. A diploid zygote is formed by the fusion of gametes. By mitosis the zygote develops into a diploid multicellular *sporophyte*. In the sporophyte some cells undergo meiosis to produce the haploid spores. While in some species like ferns and other lower plants, alternating *sporophyte* and *gametophyte* generations are separate plants, in others one generation is reduced in size and appears like a special portion of the more highly developed generation. For example, in flowering plants the gametophytes are microscopic. They are parts of the flower that form gametes—male gametes in the pollen and egg cells in the base of the flower.

Thus in higher plants and virtually all multicellular animals, most cells are diploid, the products of mitotic divisions of the zygote. Each individual starts with an essentially unique set of alleles, except for the occasional occurrences when a single zygote produces two (identical

(A) Overall behavior of chromosomes

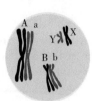

At metaphase the chromosomes line up in pairs at the spindle's equator.

At mid-prophase of the first of the two meiotic divisions homologous chromosomes are paired; each chromosome is of two chromatids.

At anaphase homologous chromosomes separate to opposite poles while the two chromatids of each chromosome stay together.

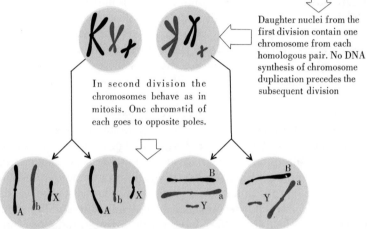

Daughter nuclei from the first division contain one chromosome from each homologous pair. No DNA synthesis of chromosome duplication precedes the subsequent division

In second division the chromosomes behave as in mitosis. One chromatid of each goes to opposite poles.

Thus, each cell entering meiosis produces four haploid daughters. Each homologous pair segregates independently of the others. For example, A enters the same daughter as B or b with equal frequency. Thus, another (Aa, Bb, XY) cell entering meiosis will produce daughters abY, ABY, aBx and AbY.

(B) Crossing over

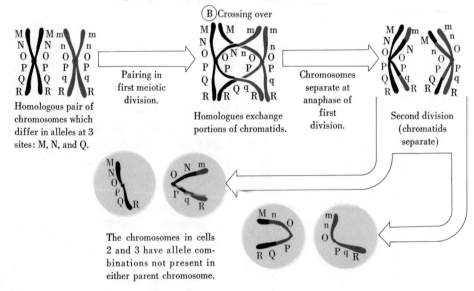

Homologous pair of chromosomes which differ in alleles at 3 sites: M, N, and Q.

Pairing in first meiotic division.

Homologues exchange portions of chromatids.

Chromosomes separate at anaphase of first division.

Second division (chromatids separate)

The chromosomes in cells 2 and 3 have allele combinations not present in either parent chromosome.

twins) or more organisms. Meiosis occurs in specialized reproductive cells and results in gametes (see Chapter 3.10). In males all four meiotic products form gametes. In females three of the cells resulting from meiosis generally degenerate and only one functions in reproduction. The cells resulting from meiosis in animals differentiate directly into sperm and ova.

4.3.2   **THE**   The diversity in detail of gamete and zygote
TRANSMISSION OF   behavior is important for understanding the
HEREDITARY   hereditary patterns of different organisms.
INFORMATION; AN   But the variety of life cycles all involve two
EXAMPLE   constant features, the reduction of diploid
cells to haploid (meiosis) and the formation
of new diploid cells with new combinations of alleles generated by fusion of haploid nuclei (sexual reproduction). This constancy produces predictable patterns of transmission of chromosomes and genes from one generation to the next, making genetic analysis possible. For example, earlier (Section 2.3.4) it was mentioned that matings of certain individuals of the toad *Xenopus* produce offspring of which one quarter are inviable due to a chromosomal defect affecting ribosome and nucleous formation. This pattern of inheritance is explainable by the proposal that the parents in the matings each contain one normal (N) and one abnormal (No) homologue in the chromosome pair responsible for formation of nucleoli. Since a normal chromosome is present, ribosome synthesis can take place. However, meiosis results in gametes half of which contain the N, and half the No, homologue. If "m" represents the homologue contributed by the male parent and "f " the homologue contributed by the female parent, then equal numbers of the following types of zygotes will formed; $mNfNo$, $mNofN$, $mNfN$, $mNofNo$. Thus one quarter of the zygotes are NN, one quarter $NoNo$, and one half NNo. The $NoNo$ category are the ones that die, since they have no normal

◄ *Fig. IV-24* *Chromosome behavior in meiosis. First meiotic prophase is usually thought of as having 5 major stages: (1)* **Leptotene,** *when the chromosomes have coiled to the point where they are visible as discrete slender threads; (2)* **Zygotene,** *when homologues begin visible pairing; (3)* **Pachytene,** *when gene-to-gene pairing is completed (see Fig. II-18); (4)* **Diplotene;** *and (5)* **Diakinesis,** *when the chromosomes coil to reach maximum thickness and although they remain associated, relax pairing so that homologues are associated only at some points, the* **chiasmata** *(Fig. IV-26). The X and Y chromosomes drawn in the diagram represnt the sex-determining chromosomes seen in many organisms (Fig. IV-6 and Section 4.3.3).*

homologue. Their survival through the early stages of development is based on ribosomes stored in the oöcyte before the completion of meiosis. At this time a normal homologue is still present in the eggs that eventually contain only the *No* homologue, since the oöcytes like the other *diploid* cells of the parents are all *NNo*.

This explanation predicts that there should be a class of surviving offspring that contains one *N* and one *No* homologue and another class with two *N* homologues. There are two lines of evidence that this is so. Among the viable offspring of the matings under discussion, some have many cells in which two nucleoli are present (in some cells the two fuse to form a single organelle), while in the cells of other individuals only one nucleolus is ever present. Also, DNA can be extracted from embryos prior to the death of the *NoNo* class and its ability to form molecular hybrids (Section 2.2.3) with purified rRNA determined. The expected three categories of individuals are found: some have essentially no DNA sites that bind rRNA, and, of the rest, one group has twice as many sites as the other. This last finding indicates that the abnormality in *No* chromosomes involves the loss (*deletion*) of the rRNA-producing DNA segments.

4.3.3   **GENETIC DIVERSITY; CROSSING OVER**   Sexual reproduction and meiosis continually generate new genetic combinations. In Figure IV-25 it is shown that a hypothetical organism with six chromosomes (three pairs) can produce gametes with eight different combinations of chromosomes. Figure IV-25 diagrams the 10 different zygote combinations that can result from matings of two individuals, each having only four chromosomes (two pairs) with the same arrangement of alleles in both individuals. These are extremely simple examples. With large numbers of chromosomes and matings, as is usual between individuals carrying many different alleles, enormous numbers of combinations are possible.

Even greater diversity of gametes and zygotes results from *crossing over* between homologous chromosomes. This depends on the reciprocal exchange of portions of chromatids during the first meiotic division (Figs. IV-24 and IV-26). Crossing over produces characteristic chromosome configurations known as *chiasmata* (Fig. IV-26) which are visible just before and during the first meiotic metaphase. It is not an occasional phenomenon, but rather an almost invariable feature of the association of chromosomes in meiosis in most organisms studied. (There are a few exceptions, such as males of the fly *Drosophila*, in which a type of pairing occurs but chiasmata do not form and crossing over is rare.) In any given crossover the two chromatids usually exchange comparable portions which may differ in alleles but which carry the same set of gene sites (*gene loci*). However, when a particular homologous pair is compared in different cells undergoing meiosis, the chromosome portions involved in crossing over vary almost at random, so that different sets of loci are exchanged in different cells. The frequency with which crossing over occurs between given loci of a chromosome is the basis of classical techniques for mapping gene locations; the further apart two loci are, the more likely a crossover will occur between them.

Sex determination can often (although not invariably) be readily demonstrated to depend on meiosis. For example, in humans, males are usually XY and females XX, where X and Y refer to special homologous sex chromosomes which pair at meiosis (Fig. IV-6), although they differ in genes and in morphology. The segregation of these chromosomes in meiosis in males (Fig. IV-24) results in equal numbers of X and Y sperm. Since female gametes all are X, half the zygotes will be XX and half XY. In other organisms somewhat different situations are found. For example, in some insects males have one X and females two; no Y chromosome is involved. In meiosis in these males, equal numbers of gametes have either one X or none; all female gametes have one X.

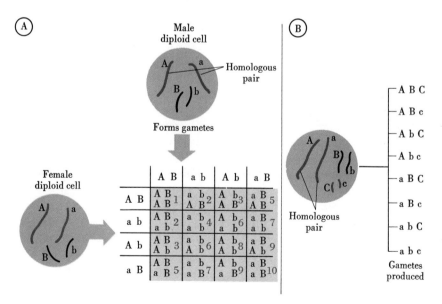

*Fig. IV-25 Some genetic consequences of meiosis. **A**. The results of a genetically simple mating. As indicated in the diagram, from the viewpoint under consideration the zygote combination (^{AB}maternal;^{ab}paternal) can be considered to be the same as (^{AB}paternal; ^{ab}maternal.) **B**. The gametes that can be produced by an individual whose diploid cells contain three pairs of chromosomes.*

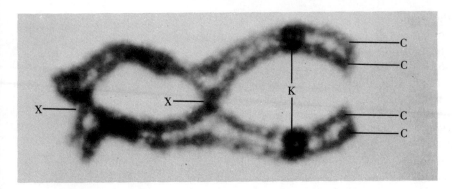

Fig. IV-26 *A pair of homologues during late prophase of the first meiotic division of a salamander spermatocyte. The four chromatids of the two chromosomes are readily seen at C, and the two kinetochores are indicated by K. At X chiasmata are present. That only one chromatid from each chromosome is involved can clearly be seen in the chiasma at the right. (Courtesy of J. Kezer.)*

Meiosis does not occur in procaryotes. The few species of bacteria that have been adequately studied are haploid in the sense that there is only a single copy of the chromosome per nuclear region. Although it is not known how widespread this is, some bacteria are capable of a form of sexual reproduction (referred to as *genetic recombination*) in which DNA from one cell is transferred to another, recipient cell. This can occur by several mechanisms, including one which involves the fusion of small regions of the cells' surfaces and the passage of the chromosome of one cell into the other cell (see Section 3.2.5 for the involvement of the *F* episomal factor). The recipient becomes "diploid" by this process, but eventually a single *recombinant* chromosome is produced that is comparable to a crossover product in higher organisms. It contains genes from both cells, and, normally, haploidy is restored. The number of nuclear regions per bacterial cell may vary under different conditions of growth, but each nuclear region apparently contains a copy of the same chromosome. The multinucleate condition probably represents a feature of cell growth and division rather than one of sexual reproduction.

4.3.4 **PROBLEMS OF MEIOSIS** There are still many unanswered questions regarding meiosis. What makes spermatogonia and oögonia, after numerous mitotic divisions, switch their mode of chromosome behavior to meiosis? Many eggs remain in the prophase or metaphase of the first meiotic division until fertilization occurs, which can be weeks or months later. What produces this arrest and how is it released at fertilization? What determines the unequal division of many animal oöcytes into a small "polar body," a cell that degenerates, and a much larger cell, the mature egg cell?

Homologous chromosomes pair gene for gene, probably with their kinetochores in some special arrangement. Is there something special about the chemistry or structure of meiotic chromosomes that permits this discriminatory mode of association? Is there any relationship between pairing of chromosomes and base pairing of nucleic acids? Are repetitive DNAs involved? Is the *somatic pairing* of homologues seen in the nonmeiotic cells of some insects (such as *Drosophila*) based on mechanisms similar to meiotic pairing? In a few instances, differences have been reported between histones of meiotic and those of mitotic cells, but the significance of this is unknown. In many species, electron microscopy of meiotic stages reveals a special structure, the synaptonemal complex (Fig. IV-27), believed to represent paired chromosomes. Further analysis of this structure may illuminate meiotic mechanisms.

In grasshopper spermatocytes it is possible to dislodge paired homologues from the meiotic metaphase spindle with a microneedle and to turn the pair around so that the chromosome that was about to segregate to one spindle pole now faces the opposite pole. The paired chromosomes reattach to the spindle and separate normally, but they segregate to the poles opposite those they would have moved to without the experimental reorientation. This experiment strongly hints that the chromosomes themselves, rather than some external system, control attachment of homologues to the spindle in the manner necessary for meiotic separation. Presumably the kinetochores are the responsible parts.

The molecular mechanisms of pairing and crossing over still are unknown. Some promising leads have emerged from studies on viruses. Genetic recombination between viral "chromosomes" (DNA molecules) occurs when different viral strains infect the same cell. A single recombinant chromosome is formed that carries hereditary information contributed by both parents; this is comparable to a chromatid resulting from a crossover.

One set of theories (*copy-choice* theories) states that this results from phenomena of DNA replication. During its synthesis a strand of DNA might start to form by alignment of nucleotides along a strand of one parental DNA molecule. After partial completion it might somehow switch and finish its growth by alignment of nucleotides along a strand of second parental DNA molecule (Fig. IV-28). For part of its length the

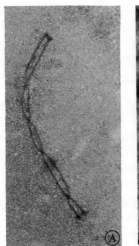
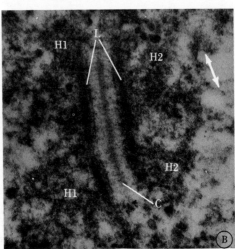

Fig. IV-27 The synaptonemal complex. A. A pair of homologous chromosomes from human spermatocytes (pachytene stage of meiosis; see Figs. II-18 and IV-24). The cells have been disrupted by placing them at the interface between a salt solution and air, and then the chromosomes have been picked up on a thin piastic film for viewing in the electron microscope without sectioning. The paired structures seen in the micrograph are the synaptonemal complex at which the two homologues are associated. B. shows a thin section of part of a synaptonemal complex of a spermatocyte of a rooster. The two chromosomes are represented by the irregular fibrous regions at H-1 and H-2. The long axes of the chromosomes are indicated by the double-headed arrow. The synaptonemal complex is seen at the surfaces where H-1 and H-2 are associated. The complex consists of the pair of dense bands (called lateral elements) indicated by L and the single central element (C) found between them. Faintly visible filamentous material runs from the central element to the lateral ones. It is the lateral elements that are visible in A. A × 9,000; B × 50,000. (From M. Moses. S. J. Counce, and D. F. Paulson, Science, *187: 363-365, Jan. 31, 1975. Copyright 1975 by the American Association for the Advancement of Science.)*

new strand would contain information specified by the base sequences of one parental DNA, while for the rest of its length the information would have been specified by the other parental DNA.

An alternative theory is based on breakage and rejoining; two parental DNA molecules might break, and portions of each join with portions of the other. An experiment was designed to determine whether or not this is a mechanism of recombination (Fig. IV-28). Viral DNA can be labeled with *density markers* (such as C^{13} and N^{15}, isotopes of carbon and nitrogen), which permit separation from nonlabeled viral material by density gradient centrifugation (see Section 1.2B). The higher the proportion of markers in the DNA, as compared with C^{12} and N^{14} which normally predominate, the greater the DNA density. When recombination takes place between two different strains of viruses, both equally labeled, some of the recombinant chromosomes with genetic information from both parents have DNA essentially as dense as the parental DNAs. This means that breakage and rejoining can produce recombinant DNA and that extensive new DNA synthesis is not necessary for recombination. Under the conditions of the experiment newly synthesized DNA would contain no label and its presence in recombinant chromosomes would lead to a decrease in density to levels below that of the parental chromosomes. However, this does not imply that extensive DNA synthesis is never involved in recombination. Much new DNA may be made as part of the processes by which the segments of DNA molecules participating in recombination are joined together. The important point is that the initial events of recombination involve physical rearrangement of preexisting DNA, rather than the synthesis of an entirely new strand.

That breakage and rejoining of chromosomes is also likely to be involved in crossing over in eucaryotes is indicated by several observations. In some species slight morphological differences occur between members of homologous pairs, and recombinant chromosomes are produced that are visibly combinations of both. If chromosomes are labeled with tritiated thymidine before meiosis, subsequently autoradiography can be used to demonstrate exchanges of DNA between homologues. Since crossing over requires that parts of different chromosomes be close to one another, crossing over and pairing probably are related processes. But there is no unequivocal evidence that crossing over must occur *after* the chromosomes are visibly paired in prophase; it might be part of the pairing process. It has even been suggested that crossing over actually takes place during the S period of premeiotic interphase, when most of the DNA replication occurs. (This might be expected were copy choice the predominant mechanism.) However, several lines of investigation suggest that crossing over occurs later. Thus in the unusual life cycle of certain fungi and algae, meiosis, presumably including genetic recombination, takes place *immediately after* two parental nuclei are brought together in a zygote. In several of these species, premeiotic DNA replication occurs in the separate parental nuclei before the chromosomal complements come together to participate in meiotic events. On the other hand, findings made initially on plant material and then extended to some animals indicate that there may be one or more periods of limited DNA synthesis during meiotic prophase. Although only 1 percent or less of the total DNA is involved, it may be quite significant. For example, exposure of meiotic prophase cells to deoxyadenosine, an inhibitor of DNA synthesis, can result in fragmentation of chromosomes and interruption of pairing and continued meiosis. Is the DNA made during prophase important in joining separate DNA segments or in other events of pairing or crossing over? Interruption of

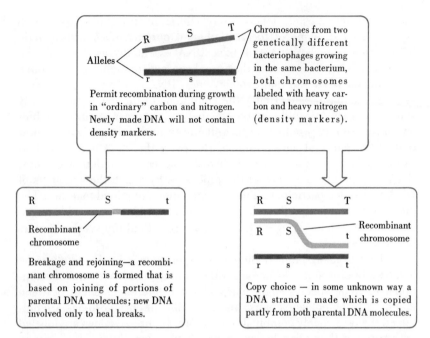

Fig. IV-28 *Alternative hypotheses to explain recombination in bacteriophages. The finding that some of the chromosomes (DNA double helices) that can be shown to be genetically recombinant are essentially as dense as the parent molecules indicates that recombination can take place by breakage and rejoining; extensive DNA replication is not required. However, the experiment is not designed to establish that this is the only mechanism by which recombination is possible. (From the work of J. J. Weigle and M. Meselson.)*

protein synthesis during meiotic prophase also leads to abnormal chromosome behavior, and it is likely that the chemical events underlying visible pairing and crossing over are complex.

Exceptions to the general rules of meiosis are known; the exceptions indicate the extent to which even fundamental cellular processes have undergone evolutionary change. During sperm formation in the insect *Sciara*, segregation occurs between *maternal* and *paternal* chromosomes, that is, between the chromosomes derived from those originally contributed by the male and female parents to the zygote that produced the individual under study. At the first meiotic division, the maternal chromosomes move to one pole and their homologues of paternal origin move to the other. Only cells receiving the maternal chromosomes form functional sperm. The paternal chromosomes are not transmitted to the next generation. Accordingly, the patterns of heredity are quite unlike those generally found. In most species maternal and paternal homologues separate at random to the poles of the first meiotic division, and therefore all gametes usually contain some chromosomes derived from those contributed by both parents. (Crossing over additionally mixes maternal and paternal chromosomal material.) The recognition devices responsible for the unusual behavior of *Sciara* chromosomes are not known.

In the brine shrimp *Artemia*, development is by *parthenogenesis*, the development of an egg without fertilization. The first division of oöcyte meiosis is normal, but in the second the chromatids separate without cytoplasmic division, and a single nucleus eventually forms with both chromatids of each chromosome. Consequently, the egg is diploid and, although there is no contribution of chromosomes by sperm, the egg produces a diploid organism.

4.3.5   ***SOME***   The variation in genetic constitution result-
       CONSEQUENCES   ing from mutation, meiosis, and sexual re-
       OF GENETIC   production is the raw material upon which
       DIVERSITY   selection operates in evolution. The pattern
                               of interaction of gene and environment
differs somewhat in organisms with different types of life cycles. For example, mutations in diploid organisms are less rapidly, or less directly exposed to selective "testing" by the environment than those of haploid organisms. In a haploid organism a mutation that results in the production of a defective form of an important enzyme will result in an inviable cell. In diploid organisms this is not necessarily so, because each gene is represented at least twice in each cell. Thus the presence of a newly mutated allele on one chromosome may be "masked" (the allele acting as a *recessive*) by the presence on the homologous chromosome of a different (*dominant*) allele that can support the synthesis of adequate amounts of "normal" enzyme. In addition, the presence of two different alleles (*heterozygosity*) in an individual may be more advantageous than the presence of the same allele on both homologues (*homozygosity*). Also, the extent to which a gene is advantageous or disadvantageous often depends on the environment. Thus heterozygous individuals with the allele for sickle cell anemia (p. 332) on one chromosome and an allele for normal hemoglobin on the other homologue may have little or no difficulty (beyond a slight anemia) at low altitudes in temperate climates. However, at high altitudes where the oxygen pressure is low, their red-blood cells will assume the abnormal shriveled shape. But heterozygosity also confers resistance to forms of malaria, a disease prevalent in the tropics. Individuals homozygous for the sickle cell allele usually die from severe anemia. The net result of factors such as these is that populations of diploid organisms tend to have a much more complex *gene pool* (the total number of genes and alleles in all individu-

als) than do populations of haploid organisms. The preservation in diploid populations of alleles that may be selectively disadvantageous
under some conditions provides a source of evolutionary variability not
available to haploids. If the environment changes, these alleles may
become advantageous. On the other hand, the large populations and
rapid division rates of many haploid unicellular organisms permit rapid
evolution by environmental selection among mutants that continually
arise at random within all populations.

4.3.6   ***CYTOPLASMIC***   In one type of cytoplasmic inheritance,
          INHERITANCE   *maternal inheritance*, characteristics of
                           the offspring are determined solely by the
female parent. As in cytoplasmic inheritance generally the transmission
patterns of the relevant hereditary information are not those expected
for genes on chromosomes governed by meiosis and mitosis. In most
cases DNA molecules in plastids or mitochondria seem to be the sites
of the genetic information, and it is the pattern of their transmission that
is reflected. In many species paternal cytoplasmic genetic factors are
not passed on to the offspring. For higher organisms this may be due, in
part, to the gross disparity between the sperm's and the egg's contribution of cytoplasm to the zygote. However, uniparental cytoplasmic inheritance also occurs in unicellular organisms, such as *Chlamydomonas*
(Fig. II-46) where sexual reproduction involves fusion of two cells of
similar size. Apparently there are factors, still unknown, that differen-
tially suppress the perpetuation of cytoplasmic genes from one parent
cell.

Both spontaneously occurring and experimentally induced mutations of cytoplasmic genes have become important experimental tools.
In yeast, "petite" mutants possess mitochondria that are abnormal in
structure and function. Aerobic metabolism is disrupted and the cells
grow slowly since they must rely on the less efficient anaerobic
metabolism (Section 1.3.1). Some "petite" yeast strains result from
mutations in nuclear genes, others involve changes in mitochondrial
DNA including marked alterations in base composition. With the fungus
Neurospora, cytoplasm from a strain with mutant mitochondria has
been injected into a normal strain, and the successful "colonization" of
the recipient by proliferating abnormal mitochondria has been observed. Such studies help confirm the concept of semiautonomous
mitochondrial duplication outlined earlier (Section 2.6.3). Detailed
analyses of the effects of nuclear and cytoplasmic mutants also help
clarify the origins of different organelle components. For example, recent studies on *Chlamydomonas* mutants have led to the surprising
hypothesis that some proteins of chloroplast ribosomes may be coded

for by nuclear genes while the rest are coded for by chloroplast DNA. If this is borne out by subsequent studies, it implies an even greater degree of cooperation between nuclear and chloroplast genomes than has hitherto been suspected. Efforts are also underway to detect and analyze mutations affecting centriole or basal body structure and behavior. The findings may help settle the controversy over the possible presence of nucleic acids in these organelles (Section 2.10.5), since it should be possible to demonstrate whether or not cytoplasmic mutants occur. We have already outlined evidence that the patterns of aligned basal bodies in protozoa can be inherited independently of the nucleus (Section 3.3.3) but this need not depend upon genetic autonomy of the basal bodies.

Cytoplasmic mutants of chloroplasts that affect plastid structure or function (Section 2.7.4) often yield plants with varying patterns of groups of cells that are green or white (absence of chlorophyll). The patterns depend upon the details of plastid transmission during sexual reproduction and during cell divisions in the embryo. Some ornamental plants owe their origin to such cytoplasmic inheritance, and there also are a few phenomena of this type that are important in agriculture.

Especially convenient for genetic analyses are plastid mutations leading to altered resistance to experimentally administered drugs. Through use of such mutants in *Chlamydomonas*, and of comparable mitochondrial mutants in yeast, it has been found that under some experimental conditions sexual reproduction can lead to patterns of inheritance suggesting genetic recombination between cytoplasmic DNAs from both parents. Do the DNAs interact when organelles contributed by the parents fuse? Another line of investigation concerns the possibility that a given organelle possessing several DNA molecules may have different information in the different DNAs (Section 2.6.3). Thus far the patterns of inheritance detected suggest that such heterogeneity is not very marked, if it occurs at all.

c h a p t e r **4.4**

DIVERGENCE OF CELLS IN EMBRYONIC DEVELOPMENT

4.4.1 DEVELOPMENT AND DIFFERENTIATION Figure IV-29 illustrates the major stages in the development of eggs of higher animals. The fertilized egg divides by mitosis to produce the *blastula*. As mentioned earlier (Section 4.2.3), the divisions are referred to as *cleavages*. Little cell

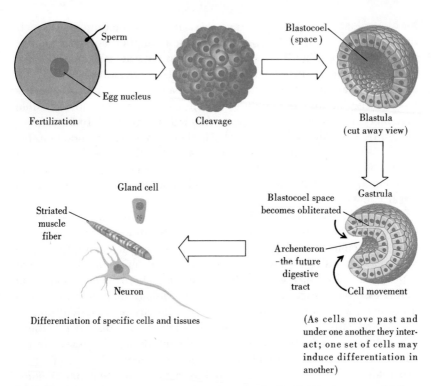

Fig. IV-29 A schematic outline of early development in some higher animals.

growth occurs, and thus sequential divisions separate the fertilized egg into successively smaller cells. Many eggs have special localized cytoplasmic regions that become included in only a few of the cells resulting from the cleavages. The blastula forms a *gastrula* by kinds of cell movement that vary according to the species, and during this process some cells are brought into new spatial relations with others. Subsequently, *differentiation* of cell types occurs, and the embryo increases in size and mass by cell growth and division. The juxtaposition of cells in new geometric relations during gastrulation is accompanied by the *induction* of differentiation in groups of cells through interaction with other cells that have come to lie close to or in contact with these groups. Cells acquire characteristics enabling them to "recognize" one another (Section 2.1.6). Additional patterns of cell growth and migration result in the establishment of other specific associations of different cell types, for example, the complex relationships among neurons, sensory receptors, and muscle cells.

Eventually embryologists will integrate detailed descriptions of these many processes into a coherent description of molecular changes.

Emphasis here is chiefly on the nuclear and chromosomal aspects of differentiation and on the specific activation or repression of genes. Increasing evidence supports the proposition that in different cell types, different genes are switched on or off so that cells come to differ in their structural and chemical characteristics although they possess the same genetic constitution. When different genes are switched on, different specific mRNAs are produced from different DNA templates. Consequently, cell-specific proteins (such as enzymes, secretory materials, or contractile proteins) are synthesized. A point to note is that there must also be some DNA templates that function in virtually all cell types, such as templates for ribosomal or transfer RNAs.

4.4.2   **DIFFERENTIA-** In the few eggs that have been thoroughly
TION IN EMBRYOS studied (sea urchins and amphibia) it appears that protein synthesis in the initial development of the egg depends chiefly on ribosomes and on mRNA that is stored during oögenesis and activated at fertilization (Section 3.10.2). About the time of gastrulation, RNA synthesis and ribosome formation increase dramatically; in many embryos, nucleoli first become prominent at gastrulation. The differential synthesis of specific mRNAs in different cells presumably becomes more and more pronounced following gastrulation.

Experiments with the sea urchin eggs have shown that cleavages up to the blastula stage can take place in the absence of a nucleus. Differentiation, however, depends on the presence of a nucleus and on the interaction of nucleus and cytoplasm. In some developing amphibian embryos, nuclei can be taken from cells at progressively later stages of development and substituted, by microsurgery, for the nucleus of a zygote that has not yet begun development. The purpose of such experiments is to determine whether the nucleus undergoes irreversible changes during development that are related to the differentiation of cells. Do the offspring of the zygote nuclei retain the ability to give rise to the nuclei of all the varied cells of the organism, or do they change to "specialized" states in which they can no longer support development of many different cell types?

With eggs of the frog *Rana*, nuclei taken from cells up to the blastula stage can support the normal development of eggs. Nuclei taken from most kinds of cells at later stages cannot. With such nuclei the egg starts to develop but the embryos become abnormal and usually die. This might suggest that nuclear changes do accompany differentiation. But interpretation is complicated by the possibility that these changes are not so much alterations in the genetic machinery related to differentiation as they are changes in the response of nuclei to the abnormal experience of transplantation. Perhaps nuclei from later stages are increasingly more easily injured or less able to adapt to the rapidity of early embryonic divisions. That this is likely is suggested by the observation of abnormalities (for example, the presence of ringshaped chromosomes) in the nuclei of some transplant embryos. The frequency of these abnormalities is much greater than in ordinary embryos.

If embryos of the toad *Xenopus* are studied, the findings differ significantly from those on *Rana*. The nuclei from advanced stages sometimes support normal development; adult *Xenopus* have been grown from eggs with nuclei transplanted from intestine cells of tadpoles. It was mentioned earlier (Section 4.3.2) that there are some individuals in appropriate strains of *Xenopus* whose nuclei contain only one nucleolus and other individuals whose nuclei contain two nucleoli. The number of nucleoli is a permanent characteristic for the nuclei of each type of individual. Thus by counting nucleoli after the proper choice of donor and recipient cells for nuclear transplantation, it is possible to demonstrate directly that the nuclei introduced experimentally have given rise to the nuclei of all the cells of the developing individuals. It may be concluded from these experiments that, while nuclei may change in development, the changes are not invariably irreversible. The nuclei of at least some differentiated cells can give rise to all the nuclei of an organism. For these cells differentiation does not involve the permanent loss of genetic information or the inactivation of a portion of the genetic material in a way that makes reactivation impossible. Some of the experiments on plant cells mentioned in Chapter 3.11 lead to a similar conclusion.

4.4.3 ***INTERACTIONS*** In frogs, nuclei from one species, capable
OF NUCLEUS AND of supporting normal development in eggs
CYTOPLASM of that species, cannot do so when transplanted into eggs of a different though related species. Some features of nucleo-cytoplasmic interactions are abnormal.

Normally inactive nuclei of chicken red blood cells (birds, unlike mammals, have nucleated red blood cells) can be induced to metabolic activity, such as RNA synthesis, when placed in the cytoplasm of mammalian tissue culture cells. As outlined in Chapter 3.11, this may be done by virus-induced fusion of the red blood cells with HeLa cells, a strain of human tissue-culture cells. The red blood cell nuclei are normally small; their chromatin is densely packed and shows little or no uptake of radioactive RNA precursors. In the fused cells the chicken nuclei enlarge, the chromatin spreads out, and much RNA synthesis is detectable by autoradiography. Apparently the HeLa cell cytoplasm has "activated" the red blood cell nucleus. DNA synthesis also may be initiated in the red blood cell nuclei. Normally red blood cells do not divide.

These experiments must be interpreted cautiously, since they involve abnormal conditions and drastic manipulation of cells. However, they provide direct evidence that the interaction of nucleus and cytoplasm is a reciprocal process.

Another example of such reciprocity occurs during the development of a number of species of insects. Like those of some other organisms, the eggs of these species contain microscopically distinguishable cytoplasmic regions. One region (the *germinal pole*) contains large microscopically distinctive granules in some insects, and it is rich in RNA. In early development, the nuclei undergo mitosis without cytoplasmic division (Section 4.2.3), and they become distributed through the egg cytoplasm. Later cytoplasmic division separates the egg into uninucleate cells. Only the nuclei that become associated with the germinal pole cytoplasm do *not* undergo an unusual series of mitoses in which a specific group of chromosomes fails to move on the spindle. The chromosomes are eliminated from the nuclei and disintegrate in the cytoplasm. The cells in which the full chromosome complement is retained are those which later produce gametes, the *germ-line* cells. Thus the germ-line cells have a different chromosome complement from the other (*somatic*) cells of the organism. It is not known what mechanisms underlie these striking differences in nuclear behavior that are apparently dependent upon the cytoplasm.

4.4.4 SALIVARY GLAND CHROMOSOMES

Some tissues of insects such as *Drosophila*, the fruit fly, and *Chironomus*, the midge, show interesting chromosomal behavior. This has been studied most intensively in the salivary glands. The cells of these glands are extremely large and their nuclei are correspondingly large. Although the cells are nondividing, the chromosomes are clearly visible. Measurements of the DNA content indicate that the nuclei may contain multiples of the normal amount of DNA as great as 1000 times or more. Cytochemical and autoradiographic studies of the cells as they enlarge indicate that this DNA accumulates by repeated replication of most of the chromosomal DNA. Some of the very highly repetitive sequences seem not to take part in this, and thus they are "underrepresented" in the mature nucleus. Histones and some of the other nuclear proteins are present in elevated amounts paralleling the DNA. The number of chromosomes is normal (four pairs in species of *Drosophila*), and homologous chromosomes are usually paired gene for gene. However, each chromosome consists of a great many parallel fibers (such chromosomes are referred to as *polytene*). The chromosomes are several hundred micrometers in length and several micrometers thick, in contrast to the ordinary chromosomes of diploid cells of insects, which are a few micrometers long and less than a micrometer thick. The most reasonable interpretation is that the chromosome has undergone uncoiling and has repeatedly replicated without separation of the daughters. It is thought by proponents of the "unineme" view of chromosome structure (Section 4.2.5) that one fiber of a salivary gland chromosome is comparable to the morphological and functional unit of an ordinary chromosome.

Study of these giant chromosomes has afforded a unique opportunity to correlate cytological, genetic, and developmental information. Figure IV-30 shows a polytene chromosome from *Chironomus*. The striking pattern of transverse bands is evident. Although DNA and protein probably are present throughout the entire length of the chromosome, the bands show a high concentration of DNA and histone. This results from the alignment in the bands of tightly folded or coiled regions (*chromomeres*) of the parallel fibers. The size, appearance, and arrangement of the bands along a given chromosome are different in nonhomologous chromosomes, but identical in the two homologous chromosomes. Several tissues other than the salivary glands also have polytene chromosomes. The banding pattern of a given chromosome is the same in all tissues, and thus there is a parallelism between bands (or more precisely the chromomeres comprising the bands) and genes. Genes also are linearly arranged along chromosomes, and they are the same in the two homologous chromosomes but different in nonhomologous chromosomes. Different cells of an individual have identical endowment. By a variety of techniques it is possible to map the chromosomes, that is, to establish the location of specific genes controlling different characteristics (eye color, wing morphology, and many others) at specific bands. The techniques used include observation of correlation between altered characteristics of the organism and loss (deletions; Section 4.5.1) of specific chromosome portions identifiable by examination of the salivary gland chromosomes.

At certain times some bands show modifications called *puffs*. (Fig. IV-30) which result in part from uncoiling of the chromomeres. These puffs are rich in RNA and are seen by autoradiography to incorporate rapidly radioactive precursors of RNA. In some species there is a localized increase ("amplification") of the amount of DNA in certain puffs. Of great interest is the fact that different bands form puffs in different tissues; differences in location of the puffs are seen even

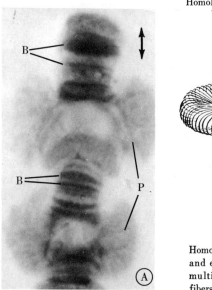

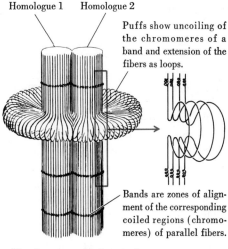

Homologue 1   Homologue 2

Puffs show uncoiling of the chromomeres of a band and extension of the fibers as loops.

Bands are zones of alignment of the corresponding coiled regions (chromomeres) of parallel fibers.

Homologues are closely paired and each chromosome is of multiple, parallel replicate fibers.

(A)

(B)

*Fig. IV-30   Salivary gland chromosomes. **A.** Portion of a chromosome from the salivary gland of the insect* Chironomus. *The double-headed arrow indicates the direction of the long axis of the chromosome. The cross-banding (B) characteristic of these chromosomes is readily visible as are two puffs (P) where the chromosome bands have been altered as indicated in part **B** of the diagram. × 1900. (Courtesy of U. Clever.) **B.** Schematic representation of a* polytene *chromosome like the one shown in **A**. What appears at first glance to be a single thick chromosome is actually two homologues closely paired, gene for gene, and each consisting of many parallel longitudinally arranged fibers.*

among cells of the salivary glands that produce different secretions. Furthermore, specific puffs appear and disappear at specific times of development. From these observations it has been hypothesized that puffs are sites where particular genes are especially active. Puffs show the expected differences from cell to cell and also the high levels of RNA synthesis one would expect of such gene sites. Polytene chromosome puffing is considered a morphological manifestation of a fundamental molecular and developmental process—the activation of specific genes.

Puffing patterns in salivary gland cells are also influenced by materials originating elsewhere in the body. This is experimentally demonstrated by injection of *ecdyson*, a hormone that plays a key role in the control of the developmental cycle of molting (skin shedding) and co-

coon building. Injection of the hormone into young larvae induces a puffing pattern that is not normally seen in the salivary chromosomes until later in development, at the time when the organism's own ecdyson normally acts. Interruption of part of the circulatory system during normal development, so that ecdyson reaches only some of the salivary gland cells, will limit puffing to these cells; cells not reached by the hormone do not puff. Thus intracellular events at the chromosomal level in the cells of an organ (the salivary gland) are controlled by hormones made by cells of another organ (the prothoracic gland). Such cell interactions during development are difficult to study in most organisms because they lack the convenient specializations, giant chromosomes.

One particular puff in *Chironomus* has been studied extensively by microscopy and by biochemical techniques. The puffs are large enough to be isolated by microdissection for detailed analysis. As it is synthesized, the puff RNA seemingly takes the form of ribonucleoprotein granules attached laterally to the DNA fibers of the puff in a manner reminiscent of Figure IV-18. The RNA molecules are unusually large (75 S; molecular weight of at least 10 million), contain repetitive information, and seem to pass into the cytoplasm without substantial diminution in size. Little is yet known of the protein molecules for which they code, although one of the potential candidates may be 3000 amino acids long, which is big enough to require a "giant" RNA for its message. Further work on this and other puffs should contribute much useful information about transcriptional processes and the organization and functioning of mRNAs.

The chromosome set in a nucleus of a *Drosophila* salivary gland cell contains on the order of 5000 distinguishable bands. Since in general a given band seems to be associated with a single genetic function, this has led to the estimate that *Drosophila* possesses roughly 5000 distinguishable "genes" (Section 4.2.6). However, matters may not be quite this simple. The DNA present at a single band may contain several thousands to tens of thousands or more base pairs. From rough calculations of the quantities of DNA required to code for proteins (Section 4.1.3) it appears that each fiber at a band contributes several to many times more DNA than is needed to specify typical proteins. In other words as with chromosomes generally (Section 4.2.6) there is an apparent excess of DNA that has yet to be explained.

4.4.5   ***LAMPBRUSH*** In many species the chromosomes of the
     CHROMOSOMES initial stages of meiotic prophase seem to
                              show chromomeric organization when
viewed in the light microscope. That is, they show longitudinal patterns of darker and lighter regions, probably corresponding to tighter and

looser coiling (hints of this are visible in Fig. II-18). The chromomere pattern differs characteristically among the different chromosomes in a cell. Whether this is comparable in some way to the organization of polytene chromosome fibers is not yet clear.

In some species, especially in amphibia, the meiotic chromosomes of oöcytes have proved particularly interesting. During the first meiotic prophase the oöcytes deposit large amounts of yolk and grow enormously. At this time the chromosomes possess numerous lateral loops, thus giving them their "lampbrush" (or test-tube brush) appearance (Fig. IV-31). Each pair of homologous chromosomes has a characteristic pattern of loops. A total of several thousand loops is present in the chromosome complement.

When the ends of lampbrush chromosomes are pulled with fine needles, the loops pull out as if they are kinks in a continuous thread. Thus the loops are not separate structures attached to the chromosomes, but are specialized regions of a continuous structure running the length of the chromosome. Much RNA is present on the loops; its removal by enzymatic (RNase) treatment does not disrupt the chromosome. Proteases have similar effects. On the other hand, DNase treatment quickly fragments the chromosome into short pieces, indicating that DNA maintains the continuity of the chromosome. This is important evidence that the basic longitudinal structure of chromosomes depends on DNA and may not involve non-DNA linkers (Section 4.2.5).

The lampbrush configuration apparently reflects activation of genes. At the base of each loop the chromosome is tightly coiled (Fig. IV-31), while in the loop itself the chromosome is uncoiled. Much RNA is concentrated at the loops. By autoradiography, it can be seen that the radioactive precursors of RNA rapidly accumulate at the loops. Thus it is probable that they are sites of active RNA production. A similar apparent parallelism between uncoiling and gene activity is encountered in the polytene chromosomes (Section 4.4.4).

The polytene chromosomes of insects are found mainly in cells which are at terminal stages in differentiation; they are destined eventually to degenerate. The chromosomes do not return to "normal" or take part in mitosis. The situation is different with the lampbrush chromosomes. The oöcyte nuclei complete meiosis and do return to "normal"; the lampbrush chromosomes change back to "ordinary" chromosomes.

4.4.6   GENE CONTROL   Several interacting mechanisms are thought to contribute to the regulation of the expression of genetic information. Most work has been done on *transcriptional controls*, that is, those that govern the rates at which particular

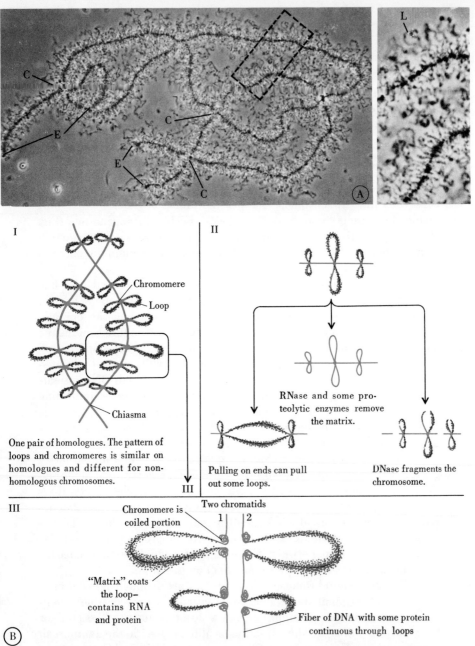

I

Chromomere

Loop

Chiasma

One pair of homologues. The pattern of loops and chromomeres is similar on homologues and different for non-homologous chromosomes.

II

RNase and some pro-teolytic enzymes remove the matrix.

Pulling on ends can pull out some loops.

DNase fragments the chromosome.

III

Two chromatids

Chromomere is coiled portion

1   2

"Matrix" coats the loop—contains RNA and protein

Fiber of DNA with some protein continuous through loops

Fig. IV-31 Lampbrush chromosomes. A. A pair of homologous chromosomes from an amphibian (Triturus) oöcyte nucleus (first meiotic prophase). The four ends of the two chromosomes are seen at E and chiasmata holding the homologues together are seen at C. Projections (loops) along the entire length of the chromosomes give them a brushlike appearance. The outlined area is enlarged in the photograph at the right to show the lateral loops (L.) Approx. × 400; right × 800. (Courtesy of J. Gall.) B. Experiments to reveal the structure of lampbrush chromosomes and the location of the type shown in II. (After the work of Callan, MacGregor, Gall, and others.)

DNA templates are used to generate RNAs. However, we have already mentioned some others. For example, there are instances in which the amounts of DNA in the nucleus are increased with corresponding increases in pertinent RNAs (Sections 3.3.1, 3.10.1, and 4.4.4). In a few special cases loss of DNA takes place (Sections 3.3.1 and 4.4.3). *Translational control* is also possible, that is, the rates at which information encoded in mRNA is translated into protein may be specially regulated. There are few unequivocal demonstrations of this, but one "extreme" case is the storage of mRNAs and ribosomes in inactive form during oögenesis and their subsequent extensive use in protein synthesis following fertilization. Degradation of macromolecules also is subject to controls. The lifespans of molecules of some enzymes and other proteins appear to vary considerably in a number of cell types exposed to different hormones and other agents. Probably the survival of mRNA molecules in the cytoplasm also varies. More speculative is the possibility that those RNAs that emerge from the nucleus into the cytoplasm are molecules that have survived or escaped from some selective intranuclear degradation. The processing of intranuclear RNAs into their functional forms (Sections 2.3.4 and 2.3.5) or the movement of RNA into the cytoplasm are also potential loci for regulation. Similarly, proteins undergo modifications and "packaging" processes (Chapters 2.4 and 2.5 and Section 3.6.1) which may vary under different conditions. And the activities of enzymes are influenced by a variety of metabolic controls, such as feedback inhibition (Section 3.2.5).

Transcriptional controls are thought to be central to cell differentiation. RNA-DNA hybridization experiments indicate the presence of different nuclear and cytoplasmic RNAs at different stages of development and in different fully differentiated cells. For various mature mammalian cells some preliminary estimates suggest that a given cell transcribes only 5–25 percent of its unique sequence DNA. (Only one of the strands of a duplex is transcribed, so these numbers should be doubled for comparison with the information that actually is potentially available.) There are inducible enzymes in eucaryotes, and it seems plausible that the corresponding genes might be subject to the same general types of regulatory mechanisms as those discussed for procaryotes (Section 3.2.5). But procaryotes undergo relatively few changes comparable to eucaryote differentiation; sporulation may be one of these. The relative complexity of the eucaryote chromosome may well involve the existence of regulatory mechanisms without direct counterparts in bacteria or viruses.

The observations on polytene chromosomes and lampbrush chromosomes described above show that gene activation can be accompanied by striking morphological changes in the chromatin. In the case of puffing this extends to selective activity of certain genes. Observations on cells with ordinary chromosomes point in a similar direction. There seem to be alternative states of chromatin, a condensed form that is inactive ("turned off") and an extended form that is active ("turned on"). (1) In sperm cells the chromosomes are densely packed and little RNA synthesis occurs. During cell division the chromosomes are coiled into compact arrays, and again there is little RNA synthesis. (2) In some strains of mice and cats when the two X chromosomes carry different alleles for coat color, females may show coats that are mosaics of the two colors. The explanation, often called the Lyon hypothesis after one of its formulators, is that during development, a given coat region is formed by the clonal progeny of a single cell and that either one of the two X's may be turned off in different progenitor cells. When the Xs of a female carry alleles for two different forms of an enzyme, tissue culture clones grown from different single cells of the individual manufacture either one or the other form of the enzyme, depending upon which X is used for transcription. These phenomena correlate with the presence of a condensed X chromosome (Fig. II-12) in many cells of female mammals. (3) In white blood cells much chromatin is coiled into dense regions. Autoradiography shows these regions to synthesize much less RNA than adjacent extended chromatin. Most cells show at least some condensed chromatin in their interphase nuclei (Fig. II-11).

As mentioned in Section 2.2.5, the term *heterochromatin* is often used to designate the coiled dark-staining condensed regions of interphase chromatin. This heterochromatin is sometimes referred to as "facultative" to indicate that it varies from cell to cell and under different conditions and to distinguish it from similarly staining chromatin of an apparently different type such as that near the kinetochore ("constitutive" heterochromatin; see Section 4.2.6). *Euchromatin* refers to the extended regions. It is conceivable but still speculative that the tight coiling of condensed chromatin helps maintain the inactive state, since it could impede penetration of substrates and enzymes to the DNA. Even if this is so, the underlying molecular mechanisms by which genes are switched on or off remain unknown. Indeed, with the exception of salivary gland chromosomes, the evidence on chromosome condensation concerns whole chromosomes or large portions of chromosomes with extensive blocks of genes. It is an open question whether this "coarse" control is an exaggeration of similar events at the level of individual genes. There may be quite different, more subtle, regulatory factors operating at the latter level.

The prime suspects as possible agents that control chromatin condensation or regulate genes at the molecular level are the chromosomal proteins. Work on bacteria discussed in Section 3.2.5 has established that specific proteins can interact with specific regions of DNA so as to turn genes on or off. For eucaryotes the chromosomal histones initially were the center of attention. With either isolated chromatin or mixtures of purified DNA and proteins, histones inhibit enzyme systems involved in DNA replication and RNA transcription from DNA templates. Removal of histones from DNA enhances these activities. Sperm often have protamines associated with their DNA. These proteins are even more basic than histones, and their binding to DNA might contribute to the inactivity of the sperm cell nucleus. However, the bacterial repressor proteins isolated thus far do not resemble histones (or protamines). Histones appear not to have the degree of specificity expected for regulators of individual genes; there are only a few distinct histone species (Section 4.2.5), and histones of different cell types and of species as disparate as calves and peas are quite similar in amino acid composition and other characteristics.

The nonhistone proteins of chromosomes are now receiving increased attention. These proteins show much more variability in both type and amount from tissue to tissue than do the histones, and changes in the nonhistone protein population are noted at different stages of cell life cycles and during development. Steroid hormones which can stimulate the synthesis of particular proteins by target cells, are thought to act by entering the cell and complexing with a protein receptor in the cytoplasm. The complex then enters the nucleus and interacts with the DNA and nonhistone proteins in a manner that is not understood, but produces an increase in synthesis of specific mRNAs. It is also reported that some of the inhibitory effects of histones on nucleic acid synthesis in mixtures of purified DNA and proteins, are diminished by addition of nonhistone proteins.

Perhaps the nonhistone proteins and the histones cooperate in gene control, forming complexes with one another as well as with DNA. Other possible regulatory factors include the selective modification of the chromosome proteins. Acetate or methyl groups can be added to the positively charged amino groups of basic amino acids, and phosphates can be attached to amino acids such as serine. These additions would weaken binding of the proteins to the negatively charged DNA phosphates. Increases and decreases have been noted in the extent of phosphorylation or acetylation of chromosomal proteins under different circumstances.

Experiments now underway on isolated chromatin should help clarify matters. There is still much controversy over interpretations. But some investigators claim that when carefully isolated chromatin is supplemented by appropriate precursors and RNA polymerases purified from bacteria, the RNAs produced resemble the ones characteristically made by the cells from which the chromatin originates. Reticulocyte chromatin, but not that from liver, yields RNAs that behave like those known to specify hemoglobin polypeptides, in RNA-DNA hybridization experiments and other tests. Through dissociation and reconstitution experiments, efforts are being made to determine which components of the chromatin generate this specificity. Initial results hint that the proteins separated from a given type of chromatin will restore transcriptional specificity when added back to the DNA.

c h a p t e r **4.5**
DIVERGENCE OF CELLS DURING EVOLUTION

4.5.1 **EVOLUTIONARY CHANGES IN CHROMOSOMES** Figure IV-32 is a diagram of the more common chromosomal changes in number or morphology that contribute to the differences in karyotype among species.

Many plant species appear to have evolved, in part, by *polyploidy*, the presence in the cell of extra *sets of chromosomes*. This is related partly to the manner by which polyploidy can overcome sterility of *species hybrids* (the products of matings between organisms of two different species). In most such hybrids, when the two species are only distantly related, the chromosomes contributed by one parent are not homologous to those from the other parent. Normal meiotic pairing cannot occur, and the gametes produced have variable numbers and abnormal combinations of chromosomes; zygotes rarely survive. If, however, the entire chromosome complement of the hybrid is doubled, or if the gametes that produce the hybrid are diploid rather than haploid, the hybrid will possess pairs of chromosomes; normal meiosis and gamete formation can take place. Such doubling may occur naturally by accidental failure of cytoplasmic separation in a dividing cell. Plant breeders induce doubling by the use of agents, such as *colchicine*, which dissolve the spindle. Chromosomes duplicate to form chromatids but, in the absence of a spindle, both chromatids end up in the same cell.

Polyploidy often results in larger cells, with nuclei enlarged in proportion to the number of extra chromosome sets. Some large polyploid cells are often present along with the diploid cells in tissues (liver, roots, and so forth) of multicellular organisms but, beyond the change in size, the significance of this is not clear.

Several human disorders are related to *aneuploidy*, the presence of abnormal numbers of one or several chromosomes of the complement, rather than of entire sets of chromosomes as in polyploidy. Among these is Down's syndrome (once called "mongolism"), a form of mental retardation. This condition is associated with the presence of three copies of one chromosome (number 21; Figs. IV-6 and IV-7) rather than the normal two copies. Thus a total of 47 chromosomes per cell is present, rather than the normal 46.

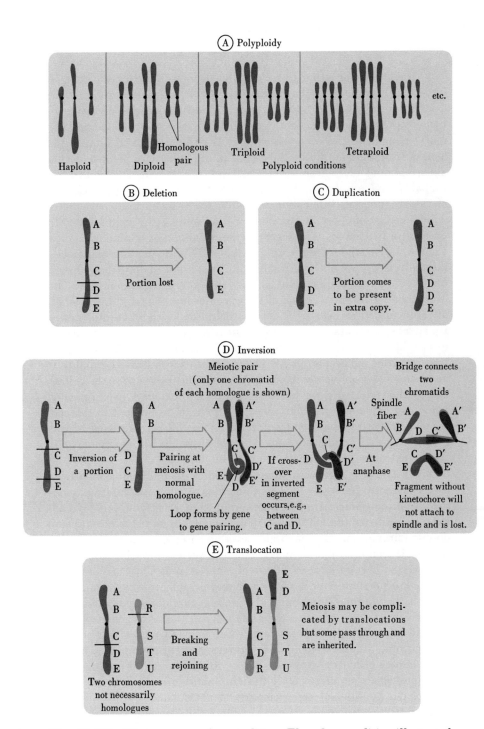

Fig. IV-32 *Chromosome abnormalities. The abnormalities illustrated occur occasionally in nature and may be induced experimentally by certain drugs, irradiation, or other means.*

Radiation and other types of damage as well as abnormalities in crossing over occasionally produce *deletions* and *duplications*. Deletions are usually harmful; for example, defective recessive genes on one homologue may be unmasked (Section 4.3.5) if the corresponding "normal" segment of the other homologue is deleted. If both homologues lack the same regions, functions vital to the cells may be absent. The abnormal nucleolar chromosomes in *Xenopus* (Sections 2.3.4 and 4.3.2) probably have undergone deletion of the DNA segments responsible for rRNA production. Duplications can sometimes result in abnormalities. Apparently some sort of balance existing among the genes normally present in two copies is upset by the introduction of additional copies. The fact that some genetic information (DNA nucleotide sequences) normally is present in many duplicate (repetitive) copies has already been mentioned. Duplications are of interest from another viewpoint. Mutations in the extra region of the chromosome may not dramatically affect the organism, since the cell is still left with two unmutated genes. By successive mutations the genes in the extra segment can diverge quite widely from their original nucleotide sequences. This can lead to the formation of proteins with new properties, such as the ability to catalyze a reaction for which no enzyme previously existed in the cell. In effect, then, a new gene has arisen from a previously existing gene, while two copies of the "old" gene continue to be present in the cell. (Evidence that this has occurred in evolution will be discussed in Section 4.5.2.) Very similar phenomena probably occur with polyploidy also when "extra" copies of genes come to be present.

Inversions may have a somewhat different significance. Crossing over constantly generates new combinations of alleles of different genes *within* chromosomes. Under some circumstances it is of selective advantage to the population that particular combinations of alleles present on one chromosome and favorable in a given environment not be disrupted. If a chromosome region becomes inverted, crossing over with a noninverted chromosome is, in effect, suppressed. Close meiotic pairing may be prevented for the region. If pairing and crossing over do take place in the chromosome region, the products are abnormal and usually will not be transmitted by viable offspring to the next generation. As seen in Figure IV-32, a crossover in the inverted region will produce chromosome fragments and also *bridges* that are connected to the kinetochores of both homologues; these bridges break in meiosis when the homologues separate. Only the chromatids in which no crossover has occurred in the inverted region will be normal, and usually only they will contribute to viable offspring. Thus the combination of alleles within an inversion tends to be inherited as a block that is not changed from generation to generation by crossing over. Different populations of a single species of the fly *Drosophila*, living at different altitudes or temperatures, show differences in the pattern of inversions on their chromosomes, an evolutionary result of such phenomena.

Since *translocations* (Fig. IV-32) often result in altered lengths of chromosomes, they probably contribute to evolutionary changes in the karyotype. One finding of interest in cells carrying translocations is that the expression of a given gene in terms of cell characteristics may differ depending on its neighbors. Gene A may behave differently if placed next to gene B than it does when next to C; this also is observed with some inversions. Although the basis for these "position effects" is not known, the findings emphasize that genes are not to be considered as totally separate and autonomous agents strung out along the chromosome. The evolution of the chromosome and gene set of a species involves not only specification of the appropriate *kinds* of metabolism, but also the balance and integration of metabolism.

If by chance one of the chromosomal breaks should occur in a nucleolar organizer region, nucleoli may be formed by both chromosomes that receive part of the organizer through translocation. This supports the other evidence that the genetic information in the nucleolar organizer is present in many similar or identical copies (Section 2.3.4). The effect of the translocation is to move some of the copies elsewhere, so that two groups of copies are present rather than one.

Other interesting questions arise with respect to the repetitive DNAs. Mutations continually occur in all DNA sequences. With unique DNAs the effects will be felt immediately, and natural selection will be a controlling factor for the frequency in the population. But when a mutation occurs in one or two of the copies of a repetitive set it may be masked by many normal copies. Thus one might expect a good deal of heterogeneity within the population of DNA molecules of a given repetitive set in an individual or species. For at least some, such as the rDNAs, this seems not to be the case. Are there presently unknown mechanisms wherein copies are somehow checked against one another and deviations "corrected"? Or do special DNA replication processes come into play? There exist mechanisms (such as the "rolling circle" mechanism) by which a DNA molecule can generate multiple copies of itself linked together in a linear array. According to some investigators such mechanisms are involved in the production of extra nucleoli in amphibian oöcytes. Could the homogeneity of other repetitive DNAs somehow reflect their production from a small number of parental sequences in each cell, or at some stages in the organism's life history? Other hypotheses rely on complex crossing-over phenomena.

Comparisons of DNAs of different species, including some closely related ones, such as different *Xenopus* species, have indicated that certain of the repetitive DNAs vary much more than do others. It appears that the DNAs coding for ribosomal RNAs, or those coding for histones, evolve at a much slower rate than do the very highly repetitive DNAs or the "spacers" (Section 4.2.6) present in some of the repetitive clusters. This might relate in part to the fact that the highly repetitive and spacer DNAs may not be transcribed into RNA or translated into proteins. Thus the roles they play may not be as drastically influenced by changes in base sequences as is true with DNAs carrying more "conventional" information.

As yet, the evolutionary effects of mutations of genes in cytoplasmic organelles have been little studied. Although such mutations represent a potential source for the variations observed in structure and chemistry in different organisms, the fact remains that most organelles of a given class are basically similar in a vast diversity of organisms. This suggests that the basic features or organelles were established relatively early in the evolution of present-day organisms. Once the organelles had arisen as highly efficient functioning units (Section 4.5.2), most major mutations tended to be deleterious and therefore were not perpetuated.

4.5.2   ***CELLULAR***   In the course of evolution, new enzymes and
EVOLUTION: metabolic pathways have arisen; diverse
SOME FACTS AND   groups of procaryotes, algae, and protozoa
SOME SPECU-   have adapted to different environments;
LATION   and specialized cells within multicellular
organisms have assumed distinctive roles. At some point the molecules of living systems must have originated from inorganic materials, cellular life must have arisen from precellular life, eucaryotes from procaryotes, diploid cells from haploid cells, and multicellular organisms from unicellular organisms.

Some early cells appear to have been preserved in the fossil record. Objects resembling present-day procaryotes have been identified in rocks formed 2–3 billion years ago (although it is not always possible to determine whether they are truly the remains of cells or merely incidental mineral deposits unrelated to life). It is easier to identify early algae and protozoa because the characteristic cell walls of the algae and the silicon- or calcium-containing exoskeletal structures of some protozoa are well preserved. Information about tissue organization (for example, bones in animals, wood of plants) may be obtained by microscopic examination of fossilized multicellular organisms. Aspects of cell chemistry can be deduced from the analysis of organic deposits such as coal or certain petroleum oils, which are formed by the transformation of living organisms and their products.

Present methods, however, do not permit deductions regarding fine structure or metabolic organization from fossils. Knowledge of cell evolution must, therefore, depend heavily on the comparative study of existing species, on speculation from known features of macromolecules and cells, on postulated features of primitive environments, and on experiments such as those in which amino acids and components of nucleic acids are formed in test-tube systems thought to simulate the earth's early environment. Nucleic acids are universally found as carriers of genetic information. A similar nucleotide-carried genetic code is universally used, and essentially similar RNAs and ribosomes are used to make protein by all species studied, procaryotes and eucaryotes alike. This suggests that the nucleic acids and present-day mechanisms of replication, transcription, and translation arose very early in the evolution of life. Geological investigations suggest that oxygen became abundant in the earth's atmosphere about 1–2 billion years ago. By then, processes similar to photosynthesis had probably evolved, since such processes are held responsible for the presence of abundant atmospheric oxygen. As the amount of atmospheric oxygen increased, organisms with aerobic metabolism evolved. Similar mechanisms for the anaerobic breakdown of sugar are found in most organisms, from bacteria to higher plants and animals. The mechanisms must have arisen early in the evolution of present-day species, probably in some ancestor common to most present-day organisms. The addition to these anaerobic mechanisms of the aerobic oxygen-dependent metabolic sequences of sugar breakdown resulted in an enormous increase in the efficiency with which energy could be obtained by cells from their nutrients (see Section 1.3.2). Presumably this permitted the expenditure of energy for "luxury" items, such as the replication and segregation of very large amounts of DNA and the maintenance of very large stable multicellular aggregates. As cells evolved more complex structures, the selective advantages in metabolic efficiency, flexibility, or adaptation to different environments must have outweighed the problems of maintenance and reproduction of more complex organization.

It is becoming possible to add some detail to the general picture presented in the preceding section. There is a growing body of evidence on the evolution of molecules and metabolic systems, and of speculation on the evolution of organelles.

Comparative studies of the amino acid sequences in the respiratory protein *cytochrome c* indicate that it arose early in the evolution of aerobic metabolism. The amino acid sequences of portions of the molecule are identical in the cytochrome *c* of yeast, invertebrates, mammals, and higher plants. Organisms as distantly related as yeast and mammals show the same amino acid sequence in almost half of the molecule (the total number of amino acids in a cytochrome *c* molecule is 104–108 in different organisms). Thus once cytochrome *c* evolved as part of a highly efficient enzyme system, the great majority of mutations that changed the amino acid sequence of certain portions of the molecule must have been selectively disadvantageous because they adversely affected the functioning of the system. Organisms carrying such mutations would be eliminated by natural selection. The portions of the molecule that are similar or identical in all species studied are presumed to include amino acid sequences of particular importance to the respiratory function of the molecule. Other proteins that have been highly "conserved" in evolution include some histones and possibly tubulins and actin.

However, part of the cytochrome *c* has changed, indicating that not every amino acid in the sequence is fixed in evolution or vital for the functioning of the protein. Under some circumstances, when selective advantages result, new enzymes can evolve from more ancient proteins. Obviously this must be based on changes in the nucleotide sequences of the appropriate genes. This probably can occur with chromosome duplications (Section 4.5.1). Mutations in an extra copy of a gene may eventually result in a protein that assumes new functions. That this has occurred in evolution is again indicated by comparative studies of proteins. For example, *hemoglobin*, the iron-containing protein of red blood cells, and *myoglobin*, an iron-containing protein found in some muscles and a few other tissues, probably evolved simultaneously along separate lines from a common ancestral protein. Hemoglobin contains four subunits, each a polypeptide chain about the size of a myoglobin molecule (which is a single polypeptide chain) and each having a three-dimensional structure similar to myoglobin. Both proteins bind oxygen and have similar amino acid arrangements in the region of the molecules responsible for oxygen binding and for three-dimensional shape. However, the two proteins have diverged considerably in amino acid sequence, indicating that the evolutionary separation must be of long standing; at some point subsequent to the separation, the evolutionary changes in hemoglobin resulted in the formation of a tetrapartite molecule. The mammalian digestive enzymes *trypsin* and *chymotrypsin* catalyze different reactions in the breakdown of proteins; however, they probably have a common ancestor since they show considerable similarity in amino acid sequence. Several different pituitary gland hormones also have similar amino acid sequences. In all these cases the similarities between distinct proteins can be most simply explained by the proposal that extra copies of genes arose and that this was followed by evolutionary divergence of the original and the extra gene. It has been speculated that the variety of transfer RNAs, each specific for a given amino acid, may have arisen by a similar process, since the few tRNAs that have been studied thoroughly show many similarities in the sequence of nucleotides (apart from the coding nucleotides involved in binding to mRNA). How protein synthesis would take place in the hypothetical organisms in which this evolution started is not known, but it is possible that inefficient mechanisms involving only a few tRNAs in the synthesis of simple proteins might have permitted survival.

Probable sequences of the evolution of organelles are less clear, and current ideas are considerably more speculative. The present plasma membrane could be the evolutionary descendant of an ancient spontaneously formed aggregate of lipids and other molecules. Multienzyme complexes might have arisen when mutations resulted in the presence of groups at the surface of two enzymes that promoted binding the two together. A sort of evolutionary self-assembly might have taken place with the selective advantages of increased efficiency leading to the preservation of useful intermolecular associations that arose by chance.

For the next step, the formation of complex organelles such as mitochondria and plastids, an interesting theory is that symbiosis was involved, as may be indicated by some of the facts outlined in the discussion of these organelles. Thus as one hypothetical sequence suggests, a primitive nonmotile photosynthetic cell containing photosynthetic enzymes and pigments bound to membranes (perhaps resembling present-day photosynthetic procaryotes) might have been phagocytosed by a primitive amebalike cell; for some reason it was not digested. The combination of motility, photosynthetic capacity, and perhaps phagocytic ability would be of mutual advantage. If the two cells multiplied synchronously, the association might be stable; in time it might become necessary for the survival of the partners. (In the evolution of *lichens*, a symbiotic relationship between algae and fungi has apparently evolved to the point where the fungi normally cannot survive without the algae.) The photosynthetic cell might, under these circumstances, evolve into a plastid and lose its ability to live independently. The evidence that symbiosis of this general type might have occurred includes the following. Mitochondria and plastids contribute to their own duplication; they possess nucleic acids and ribosomes which differ from the rest of the cell; in some cases the ribosomes resemble the ribosomes of procaryotes, and there are similarities to bacteria in responses to inhibitors of protein synthesis and in the probable use of N-formyl methionine to initiate protein synthesis. There are present-day cases of situations in which one cell type lives within another in a symbiosislike association. But while there is some evidence for origin of organelles through symbiosis, it is far from conclusive. In particular, the extensive involvement of nuclear genes in the synthesis of organellar macromolecules must be kept in mind. One can construct plausible schemes in which organellar DNAs derive, evolutionarily, from nuclear DNAs, or in which the several cellular DNAs all derive from a common ancestor.

The evolution of mitosis, a key mechanism in eucaryotes, must have occurred early, since mitosis occurs in virtually all present-day eucaryotes, despite the fact that the evolutionary separation of groups such as plant cells and animal cells probably is very ancient. The evolutionary significance of mitosis lies in the fact that it accomplishes a regular segregation of large amounts of DNA and thus provides for constancy of complex genetic combinations. The steps in the evolution of the process may be reflected in the presence of alternate division mechanisms in a few eucaryotes. For example, in some ciliated protozoa the micronucleus divides by mitosis, while the macronucleus apparently is pinched in two after duplication of its contents. The nuclear division mechanisms of dinoflagellates (Section 4.2.7) may also be of ancient origin.

At present the origins of most of the highly diversified cells of present-day organisms remain a matter of conjecture. It is hoped, however, that an understanding of them eventually will be reached, explaining evolutionary steps such as those that have led to the presence in the same organism of pigment cells producing melanin from the amino acid tyrosine and of gland cells producing the hormone epinephrine, from the same starting material. Study of evolution at the level of whole organisms started from a handful of fossils, a few unusual species, and a mass of anatomical descriptions. By now it has progressed to experimentation on the subtle effects of the environment on the frequency of particular genes in populations. It may be expected that the study of evolution at the cellular and molecular level will progress in similar manner.

FURTHER READING

Beerman, W., And U. Clever. "Chromosome Puffs." *Scientific American*, **210**: no. 4, p. 150, Apr. 1964.

Bogorad, L. "Evolution of Organelles and Eukaryotic Genomes." *Science*, **188**:891, 1975.

Britten, R. J., and D. E. Kohne. "Repeated Segments of DNA." *Scientific American*, **222**: no. 2, p. 24, April 1970.

Brown, D. D. "The Isolation of Genes." *Scientific American*, **229**: no. 2, p. 20, Aug. 1973.

Clarkson, B., and R. Baserga, Eds. Control of Proliferation in Animal Cells. Cold Spring Harbor, N.Y.: Cold Spring Harbor Press, 1974. A collection of research articles.

Cohen, S. N. "The Manipulation of Genes." *Scientific American*, **233**: no. 1, p. 24, July 1975.

Cold Spring Harbor Symposium on Quantitative Biology 38: Chromosome Structure and Function. Cold Spring Harbor, N.Y.: Cold Spring Harbor Press, 1974. A collection of research articles detailing recent progress.

Hood, L. E., J. H. Wilson, and W. B. Wood. *Molecular Biology of Eucaryotic Cells: A Problems Approach.* Menlo Park, Calif.: Benjamin, 1975, 343 pp. A useful concise summary of the organization and functioning of nucleic acids in eucaryotes.

Goodenough, U., and R. P. Levine. *Genetics.* New York: Holt, Rinehart and Winston, 1974, 882 pp. A comprehensive text.

Gurdon, J. B. *The Control of Gene Expression in Animal Development.* Oxford: Oxford Univ. Press, 1974.

Gurdon, J. T. "Transplanted Nuclei and Cell Differentiation." *Scientific American,* **219:** no. 6, p. 24, Dec. 1968.

J. Supramolecular Structure, **2:** 2-34, 1974. A collection of research articles on problems of self-assembly.

Margulis, L. *Origin of Eucaryotic Cells.* New Haven, Conn.: Yale Univ. Press, 1970, 349 pp. A stimulating speculative account of cellular evolution.

Mazia, D. "The Cell Cycle." *Scientific American,* **230:** no. 1, p. 55, Jan. 1974.

Miller, O. L. "The Visualization of Genes in Action." *Scientific American,* **228:** no. 3, p. 34, March 1973.

Padilla, G. M., Cameron, I. L., and A. Zimmerman. *Cell Cycle Controls.* New York: Academic Press, 1974, 370 pp. A collection of research articles on cell division and its controls.

Sager, R. *Cytoplasmic Genes and Organelles.* New York: Academic Press, 1972, 405 pp. A detailed discussion of cytoplasmic inheritance.

Stein, G. S., J. Swinehart-Stein, and L. J. Kleinsmith. "Chromosomal Proteins and Gene Regulation." *Scientific American,* **232:** no. 2, p. 46, 1975.

Stroud, R. M. "A Family of Proteins Cutting Proteins." *Scientific American,* **231:** no. 1, p. 74, 1974.

Structure and Function of Chromatin [CIBA Foundation Symp. 28 (new series)]. Amsterdam, Netherlands: Elsevier, 1975. A collection of research articles.

Wood, W. B., and R. J. Edgar. "Building a Bacterial Virus." *Scientific American,* **217:** no. 1, p. 60, July 1967.

Zubay, G., and J. Marmur. *Papers in Biochemical Genetics,* 2nd ed. New York: Holt, Rinehart and Winston, 1973. A collection of research articles focusing on procaryotes.

Consult also the series of annual reviews and monographs listed at the end of Part I. Many of these series contain extensive articles or books on topics pertinent to Part IV.

part 5

TOWARD
A
MOLECULAR
CYTOLOGY

chapter **5.1**
FROM WILSON TO WATSON

It is now apparent that E. B. Wilson's confidence (see p. 5) was well founded when he predicted fifty years ago that many "puzzles of the cell" would be solved as more powerful tools of analysis became available. None has yet been solved completely, but all are yielding to modern methods. Wilson's "puzzles" included the manner by which nuclear genes affect chemical reactions in the cytoplasm; the nature of cell differentiation; the continuity from one cell generation to the next of centrioles, mitochondria, and chloroplasts; and the influence of the cell's complex organization upon the behavior of macromolecules.

In Wilson's day, morphological description was the basic approach; chemical analysis was in its infancy. Today cell biologists have available a wide battery of methods, and qualitative description is now based on a variety of microscopes. Increasingly the problems defined by such descriptions are analyzed by precise quantitative measurements that are based on biochemical and biophysical procedures. Important strides are being made in taking structures apart and then reconstituting them in the test tube. Although Wilson recognized that form and function were inseparable, it has taken the expanded knowledge of biochemical events and their intracellular localizations, coupled with the extension of microscopic observations into the molecular realm, to establish this for all organelles within the cell.

The complex path of progress in cell biology is well illustrated by the history of the study of nucleoli. Clues have come from many directions. Initially it was noted by light microscopists that nucleoli are found in virtually all eucaryotic cells, but that they are particularly prominent in active cells which synthesize much protein (Chapter 2.3). Later, cytochemists found much RNA in nucleoli and correlated this with the presence of abundant RNA in the cytoplasm; the presumption of transfer of RNA from nucleolus to cytoplasm could be supported by autoradiographic studies (see Fig. I-22). The discovery of genetically distinctive "anucleolate" strains of *Xenopus* (Sections 2.3.4 and 4.3.2) established the fact that the absence of nucleoli is accompanied by the absence of the formation of ribosomes. Molecular biological work explained this by showing that nucleolar RNA gives rise to ribosomal RNA (Section 2.3.4). Large precursor molecules (45 S in HeLa cells and roughly similar size in other cell types) are formed in the nucleolus and are modified then to produce the RNA molecules (18 and 28 S) found in the subunits of the ribosomes (see Fig. II-20).

But nucleoli do not seem to be self-reproducing; they disappear as organized entities in many cells during division. How are they formed? Again, early light microscopy provided an important clue by showing that there is special chromatin associated with nucleoli during interphase and during early stages of cell division. In many organisms nucleoli are attached to special nucleolar-organizer regions of specific chromosomes (see Figs. II-18 and IV-5). When fusions between nucleoli are taken into account, the number of nucleoli parallels the number of organizer regions. RNA-DNA hybridization studies (Section 2.3.4) supported by other evidence (Section 4.3.2) have now shown that the nucleolar-organizer regions contain the genetic information for ribosomal RNA. They have revealed that this information is present in repetitive form, that is, up to many hundred copies of the information, apparently identical, may be present in each organizer. In amphibian oöcytes even more copies of the information are temporarily present under special circumstances of intensive rRNA production. Multiple replicates of the nucleolar-organizer region are produced, and each makes a nucleolus. This special situation has been used to advantage for isolation of the DNA responsible for RNA synthesis. Figure IV-18 is an electron micrograph of the DNA "caught in action"; the interspersal of transcribed rDNA regions and nontranscribed spacers are clearly seen, as is the simultaneous use of a given rDNA segment for synthesis of many RNA molecules.

Figure IV-18 also illustrates an important technical aspect of modern cell biology, the utilization of microscopes to gain information about molecules. Other examples were previously shown in Figures II-5, II-40, and II-43. Figures V-1 to V-3 show a few examples of current microscopic study of DNA, and the legends suggest the kinds of information currently obtainable.

Of central importance in elucidating cell metabolism and the gene-

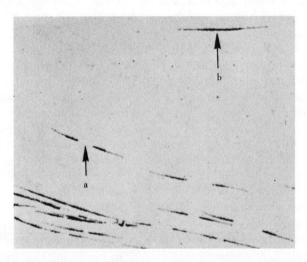

Fig. V-1   *Autoradiograms of DNA from a cultured mouse cell (L cell) that was exposed first to highly radioactive thymidine for 30 minutes and then to much less radioactive thymidine for an additional 30 minutes before the cells were broken open and their contents spread on a slide. The DNA molecules themselves cannot be seen, but the portions that have replicated during the labeling period are radioactive and produce lines or tracks of grains in the autoradiogram, seen as dark zones in the micrograph (the magnification is too low for individual grains to be visible). Each track corresponds to a replicating unit of DNA (see Figs. IV-10 and V-2; also Section 4.2.4). The unit at b commenced replication while the highly radioactive thymidine was present, and it was still replicating when the cells were switched to the less radioactive medium. Since replication begins at the middle of a unit and proceeds in both directions from there (see Fig. IV-10) the central portion of the track shows heavier radioactivity (denser accumulation of grains) than do the more lateral portions (these correspond to DNA made in the less radioactive medium). The unit at a, had started replicating before label was present; thus it shows an unlabeled central region (arrow) flanked on either side by densely labeled DNA (made after the cells were placed in the highly radioactive thymidine) and then by less heavily labeled DNA. Determinations of the length of tracks, produced with different labeling periods indicate that an end of a single growing DNA strand adds bases at a rate of approximately 1-2 thousand per minute. × 200. (Courtesy of R. Hand.)*

tic machinery that control metabolism has been the use of the procaryotic bacteria and of the noncellular viruses. Study of bacteria and viruses has led to revolutionary changes in the way the cell is viewed and in the experimental questions being asked. These changes are dramatized in *The Molecular Biology of the Gene* 2nd ed. (New York: Benjamin, 1970) by J. D. Watson who with Francis Crick first unraveled the structure of DNA. For example, in Wilson's day and long after, a mutation was an event of unknown nature, affecting a gene of unknown chemistry, and resulting in a change detectable in the organism only at a level such as the color of the eye, far removed from the primary effect. Today, largely from work on bacteria and viruses, mutation is explained as a change in base sequence of DNA, resulting in a changed mRNA and protein whose enzymatic or other properties are altered, ultimately producing the visible effect in the organism.

Much of what is known of ribosome formation and functioning comes from studies on bacteria. For example, the reconstitution experiments that bear upon ribosome assembly and functions of ribosomal proteins (Sections 2.3.2 and 4.1.6) have been done largely with organelles from *E. coli*. Work with the same organisms led to the concept that ribosome subunits dissociate once they have completed translation of the mRNA of a polysome and then reassociate with one another and with mRNAs to form new polysomes. Eucaryote ribosomes are fundamentally similar, but the impact of the presence of the nucleolus and the nuclear envelope are yet to be fully assessed. We have already mentioned that eucaryotic ribosomal proteins probably are synthesized in the cytoplasm, and evidently traverse the nuclear envelope to complex with newly formed rRNA. We also noted that the simultaneous transcription and translation of mRNA, which takes place in procaryotes, is unlikely in eucaryotes (Section 2.3.5).

Detailed information is at hand on the molecular events underlying regulation of several specific genes in bacteria. One task for the future is to determine how similar mechanisms might fit into phenomena such as embryonic differentiation in eucaryotes. Such processes seem a good deal more complex than are the enzyme inductions studied in bacteria. But the concepts of molecular biology and biochemistry are now sufficiently advanced, and we have enough information about the organelles of animal and plant cells so that direct attacks have begun upon such complex cellular phenomena as cell division, differentiation, embryonic development, and the functioning of the nervous system.

As the analysis of cell organelles and functions in molecular terms has deepened, so has appreciation of the fact that organelles constitute higher levels of integration than molecules. A mitochondrion is not a membrane-delimited sac in which respiratory enzymes, DNA, and other molecules are dissolved at random. It is a complex *organized* structure that couples phosphorylation and oxidation, transfers electrons in a

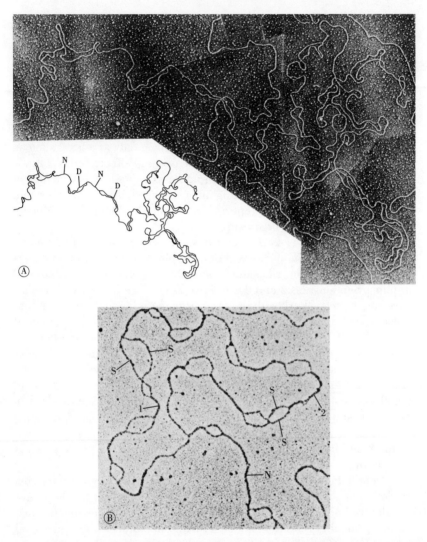

Fig. V-2 Replicating DNA isolated early in the cleavage stage of Drosophila *embryos; prepared by shadowing techniques (p. 17). The cleavages are very rapid (Section 4.2.3; interphase may last less than 5 minutes) and the corresponding rapid duplication of DNA facilitates study of replication.* **A.** *(The sketch at the lower left is a tracing of the photograph.) Replicating units of DNA arranged as would be predicted from biochemical studies of replication (see Fig. IV-10). Stretches of replicated DNA (D) alternate with stretches not yet replicated (N).* **B.** *One of the already replicated regions from a molecule like that in* **A.** *N is a portion of double helix not yet replicated; 1 and 2 are the two daughter double helices. This molecule has been partially denatured by treatment with formamide. As in the present case, it is observed that some treatments can bring about local separation of the double helix into its two constituent strands (S), while adjacent regions of the helix retain their double-stranded configuration.*

highly efficient manner, and carries out other integrated functions. Nuclear DNA in eucaryotes is not a naked template. It is part of an organized chromosome, capable of complex interactions with other chromosomes (for example, in meiotic pairing) and with other organelles (for example, the mitotic spindle).

There is great intellectual excitement and far-reaching practical importance in molecular explanations of cytological events. But there is also profound drama in cell activities at higher levels of integration: the ceaseless beating of heart muscle cells for many decades, the spectacular specializations of structure and function within a single protozoon cell, or the beautifully synchronized and precisely balanced changes in the development of an animal or plant embryo.

One reads Wilson's book, decades after its publication, with astonishment at the variety, complexity, and beauty of cells. One reads Watson's book with enthusiasm for the precision and power with which events at the molecular level can now be described.

chapter **5.2**
CYTOLOGY AND PATHOLOGY

Even while the concept of the cell as the unit of form, function, and duplication in higher organisms was being established as a principle of biology, it was extended to the study of diseased tissue. The basis for cell pathology was laid in the mid-nineteenth century by Rudolph Virchow. Since that time, pathology, cell biology, and cytology have been interdependent. With the present focus in cell study upon organelles and molecules, cell pathology is naturally moving in similar directions. Explanations of abnormal cell functions, particularly their origins (pathogenesis), are being sought in terms of organelles and molecules. Tantalizing clues have been reported—for example, changes of endoplasmic reticulum in fatty liver development, of mitochondria in riboflavin deficiency, and of lysosomes in a number of nervous system disorders. The disease that is best understood in molecular terms has

---

◄ *The regions that separate are probably those relatively rich in A-T base pairs (see Fig. II-9); A and T bind to one another by two hydrogen bonds in contrast to G and C, which form three such bonds and are thus more strongly associated. In this experiment, partial denaturation confirms the expected presence of two polynucleotide strands in the replicated DNA. In other cases this procedure is being used to map DNA molecules in terms of the locations of G-C rich and A-T rich regions. (Note: A and B are printed in opposite photographic contrast.) (Courtesy of H. J. Kriegstein and D. S. Hogness.)*

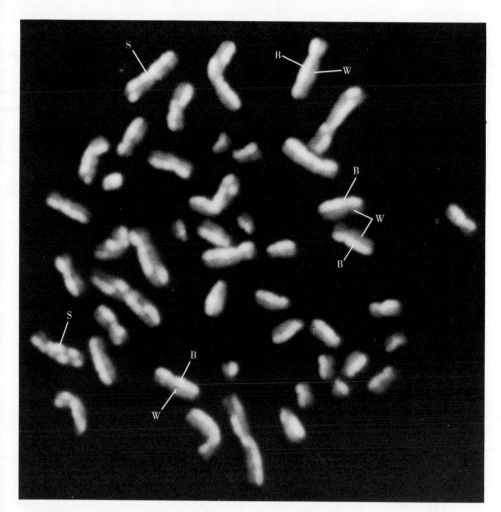

Fig. V-3 *Study of DNA replication by fluorescence microscopy. These methods may eventually evolve into procedures for observation on replication in living cells. Here, they have been used for an experiment like the one in Figure IV-11. Human lymphocytes were grown in BUdR (p. 304) which substitutes for thymidine (BUdR=Bromo-uracil-deoxyriboside; it is also referred to as BrdU). The growth period in BUdR was such that the cells went through two interphases. The cells were then fixed and stained with a dye (33258 Hoechst) that binds to DNA and fluoresces when irradiated with ultraviolet light, as in a fluorescence microscope. The fluorescence of this dye decreases when the DNA contains BUdR. By the same logic used in Figure IV-11, one would expect that after two rounds of replication in the presence of BUdR each metaphase chromosome will have one chromatid in which both strands of the DNA double helix contain BUdR and one in which only one strand contains BUdR. The present figure shows one piece of evidence that confirms this: each chromosome has one chromatid whose fluorescence is very weak (W), presumably with 2 BUdR strands and one*

already been mentioned. In sickle cell anemia a single amino acid change in a protein, resulting from mutation, leads to grossly abnormal red blood cells (Sections 4.3.1 and 4.3.5). And even for cancer, molecular biology studies are beginning to penetrate into hitherto mysterious areas, such as the possible involvement of viruses and alterations in cell surface glycoproteins.

Enzymatic changes in the so-called "inborn errors of metabolism" have been known for a relatively long time. These changes include the abnormal enzymes of amino acid metabolism in several rare mental disorders, the absence of an enzyme of pigment metabolism in some forms of albinism, and defects in lysosomal enzymes in storage diseases. Such abnormalities and other important ones (such as the aneuploidy characteristic of Down's syndrome; Section 4.5.1) can be detected prenatally by biochemical and cytological analysis. It is possible to culture into a large population a few cells obtained from the sacs surrounding a fetus in the uterus and to use these cultures for diagnosis of possible fetal abnormalities.

The ultimate perspective of cell pathology is to obtain sufficient insight into the basic causes and progressions of diseases so that cure or prevention is possible. For example, agents that might reduce or reverse the sickling phenomenon in red blood cells are being sought. Some of the lysosomal storage diseases may prove susceptible to therapy based on enzyme replacements. With tissue culture cells one can sometimes reduce the abnormal intracellular stores of lipids or polysaccharides by adding the enzyme that is missing or defective to the growth medium or by coculturing normal and abnormal cells (Section 2.8.5). Evidently, exogenous enzymes in active form can gain access to the defective lysosomes through endocytic entry into the cell and fusion of endocytosis vesicles with the lysosomes. Injection of missing lysosomal enzymes into patients has been tried and does have the expected effects on tissues that engage extensively in endocytosis. Unfortunately, important tissues such as cardiac muscle seem inadequately active in endocytosis, and others, such as central nervous tissue cells, are not normally accessible to blood-borne macromolecules. Thus clinical cures of storage diseases have yet to be accomplished.

Other mammals may be afflicted by abnormalities closely similar to those in humans. Sometimes this provides useful material for study.

◀ *with stronger fluorescence (B) presumably with only one BUdR strand. In some cases (S) "sister chromatid" exchanges seem to have occurred; 2 chromatids of a chromosome have apparently exchanged portions with one another. (From S. Latt, J. Histochem. Cytochem.* **22**:478, *1975. Copyright* © *1975 The Histochemical Society, Inc.)*

For example, the Chediak-Higashi syndrome is a poorly understood, genetically based, rare disease in which a spectrum of abnormalities is encountered, ranging from unusual hair pigmentation to low resistance to infection. Mutant mink, cattle, and mice are known with very similar disorders. These strains can be maintained in the laboratory and studied conveniently without the special ethical problems that arise in work with human subjects. Studies on one such strain, the "beige" mutant of mouse, have been important in demonstrating that many cell types have unusually large lysosomes and related inclusions such as pigment granules. Moreover, the region of endoplasmic reticulum (GERL) responsible for generating lysosomes in hepatocytes, is remarkably enlarged in the beige mouse. Conceivably such observations point toward defects in intracellular transport and digestion, or in membrane-fusion phenomena that are significant for the pathogenesis of the disease.

The changes in lysosomes and in the ER of the beige mouse exemplify the exaggeration of structure or function that often is encountered in pathological material. Such exaggeration can make evident subtleties difficult to detect in normal material—the abnormal teaches about the normal. For example, the manner of synthesis of membranes and enzymes of smooth ER in hepatocytes is being studied in animals treated with the drug phenobarbital, since such treatment leads to the manufacture of large amounts of smooth ER (Section 2.4.4). Several drugs and carcinogens produce changes in nucleoli that are visible by microcinematography of living cells. At first there is an apparent redistribution of material within the nucleoli and also changes in their size. Subsequently the nucleoli may disintegrate. Electron microscopy of treated cells reveals that the several components of each nucleolus segregate from one another, so that the nucleoli show large separate regions of granules, fibrils, and amorphous material. Of interest is the fact that some of the effective agents have known effects on RNA synthesis (the antibiotic *actinomycin D*, for example, inhibits the synthesis of RNA). Thus it may be possible to correlate structural and functional changes and to understand better the complexities of nucleolar morphology. Drugs used to lower cholesterol levels in the blood (hypolipidemic drugs) have been found to increase the amount of hepatocyte smooth ER and also the number and size of the peroxisomes associated with the ER. The implications of this for peroxisome formation and function are now being analyzed.

Diseased tissue is often the source material for isolating biologically important substances. It was from pus cells that Miescher, almost 100 years ago, isolated "nuclein," later renamed DNA. Pus contains large numbers of white blood cells in easily obtainable form and already partially separated from the anucleate red blood cells.

chapter **5.3**
EPILOGUE

One of the extraordinary aspects of science is that each generation learns enough to warrant optimism for the future and to arouse excitement in the succeeding generation. The prospects for cytology and cell biology have never been as bright as they are today. Important first steps have been taken that foreshadow test-tube synthesis of living matter. Such synthesis of proteins from amino acids has been accomplished—the first was the hormone insulin and the second, the enzyme ribonuclease. The synthesis of complete viral RNA and DNA was achieved in mixtures containing only initial copies of the nucleic acids, enzymes, and small precursor molecules. Test-tube synthesis of biologically active nucleic acid molecules in even simpler mixtures lacking the initial templates is progressing to increasingly complex base sequences. The isolation and analysis of specific portions of the genetic apparatus of procaryotes and eucaryotes and the identification of some molecules responsible for control of genetic activity also promise future findings of great interest.

Practical application of the new concepts and techniques to medicine and agriculture is accelerating. Findings with cells in culture come with ever-increasing speed. Already the genetic mapping of human chromosomes is well underway (Chapter 3.11). The mass production has begun of viruses that may replace some of the chemicals now used to kill off agricultural pests, strains of which have arisen that are resistant to the chemical pesticides. There is active research on making such virus production practical, effective, and safe. With the new developments in practical applications of genetics and molecular biology have come new problems. Scientists and the public are apprehensive at the possibilities for accident or misuse in research like use of plasmids for large-scale production of experimentally-manipulated DNA (Section 3.2.5) or genetic engineering of humans.

Biology is in the midst of a revolutionary period, and experiments inconceivable ten years ago are everyday exercises today. Roughly forty years elapsed between the last edition of Wilson's book and the first edition of Watson's. It is probable that, forty years from today, features of the cell will be described in terms of electrons and atomic nuclei. Already, concepts of the new solid-state physics are being applied to electron transport in mitochondria and chloroplasts. Knowledge of the cell changes continuously. The excitement of studying cells derives partly from solving old problems, but each solution brings new questions that require answers.

Index